MATHEMATICS FOR BUSINESS

Second Edition

Flora M. Locke
Merritt College,
Oakland, California

JOHN WILEY & SONS
New York London Sydney Toronto

Library of Congress Cataloging in Publication Data

Locke, Flora M.
College mathematics for business.
1. Business mathematics. I. Title.
HF5691.L55 1974 513'.93 73-10090
ISBN 0-471-54321-7

Printed in the United States of America

10 9 8 7 6

COLLEGE MATHEMATICS FOR BUSINESS

Other Books by the Author

Office Calculating and Adding Machines, 4th ed.
with Dorothy Dehr
(John Wiley, New York, 1974)

Machine Accounting
with Dorothy Dehr
(Locke Publications,
Sebastopol, California, 1964)

Business Mathematics
(John Wiley, New York, 1972)

Math Shortcuts
(John Wiley, New York, 1972)

ABOUT THE AUTHOR

Mrs. Flora M. Locke is a graduate of the University of California where she received an A.B. degree in mathematics, an M.A. degree in education, a general secondary teaching credential, and a general administrative credential.

She has been associated with the Oakland Public Schools for several years, most of this time at Merritt College where she is now Professor Emeritus of Business Education and has served as chairman of the business education department for several years. She has also been a master teacher at Merritt College supervising prospective junior college instructors in business mathematics.

As an instructor Mrs. Locke has served as chairman and member of numerous committees concerned with the teaching of business courses on the high school, junior college, and employment levels. She also has served on a statewide committee for business mathematics at the college level.

Mrs. Locke has had several years of business experience as a secretary, as a calculating machine operator, and as a supervisor of calculating machine operators in statistical, accounting, and merchandising work. She also has served as the chief in charge of the statistical department for a large government agency for three years.

She has been an active member of the NBEA (National Business Education Association) and the CBEA (California Business Education Association) for many years. Other affiliations include Pi Lambda Theta (national honor and professional association for women in education) and Alpha Delta Kappa (international honor society for women in education). She is listed in the 1974–1975 National Register for Prominent Americans and in Contemporary Authors (a biobibliographical guide to current authors and their works) published by Gale Research Company.

PREFACE

This text emphasizes the application of mathematics to personal and professional business problems. Instruction in methods is stressed, with many examples and illustrations to aid you in your work. Business students will benefit from the study of basic concepts and examples in such specialized areas as accounting, banking, merchandising, credit, insurance, taxes, payroll, and data processing.

Whether a business or a liberal arts student, you will be better able to handle your own personal business affairs more adequately. In addition, your background in business terminology will be greatly enriched through an exposure to many business problems or applications. You will develop a greater appreciation and understanding of number relationships and improve your ability to perform the fundamental operations of arithmetic.

As stated in the preceding edition, it is assumed that you have completed courses in arithmetic and basic mathematics before starting this course of study. The first chapter is a review of fundamental processes with an emphasis on means by which speed and accuracy may be improved. A section on estimating has been added to give you more opportunity to develop a keener sense of accuracy for the answer to any computation.

Many additional examples, problems, reviews, and summaries are included in this edition to offer you a greater opportunity for review and testing, and to enrich your understanding of business procedures. The chapter on sales and property taxes includes new tables and more problems. All payroll tables and rates have been updated with the most recent information available. More problems have been added on notes and drafts with a review and summary at the end of the chapter. A review has also been added to the chapter dealing with installment purchases and loan payment plans. As in the first edition, the last chapter is intended for additional work if needed for any of the subject areas covered in the text or for testing.

New features of this edition include a chapter on the preparation of federal income tax returns, computer number systems, and the metric system of weights and measurements. Chapter 8 offers you an opportunity to prepare simple federal income tax returns on either the long or short form.

Since computers are used for many purposes and involve all of us to some degree, in either a personal or a business way, an introduction to computer number systems is offered. If you are a business student and plan to go into computer or data processing work, this chapter will be of special interest to you.

The metric system is used, wholly or in part, by all major countries of the world. Since it is expected to replace our present system of weights and measurements in time, this chapter will assist you in making this transition and in learning to think in metric terms and to work with metric units.

The answers to odd-numbered problems are provided to help you develop confidence in your ability to analyze and solve problems.

A workbook is available if desired. For those using this workbook, you will find guidelines and suggestions to help you in completing your assignments, and assistance in analyzing and solving word problems.

The materials, information, and suggestions contributed by educators and business men in the preparation of this text is greatly appreciated and acknowledged. I am especially grateful to Mr. Alex Pappas, Laney College, Alameda, California; the California Automobile Association; and Insurance Services, San Francisco, California for their valuable assistance.

Sebastopol, California *Flora M. Locke*

CONTENTS

Objectives

This chapter is designed, through review and the use of shortcuts, to improve one's ability to work with numbers—to develop speed and accuracy in performing the fundamental operations of arithmetic as applied to whole numbers, decimals, and fractions and to use this ability in solving practical applications.

A unit dealing with estimating answers is designed to assist in evaluating and judging the results of any calculation. Guidelines are also included as an aid in identifying and locating errors, especially as applied to cross-footing.

The successful completion of this chapter represents adequate preparation in the basic functions to work with numbers successfully in solving everyday business problems.

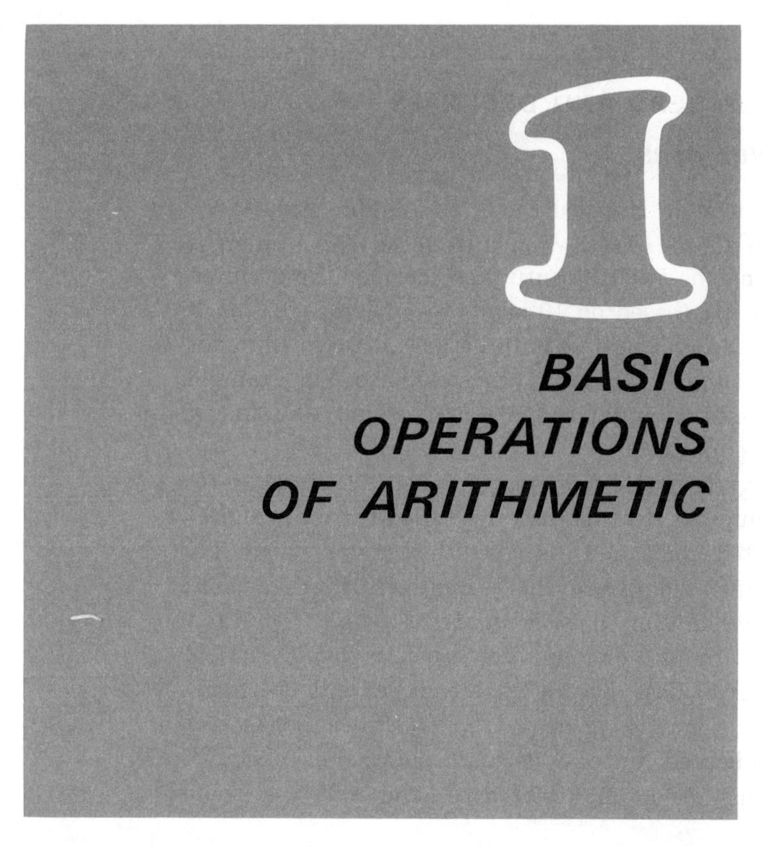

1 BASIC OPERATIONS OF ARITHMETIC

The ability to perform the fundamental processes of arithmetic with competence is necessary to succeed in the study of business mathematics. Despite the wide use of calculating machines, there is always a need for basic arithmetic calculations with pencil and paper.

The development of any skill requires practice. However, practice in itself is not enough. The best methods must be used whereby a reasonable level of speed and accuracy may be achieved. Proficiency in arithmetic computations is based on making use of one or more methods that best suit the problem. Since a complete treatment of shortcuts represents a study in itself, only those in common usage are discussed and illustrated in this book.

Always check an answer. When possible, verify the solution to any problem by some method other than the one used to solve it. When this is not possible, check the calculations as illustrated in subsequent pages.

WHOLE NUMBERS AND DECIMALS

READING NUMBERS

Our number system is called the Hindu-Arabic system. We are indebted to the Arabs who introduced this system to Europe during the twelfth century. It is called the decimal† system since it is based on 10 symbols or digits—0, 1, 2, 3, 4, 5, 6, 7, 8, and 9. The base is 10 because it takes 10 in any one place to equal 1 in the next higher place. For example, it takes 10 ones to equal 10, 10 tens to equal 100, and 10 hundreds to equal 1000.

A number consists of 1 or more digits. The number 765 is a 3-digit number. In this case, 5 is called the units digit, 6 is called the tens digit, and 7 is called the hundreds digit. The decimal point separates whole numbers from decimal numbers. By definition, a *whole number* is a digit from 0 to 9 or a combination of 2 or more digits such as 235 or 1237. A *decimal* is part of (less than) a whole number with a denominator of 10 or some power of 10 (10, 100, 1000, 10,000, etc.) which is indicated by the decimal point. For example: .5 = 5 tenths, .52 = 52 hundredths, and .005 = 5 thousandths.

Digit positions to billions (whole numbers) and to billionths (decimal numbers) are illustrated below.

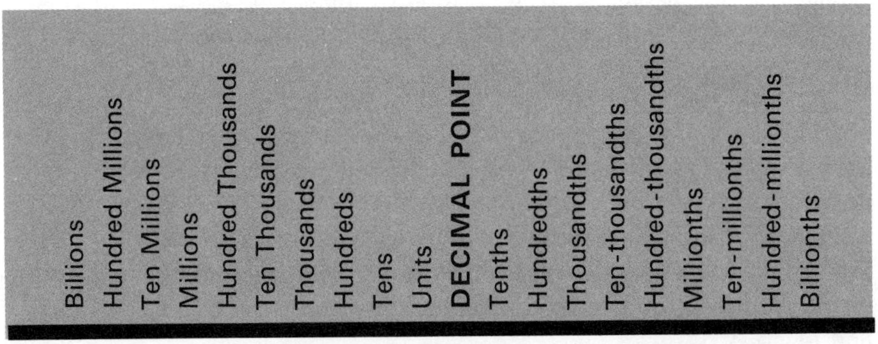

Billions	Hundred Millions	Ten Millions	Millions	Hundred Thousands	Ten Thousands	Thousands	Hundreds	Tens	Units	**DECIMAL POINT**	Tenths	Hundredths	Thousandths	Ten-thousandths	Hundred-thousandths	Millionths	Ten-millionths	Hundred-millionths	Billionths

There is no difficulty in reading small numbers. For example, 7 is read as "seven," 63 is read as "sixty-three," and 426 is read as "four hundred twenty-six." Readers often tend to add the word "and" after the word "hundred" but

†From the Latin *decem* which means ten.

this should not be done. The word "and" should only be used to indicate the decimal point for numbers that include digits to the right of the decimal. Therefore, 2.5 is read "two and five tenths," while 309.25 is read "three hundred nine and twenty-five hundredths."

A common practice in business offices when reading numbers involving both whole numbers and decimals is to identify the decimal point as a point; 13.465 then is read "thirteen point four, six, five."

Large numbers are generally easier to read when the comma is used to group hundreds, thousands, millions, etc. For example, the number 3678458 is easily and quickly read when written as 3,678,458, i.e., "three million, six hundred seventy-eight thousand, four hundred fifty-eight."

examples: Read the following orally.

2.04	8,916.2000
13.216	0.146
.079	206.12
416.2	3,000.05
42.006	400.50
3,004,162	.0300
14,000,914	2,062,526,413
10,906.07	.3
.03	1.007
100.06	90.106

WRITING NUMBERS

It is very important that all numbers be legible, not only for the sake of appearance but, more important, to avoid errors in reading. A carelessly written 9 may be mistaken for a 7 or vice versa; a zero (0) that is not closed may look like a 6; the digit 2 may be mistaken for a checkmark; a poorly written 5 may be mistaken for the letter s. When identifying numbers include letters it is especially important that both letters and numbers be clearly formed. Examples are 6K78083, 51 YC 2803, 51 SF 0931, 24C-60 all of which represent catalogue or model numbers found in industry.

ADDING NUMBERS

Addition is the means by which two or more numbers are combined and expressed as a single quantity which is called a sum or total. The plus (+) sign is used to denote addition when not stated otherwise.

When a group of numbers are to be added, units are added to units, tens are added to tens, hundreds are added to hundreds, etc. Therefore, they should be aligned so that whole numbers and decimals are properly placed in relation to each other.

examples:

36	3.06
317	23.3
29	.005
31,667	5.106
52	119.25
32,101	150.721

Developing Speed and Accuracy

Grouping by tens: Look for and combine two or more numbers that add to 10. In the first example below, starting from the top, notice that 7 and 3 add to 10 and also that 6 plus 4 equals 10 in the units column. This greatly simplifies the problem; by observing that these 4 digits add to 20, it is only necessary to add to this sum the remaining digits in that column. The total is 24. The digit 2, which is carried to 10's position, combines with 8 to equal 10 to which other digits in that column are added.

(1)	(2)
37	83.6
53	2.9
46	20.3
91	19.1
84	6.5
93	22
2	132.4
404	

With practice, considerable speed can be acquired in adding by this method. Other number combinations that tend to repeat themselves may be used in the same manner.

Using number combinations: The sum of any two digits should be automatic. If number combinations are thoroughly known, it is possible to add a column of figures rapidly by adding combinations of numbers instead of the individual numbers.

example:

$$\begin{array}{r} 3 \\ 7 \\ 9 \\ 5 \\ 6 \\ 8 \\ 4 \\ \underline{2} \\ 44 \end{array}$$

Notice that the bottom 4 numbers add to 20. In other words, instead of saying to yourself 2 + 4 = 6, 6 + 8 = 14, 14 + 6 = 20, combine these numbers at once saying 20 + 5 = 25, 25 + 9 = 34, etc., or possibly 20 + 14 (5 + 9) = 34, 34 + 10 = 44.

Using subtotals: When a long column of numbers is to be added, it is advantageous to: (1) add part of the column and record to the side and (2) add these partial totals (subtotals) for a final total. This method is also helpful if you are interrupted since the whole column need not be readded.

example:

$$\begin{array}{rr} 328 & \\ 422 & \\ 316 & \\ 989 & \\ 514 & 2569 \\ \hline 714 & \\ 158 & \\ 806 & 1678 \\ \hline 475 & \\ 814 & \\ 515 & \\ 115 & \\ 171 & 2090 \\ \hline 6337 & 6337 \end{array}$$

Checking Answers

The most common and satisfactory procedure for checking answers in addition is the reverse-order method. This simply means that if you have added a column of numbers from top to bottom, then readd in the opposite direction; that is, begin at the bottom and add upwards.

EXERCISE 1

Add each problem. Use shortcuts when possible. Prove each answer by reverse-order check.

1.
1878
5306
3582
2074

2.
77.12
43.79
62.47
5.66

3.
5460
568
823
6532

4.
63.63
31.38
82.14
56.55

5.
87.87
.89
52.30
9.19

6.
8452
7970
4574
9589

7.
798
3986
5632
981

8.
223.41
47.86
62.53
127.79

9.
.0789
.4245
.8164
.5694
.0078

10.
78.30
43.87
76.42
102.21
81.87

11.
74,390
8,665
13,187
11,215
27,898

12.
876.41
232.01
683.17
36.15
362.44

Crossfooting—Vertical and Horizontal Addition

Many business reports require both vertical and horizontal addition. Data are so arranged as to provide more than one kind of information besides including a check against totals.

HIGHLAND CATERING SERVICE

Sales Report Week Ending May____, 19____

District	Mon.	Tues.	Wed.	Thurs.	Fri.	Sat.	Sun.	Total
A	$ 23.40	$ 82.95	$ 68.79	$ 96.38	$ 92.35	$115.50	$101.70	$ 581.07
B	77.15	76.65	78.25	84.96	81.00	89.25	90.20	577.46
C	34.65	53.74	49.00	57.75	48.85	60.50	65.00	369.49
Total	$135.20	$213.34	$196.04	$239.09	$222.20	$265.25	$256.90	$1528.02

example: This report provides: (1) the total sales for each district for the week and (2) the total sales for each day of the week. It is also a convenient means of making comparisons between districts or days and other weekly periods during the year.

The total of the line or district totals ($1528.02) is also the total of the column or daily totals and is referred to as the *grand total*. No reliable office worker would consider this report complete unless the column and line totals checked out to the same grand total. If they do not agree, an error has been made which must be found and corrected.

Identifying and Correcting Errors

A great deal of time can be saved if you know how to locate an error. The possibilities depend on the type of problem. In a problem as illustrated above, proceed as follows.

1. Find the difference between the total of the district totals and the total of the daily totals.
(a) If the difference appears in the cents column, then readd the cents column only; if it appears in the units column (first column to left of the decimal point), then readd the first three columns of numbers only, etc.
(b) If the difference equals any one number in the report, it may be assumed that it was omitted in adding, either horizontally or vertically.

2. If the error cannot be localized, then readd the total column (district totals) and the column totals (daily totals) across.

3. As a last resort, check each column total and then check each line total.

EXERCISE 2

	Column					
	A	B	C	D	E	Total
Line A	3.61	72.09	271.13	10.19	34.97	
Line B	4.76	48.23	67.76	49.36	11.22	
Line C	23.34	9.37	20.35	28.78	72.09	
Line D	4.40	8.00	6.69	89.31	78.12	
Line E	91.46	27.90	5.80	149.25	30.08	
Total						

1. Copy† this report and: (a) add each column, (b) add each line, (c) add the total column, and (d) add the totals of the columns. If the same total is obtained for (c) and (d), the report may be assumed to be correct.‡

2. If the grand totals differ, check your work. All errors must be found and corrected if the report is to be completed properly.

SUBTRACTING NUMBERS

Subtraction is the process of finding the difference between numbers. Unless otherwise stated the minus (−) sign is used to indicate subtraction. To check answers, add mentally the difference (answer) to the subtrahend which should, of course, equal the minuend.

†One common fault in copying numbers is to transpose digits—to reverse their order. When this occurs the difference equals a multiple of nine when checked against original reports or source data control figures.
‡It is possible but rare to have a compensating error.

example:

	49.26	minuend
	−17.65	subtrahend
	31.61	difference or remainder

EXERCISE 3

Subtract. Check results mentally.

1.	498	**5.**	105.68	**9.**	10,016	**13.**	98.63	**17.**	31,477
	316		40.81		6,799		79.36		18,788
2.	5,896	**6.**	50,467	**10.**	21,063	**14.**	51.99	**18.**	19,318
	399		32,895		19,001		35.69		7,534
3.	51.63	**7.**	.00683	**11.**	2356	**15.**	48.56	**19.**	635.95
	49.02		.00491		498		32.05		349.83
4.	90.07	**8.**	9268	**12.**	2385	**16.**	59,531	**20.**	504.28
	81.96		5799		2064		16,483		457.19

MULTIPLYING NUMBERS

Multiplication is a short method of repeat addition. For example, 3 multiplied by 4 equals 12 which is the same as 3 added 4 times (3 + 3 + 3 + 3) or 4 added 3 times (4 + 4 + 4).

Knowledge of the multiplication tables should be automatic. When this is not the case, further study is essential for success in multiplication work.

The result of multiplying two numbers (called factors) by each other is called the *product.* Since the result is the same regardless of which factor is used as the multiplier, use the smaller number.

Check answers in multiplication work by either: (1) reversing the factors or (2) dividing the product by either factor to obtain the other. Because multiplication is an easier process than division, the first method is usually preferred.

examples: 1. Whole numbers—multiply 365 by 54.

```
  365   multiplicand
   54   multiplier
 ----
 1460
1825
-----
19710   product (answer)
```

2. Decimal numbers—multiply 3.605 by 1.27.

```
  3.605   multiplicand (3 decimals)
   1.27   multiplier (2 decimals)
 ------
  25235
  7210
 3605
 ------
 457835   product
4.57835   answer (5 decimals)
```

Mark off as many decimal places in the answer as there are in both factors added together

Multiplying by Any Power of 10 (10, 100, 1000, etc.)

Whole numbers: If a number is to be multiplied by 10, 100, 1000, etc., it is only necessary to add to the multiplicand the number of zeros in the multiplier.

examples:

$$96 \times 10 = 960$$
$$96 \times 100 = 9600$$
$$96 \times 1000 = 96{,}000$$

Decimals: If the multiplicand is a decimal, move the decimal point to the right one place for each zero in the multiplier.

examples:

$$3.167 \times 10 = 31.67$$
$$3.167 \times 100 = 316.7$$
$$3.167 \times 1000 = 3167$$

Multiplying by Any Factor of 10, 100, 1000, etc., Such As 5, 25, 50, 16⅔, and 125

1. 629×5

$629 \times 10 = 6290$
$6290 \div 2 = 3145$ (answer)

Multiply by 10 and then divide the result by 2 since 10 is 2 times as great as 5.

2. 45×25

$45 \times 100 = 4500$
$4500 \div 4 = 1125$ (answer)

Multiply by 100 and then divide the result by 4 since 100 is 4 times as great as 25.

3. 562×50

$562 \times 100 = 56{,}200$
$56{,}200 \div 2 = 28{,}1000$ (answer)

Multiply by 100 and then divide the result by 2 since 100 is 2 times as great as 50.

4. $36 \times 16\frac{2}{3}$

$36 \times 100 = 3600$
$3600 \div 6 = 600$ (answer)

Multiply by 100 and then divide the result by 6 since 100 is 6 times as great as $16\frac{2}{3}$.

5. 912×125

$912 \times 1000 = 912{,}000$
$912{,}000 \div 8 = 114{,}000$ (answer)

Multiply by 1000 and then divide the result by 8 since 1000 is 8 times as great as 125.

Multiplying by 99, 101, 90, and 110

Such numbers as these (there are many others) can be multiplied quickly by some other number as follows: Any number may be expressed as the sum or difference of 2 numbers. For example:

99 has the same value as $100 - 1$
101 has the same value as $100 + 1$
90 has the same value as $100 - 10$
110 has the same value as $100 + 10$

Therefore, the following examples may be solved as illustrated.

1. $546 \times 99 = 546 \times (100 - 1)$ or $546\ (100 - 1)$
$= 54{,}600 - 546$
$= 54{,}054$ (answer)

2. $214 \times 101 = 214\ (100 + 1)$
$= 21{,}400 + 214$
$= 21{,}614$ (answer)

3. $763 \times 90 = 763\ (100 - 10)$
$= 76{,}300 - 7630$
$= 68{,}670$ (answer)

4. $3.54 \times 110 = 3.54\ (100 + 10)$
$= 354 + 35.4$
$= 389.4$ (answer)

EXERCISE 4

Work the following problems as quickly as you can. Use shortcuts where appropriate. Check answers.

1. 43×50
2. 219×25
3. 37.5×101
4. 1.068×100
5. 976×50
6. 51.42×10
7. 800×125
8. 2.167×100
9. 73×10
10. 90×86
11. 76×35
12. 38×72
13. 57.46×80.2
14. $.198 \times 110$
15. $27.5 \times .06$
16. $.0375 \times 24$
17. 75×3160
18. 31.56×125

DIVIDING NUMBERS

Division is a short method of repeat subtraction. It is the inverse operation of multiplication. For example, 20 divided by 5 is the same as saying: "How many times can the number 5 be subtracted from 20?" In this case, the answer is 4.

Ability to divide quickly and accurately is dependent on a thorough knowledge of the multiplication tables. Any division problem may be expressed as follows: $362 \div 15$, $\frac{362}{15}$, or as illustrated.

Whole numbers: Divide 362 by 15.

$$\begin{array}{lrl} & 24 & \text{quotient} \\ \text{divisor} & 15\overline{)362} & \text{dividend} \\ & \underline{30} & \\ & 62 & \\ & \underline{60} & \\ & 2 & \text{remainder} \end{array}$$

Decimals: After the decimal point has been determined, proceed as with whole numbers.

When the number of decimals in the divisor is equal to or less than those in the dividend, mentally move the decimal point in the dividend as many decimal places as there are in the divisor to locate the decimal point in the quotient.†

examples:

1. $\begin{array}{r} 3. \\ 4.001\overline{)12.003} \end{array}$

2. $\begin{array}{r} 20.6\ + \\ 136\overline{)2805.1} \end{array}$

3. $\begin{array}{r} 5.63 \\ .05\overline{).2815} \end{array}$

4. $\begin{array}{r} .09 \\ .009\overline{).00081} \end{array}$

5. $\begin{array}{r} .05 \\ .018\overline{).00090} \end{array}$

When the number of decimals in the divisor is greater than in the dividend, add decimal places in the dividend to equal those in the divisor.‡

†See fractions, page 23. The dividend and divisor or, in fraction form, the numerator and denominator may be multiplied by any number without changing its value; that is,

$$\frac{.2815 \times 100}{.05 \times 100} = \frac{28.15}{5} = 5.63$$

‡Zeros added to any number after the decimal point do not change its value.

examples: 1. $1026 \div .15$

$$.15 \overline{)1026.00} = 6840.$$

2. $80.25 \div 2.016$

$$2.016 \overline{)80.250} = 39.+$$

Checking Answers

Multiply the quotient by the divisor and add the remainder (if any). The result should equal the dividend.

example: $953 \div 22 = 43$ with a remainder of 7

check: $43 \times 22 = 946$; $946 + 7 = 953$

Dividing by Any Power of 10 (10, 100, 1000, etc.)

Whole numbers: Mark off as many decimal places as there are zeros in the divisor.

examples:

$$54 \div 10 = 5.4$$
$$216 \div 100 = 2.16$$
$$29 \div 1000 = .029$$

Decimals: Move the decimal point to the left as many places as there are zeros in the divisor.

examples:

$$2.19 \div 10 = .219$$
$$436.54 \div 100 = 4.3654$$
$$92.7 \div 1000 = .0927$$

Dividing by Any Factor of 10, 100, 1000, etc.

Such as 5, 25, 50, $16\frac{2}{3}$, and 125.

1. $962 \div 5$

$962 \div 10 = 96.2$
$96.2 \times 2 = 192.4$ (answer)

Divide by 10 and then multiply the result by 2 since 10 is 2 times as great as 5.

2. $3620 \div 25$

$3620 \div 100 = 36.20$
$36.20 \times 4 = 144.80$ (answer)

Divide by 100 and then multiply the result by 4 since 100 is 4 times as great as 25.

3. $854 \div 125$

$854 \div 1000 = .854$
$.854 \times 8 = 6.832$ (answer)

Divide by 1000 and then multiply the result by 8 since 1000 is 8 times as great as 125.

EXERCISE 5

Work the following problems, using shortcuts when appropriate.

I. Show remainder for those problems with uneven answers to 3 places.

1. $516 \div 25$
2. $9090 \div 35$
3. $735 \div 50$
4. $5106 \div 10$
5. $375 \div 1000$
6. $2125 \div 125$
7. $4.205 \div 20$
8. $21.65 \div 3.55$
9. $.0076 \div .42$
10. $3278 \div .006$

II. Record all answers correct to 2 decimals.†

1. $.24 \div 2.7$
2. $72 \div 26.4$
3. $9.8 \div 8.4$
4. $8.8 \div .95$
5. $32.68 \div 75$
6. $74 \div .03$

III. Record all answers correct to the nearest thousandth.‡

1. $.90 \div 3.2$
2. $100 \div 2.85$
3. $24 \div .08$
4. $4.661 \div .25$
5. $.324 \div 4.86$
6. $49 \div 375$

†Division must be carried to 1 more decimal than is required in the answer. If the third decimal is less than 5, drop it; if it is more than 5, increase the second decimal by 1. For example (1) $17.428 \div .24 = 72.616$ (to 3 decimals) or 72.62 correct to 2 decimals. (2) $5.6 \div 92 = .060$. Record as .06.

‡Carry quotient to 4 decimals and increase or drop the third decimal as illustrated: (1) $48 \div 58 = .8275$. Record as .828. (2) $6.795 \div 12 = .5662$. Record as .566.

ROUNDING NUMBERS

The number of significant figures desired is dependent on their use. If the data collected are to be reported correct to the nearest hundreds, thousands, millions, etc., then all numbers are rounded to these values accordingly. If the part of the number that is to be dropped begins with 5 or more, add 1 to the preceding digit; if the number to be dropped begins with some digit less than 5, drop it and record preceding digits unchanged.

The following numbers illustrate this principle.

Rounded to Hundreds	Rounded to Thousands
863 to 900	216,862 to 217,000
450 to 500	37,500 to 38,000
319 to 300	9,499 to 9,000

Rounded to Units	Rounded to Hundredths
32.50 to 33	0.7349 to 0.73
41.25 to 41	0.10409 to 0.10
70.92 to 71	3.5 to 3.50
6.499 to 6	44.0653 to 44.07

ESTIMATING OR APPROXIMATING ANSWERS

An approximation is a value that is nearly but not exactly correct. There are many occasions when exact figures are neither desirable nor necessary. Approximations may be used for general reporting, for comparative studies, for decision making, and for checks against gross errors in computational work. The report of stock market sales for any day may be 12 million shares when actually 12,426,429 shares may be sold. The approximate figure is more meaningful to the general public and it becomes even more so when compared to sales, say of 10 million, made during the preceding day.

Approximate cost figures are often used to determine the feasibility of initiating a building program, the expansion of services, or other action of importance to the best interest of a business.

The ability to approximate the answer to a problem in-

involving computations can prevent serious errors. For example, multiplying a 3-digit number by another 3-digit number results in a 5-digit number at least and a 6-digit number at most. With a little practice errors can often be noticed by asking yourself if the answer seems reasonable. Rounding numbers is a very convenient means of arriving at estimates to answers as we shall see in the following examples.

Estimating Answers in Addition

Suppose we wanted to know the approximate total of the following list of numbers: 367, 125, 820, and 578.

Since all of these numbers are 3-digit numbers, we round off to the nearest hundreds. Then

$$400 + 100 + 800 + 600 = 1900 \text{ estimated total.}$$

And

$$367 + 125 + 820 + 578 = 1890 \text{ exact total.}$$

If the numbers to be added differ considerably in value such as 23, 1096, 603, 89, and 16 it is best to group them and round to the most common multiple of 10, i.e., 100, 1000, etc.

example:

exact	estimate		exact	estimate
22			22	
368	400		368	400
109	100		109	
8			8	
19			19	100
625	700	OR	625	
35			35	700
82			82	
192	300		192	300
1460	1500		1460	1500

Estimating Answers in Subtraction

Mr. Jones wanted to remodel his kitchen. One company offered to do the job for $876 while another company wanted $1052 or a difference of approximately $200 i.e. ($1100 − $900). The actual difference is $176.

exact		estimate
$1052	minuend	$1100
−876	subtrahend	−900
$ 176	difference	$ 200

In estimating differences, always consider the smaller number (subtrahend) and round to the nearest multiply of 10. Then round the larger number (minuend) to the same degree.

examples:

exact	estimate	exact	estimate
9836	9800	437	440
−126	−100	−26	−30
9710	9700	411	410

To arrive at a closer estimate when both numbers are of the same magnitude, round to the second digit from the *left* rather than the first. For example, in the problem below, round to the nearest 100 rather than 1000.

example:

exact	estimate
2683	2700
−1421	−1400
1262	1300

When adding or subtracting decimals, drop the decimal part and treat as whole numbers.

examples:

Addition		Subtraction	
exact	estimate	exact	estimate
23.15	20	521.60	520
8.92	10	−32.12	−30
16.45	20	489.48	490
236.09	240		
284.61	290		

Estimating Answers in Multiplication

As in addition and subtraction it is easy, with a little practice, to estimate answers in multiplication. An estimate tells us if an answer is approximately correct or, in some cases, if it is incorrect. The closer the estimate is to the true answer, the better it is in checking errors.

example: $832 \times 32 = 26{,}624$ *exact*

solution: Round each factor to the *same number of digits*. In this problem, round the number 32 first and then round 832 to the same place (tens in this case). Then the problem becomes $830 \times 30 = 24{,}900$, *estimate*.

example: $836 \times 403 = 336{,}908$ *exact*

solution: If both are rounded to hundreds, then $800 \times 400 = 320{,}000$, *estimate*. A closer estimate could be made by rounding 836 to 840 and 403 to 400, then $840 \times 400 = 336{,}000$.

When multiplying decimals, round to the nearest whole number and proceed as explained previously

example: $32.56 \times 5.80 = 188.848$ *exact*

solution: Treat as 33×6, then $30 \times 6 = 180$, *estimate*.

Estimating Answers in Division

It is more difficult to get a good approximation to the exact answer in division than in multiplication. Estimates are generally more accurate if both the divisor and dividend are rounded up (example A) or both are rounded down (example B).

Round the dividend to *one more digit* than the divisor. If the divisor is rounded to tens, the dividend is rounded to hundreds; if the divisor is rounded to hundreds, the dividend is rounded to thousands.

example A: $38\overline{)4978}$ = 131   **example B:** $230\overline{)3220}$ = 14

estimate: $40\overline{)5000}$ = 125   *estimate:* $200\overline{)3000}$ = 15

EXERCISE 6

1. Sales for the various departments of the Armco Company during the past week were \$346.50, \$227.00, \$1096.00, \$89.00, and \$589.70. What were the actual and estimated sales for the week?

2. The Howard Company produced the following units during the week: 2200, 1650, 2549, 1680, and 1980. What is the estimated number of units produced?

3. If Mr. James deposited \$3146.75 in his checking account this month and spent \$2036.50 during the same period, how much was his approximate balance?

4. Sales for the Hampton Company were \$49,500.67 last year as compared to \$57,804.16 this year. What is the approximate increase in sales?

5. Mrs. Harmon ordered carpet for her living room. If it cost \$9.75 a square yard and her living room contained 136 sq yd, what was the approximate cost?

6. What is the approximate cost of the following purchase?

   $11\frac{7}{8}$ yd of cotton at \$4.35 a yard
   16 yd of wool at \$10.50 a yard
   5 yd of lining at \$1.15 a yard

7. The employees for the Lane Manufacturing Company collected \$6102 for a company benefit. If there were 678 employees, what was the approximate average amount collected from each employee?

8. What is the approximate length of a room that occupies 504 sq ft and is 18 ft wide?

COMMON FRACTIONS

A fraction is a part of a whole unit or quantity. It may be expressed as a number (called the numerator) written above another number (called the denominator) such as $\frac{1}{2}$ and $\frac{3}{4}$ or in decimal form such as .50 and .75.

A *proper fraction* has a value less than 1 because, by

definition, it has a numerator that is less than the denominator (also called the divisor). Some examples are $\frac{3}{8}$, $\frac{5}{6}$, and $\frac{11}{12}$.

An *improper fraction* has a value equal to or greater than 1 because, by definition, it has a numerator that is equal to or greater than the denominator. Some examples are $\frac{5}{4}$, $\frac{7}{6}$, $\frac{4}{3}$, and $\frac{8}{8}$.

A *mixed number* is a number containing a whole number and a fraction. Some examples are $3\frac{1}{4}$, $5\frac{2}{7}$, and $8\frac{5}{8}$.

A *simple fraction* is one in which both the numerator and denominator are whole numbers such as $\frac{9}{13}$, $\frac{5}{8}$, $\frac{4}{9}$.

A *complex fraction* is one in which the numerator or denominator or both is a mixed number. Some examples are:

$$\frac{5\frac{1}{4}}{8\frac{1}{6}} \qquad \frac{4}{2\frac{1}{3}} \qquad \frac{10\frac{1}{2}}{5}$$

Rule of fractions: When the numerator and denominator of any fraction are either multiplied or divided by the same number, it does not change its value.

REDUCING FRACTIONS

To reduce a fraction to its lowest terms, divide the numerator and denominator by the largest whole number common to both. If the largest common divisor or factor is not recognized, the same result will be obtained by dividing both the numerator and denominator of the fraction by any common divisor, repeating the process until the fraction is reduced to its lowest terms. In the example given below, 3 is a common divisor of both 12 and 18, in which case $12 \div 18 = 4 \div 6$. Since 4 and 6 may both be divided by 2, then $4 \div 6 = 2 \div 3$. In this case an additional step in the process is required but the final result is, of course, always the same.

example: Reduce $\frac{12}{18}$ to its lowest terms. Since 6 is the largest divisor of 12 and 18, divide each by 6. Then $\frac{12}{18} = \frac{2}{3}$.

CHANGING AN IMPROPER FRACTION TO A WHOLE OR MIXED NUMBER

Divide the numerator by the denominator and write the remainder, if any, as a fraction.

examples: $\frac{12}{7} = 1\frac{5}{7}$ $\frac{23}{8} = 2\frac{7}{8}$ $\frac{15}{5} = 3$

CHANGING A MIXED NUMBER TO AN IMPROPER FRACTION

Multiply the whole number by the denominator of the fraction; add the numerator of the fraction and place over the denominator.

examples: $4\frac{3}{4} = \frac{19}{4}$ $5\frac{1}{6} = \frac{31}{6}$

CHANGING A FRACTION TO HIGHER TERMS

Follow the procedure as illustrated.

examples: $\frac{1}{4} = \frac{?}{20}$ $\frac{3}{8} = \frac{?}{24}$

procedure: $20 \div 4 = 5$ $24 \div 8 = 3$

$\frac{1 \times 5}{4 \times 5} = \frac{5}{20}$ (answer) $\frac{3 \times 3}{8 \times 3} = \frac{9}{24}$ (answer)

CHANGING A COMPLEX FRACTION TO A SIMPLE FRACTION

See page 28 under multiplication and division of fractions.

ADDING AND SUBTRACTING FRACTIONS

When the Denominators Are the Same

example:

$4\frac{1}{4}$
$5\frac{2}{4}$
$12\frac{3}{4}$

$22\frac{2}{4}$ or $22\frac{1}{2}$ (answer)

procedure: Addition—add the numerators of the fractions. Result is $\frac{6}{4}$ or $1\frac{2}{4}$. Carry the "1" to the units column and add. Reduce the fraction to its lowest terms.

example:

$$\begin{array}{r} 12\frac{4}{9} \\ -5\frac{1}{9} \\ \hline 7\frac{3}{9} \end{array} \text{ or } 7\frac{1}{3} \quad \text{(answer)}$$

procedure: Subtraction—Subtract $\frac{1}{9}$ from $\frac{4}{9}$, and reduce. Subtract 5 from 12.

example:

$$\begin{array}{r} 10\,{}_{17} \\ \not{1}\not{1}\frac{\not{5}}{12} \\ -3\frac{7}{12} \\ \hline 7\frac{10}{12} \end{array} \text{ or } 7\frac{5}{6} \quad \text{(answer)}$$

procedure: Subtraction—Since 7 is greater than 5, borrow 1 whole number from the minuend, change to 12ths, and add to $\frac{5}{12}$ths. Proceed as illustrated.

When the Denominators Are Not the Same

Fractions with unlike denominators cannot be added or subtracted until they are reduced or changed to a common denominator.

example:

$$\begin{array}{r|l} 5\frac{1}{6} & \frac{4}{24} \\ 20\frac{3}{8} & \frac{9}{24} \\ 15\frac{3}{4} & \frac{18}{24} \\ \hline 41\frac{7}{24} & \frac{31}{24} = 1\frac{7}{24} \end{array}$$

procedure: Addition—First find the smallest number that can be divided by 6, 8, and 4. This is called the least common denominator (LCD). In this case, it is 24. Then change each fraction to 24ths and add as illustrated.

example:

$$\begin{array}{r|l} 90\frac{5}{6} & \frac{15}{18} \\ 27\frac{1}{9} & \frac{2}{18} \\ \hline 63\frac{13}{18} & \frac{13}{18} \end{array}$$

procedure: Subtraction—The LCD of 6 and 9 is 18. Therefore, change $\frac{5}{6}$ and $\frac{1}{9}$ to 18ths and proceed as illustrated.

FINDING THE LEAST COMMON DENOMINATOR FOR A SERIES OF NUMBERS

When the denominators are *prime*† to each other, the least common denominator is their product.

example: Find the LCD for the following fractions: $\frac{5}{6}$, $\frac{1}{7}$, and $\frac{4}{5}$.

solution: LCD $= 6 \times 7 \times 5 = 210$ (answer)

When one denominator can be divided exactly by the other, the LCD can be found by inspection.

example: $\frac{2}{3}$ and $\frac{5}{6}$—LCD $= 6$ since it can be divided exactly by 3 and 6.

$\frac{1}{3}$, $\frac{2}{9}$, and $\frac{1}{18}$—LCD $= 18$ since it can be divided exactly by 3, 9, and 18.

When the denominators are neither all prime to each other or factors of each other, follow the procedure outlined below.

example: $\frac{4}{5}$ $\quad \frac{7}{12}$ $\quad \frac{2}{9}$ $\quad \frac{5}{8}$

procedure: 1. Arrange denominators in a row as illustrated.

$$\begin{array}{r|rrrr} 2 & 5 & 12 & 9 & 8 \\ \hline 2 & 5 & 6 & 9 & 4 \\ \hline 3 & 5 & 3 & 9 & 2 \\ \hline & 5 & 1 & 3 & 2 \end{array}$$

2. Divide by the smallest number that can be divided into 2 or more denominators and

†Numbers are said to be *prime* to each other if they have no common denominator except 1.

bring down to the next row with any denominators that were not divided.

3. Continue this process until there are no divisors.

4. Multiply all divisors and the numbers in the last row. Thus, in example illustrated, $2 \times 2 \times 3 \times 5 \times 1 \times 3 \times 2 = 360$ (answer).

MULTIPLYING AND DIVIDING FRACTIONS

Multiplication

examples:

1. $\frac{9}{10} \times \frac{3}{7} = \frac{27}{70}$

2. $\frac{4}{5} \times \frac{7}{8} = \frac{28}{40} = \frac{7}{10}$ *or* $\frac{\overset{1}{\cancel{4}}}{5} \times \frac{7}{\underset{2}{\cancel{8}}} = \frac{7}{10}$

3. $6 \times \frac{5}{9} = \frac{30}{9} = 3\frac{1}{3}$ *or* $\overset{2}{\cancel{6}} \times \frac{5}{\underset{3}{\cancel{9}}} = \frac{10}{3} = 3\frac{1}{3}$

4. $2\frac{1}{3} \times 5\frac{2}{5} = \frac{7}{\underset{1}{\cancel{3}}} \times \frac{\overset{9}{\cancel{27}}}{5} = \frac{63}{5} = 12\frac{3}{5}$

procedure:

1. Multiply the numerators by each other.
2. Multiply the denominators by each other.
3. Reduce result to its lowest terms.

Or, before multiplying the numerators and denominators, divide by common factors if there are any (see Nos. 2, 3, and 4) and proceed as illustrated.

Division

examples:

1. $\frac{1}{5} \div \frac{1}{3} = \frac{1}{5} \times \frac{3}{1} = \frac{3}{5}$

2. $\frac{9}{10} \div \frac{1}{3} = \frac{9}{10} \times \frac{3}{1} = \frac{27}{10} = 2\frac{7}{10}$

3. $\frac{11}{12} \div \frac{7}{4} = \frac{11}{\underset{3}{\cancel{12}}} \times \frac{\overset{1}{\cancel{4}}}{7} = \frac{11}{21}$

4. $16 \div \frac{86}{9} = \overset{8}{\cancel{16}} \times \frac{9}{\underset{43}{\cancel{86}}} = \frac{72}{43} = 1\frac{29}{43}$

5. $21\frac{1}{2} \div 4 = \frac{43}{2} \times \frac{1}{4} = \frac{43}{8} = 5\frac{3}{8}$

procedure: Invert the divisor and proceed as in multiplication.

Complex Fractions

A complex fraction is one in which either the numerator or denominator is a fraction or mixed number or both are. Therefore, any complex fraction may be written as follows and solved as illustrated above.

examples:

$$\frac{5\frac{1}{2}}{6} = 5\frac{1}{2} \div 6$$

$$\frac{2\frac{1}{3}}{8\frac{1}{5}} = 2\frac{1}{3} \div 8\frac{1}{5}$$

$$\frac{10}{3\frac{5}{8}} = 10 \div 3\frac{5}{8}$$

$$\frac{\frac{1}{4}}{6} = \frac{1}{4} \div 6$$

In some cases, the use of a common multiple or denominator may be used to advantage, as shown in the following examples.

examples: Use of a common multiple:

$$\frac{4\frac{1}{2}}{10} = \frac{4\frac{1}{2} \times 2}{10 \times 2} = \frac{9}{20}$$

$$\frac{5\frac{1}{3}}{18} = \frac{5\frac{1}{3} \times 3}{18 \times 3} = \frac{16}{54} = \frac{8}{27}$$

$$\frac{4\frac{1}{2}}{6\frac{1}{4}} = \frac{4\frac{1}{2} \times 4}{6\frac{1}{4} \times 4} = \frac{18}{25}$$

Use of common denominator:

$$\frac{14}{3\frac{1}{2}} = \frac{28}{2} \div \frac{7}{2} = 4$$

(In this case, think of 14 as $\frac{28}{2}$.)

EXERCISE 7

1. $7\frac{1}{6} + 12\frac{1}{3} + 8\frac{5}{6}$

2. $36\frac{1}{3} + 37\frac{2}{3}$

3. $5 + 2\frac{1}{4} + 3\frac{1}{8} + 14$

4. $1\frac{5}{8} + 3\frac{1}{2} + 6\frac{3}{8}$

5. $24\frac{1}{2} + 9\frac{3}{4} + 11\frac{1}{6} + \frac{1}{3}$

6. $14\frac{2}{3} - 4\frac{1}{6}$

7. $42\frac{5}{8} - 13$

8. $7\frac{2}{3} - 2\frac{1}{6}$

9. $14\frac{3}{4} - 10\frac{1}{6}$

10. $12 - 5\frac{1}{6}$

11. $10\frac{1}{2} \times 15\frac{2}{5}$

12. $66 \times 2\frac{1}{3}$

13. $14\frac{1}{2} \times 4\frac{1}{2}$

14. $4\frac{1}{5} \times 60$

15. $1\frac{2}{3} \times 3\frac{1}{4}$

16. $120 \div 6\frac{1}{3}$

17. $28 \div \frac{1}{7}$

18. $12\frac{3}{4} \div 3$

19. $7\frac{1}{2} \div 3\frac{3}{4}$

20. $30\frac{1}{2} \div 32$

DECIMALS OR DECIMAL FRACTIONS

A *decimal*† is another way of expressing a common fraction. Decimal fractions and mixed decimals (decimals including a whole number) are usually referred to as decimals.

†See page 2.

CHANGING A COMMON FRACTION OR MIXED NUMBER TO THE DECIMAL FORM

Any common fraction or mixed number can be changed to the decimal form by dividing the numerator by the denominator to as many decimal places as desired.

examples: 1. $\frac{3}{8} = .375$
2. $\frac{12}{5} = 2\frac{2}{5} = 2.4$
3. $5\frac{1}{6} = 5.166\frac{2}{3}$ *or* 5.167 rounded or corrected to the third decimal

CHANGING A DECIMAL TO A COMMON FRACTION OR MIXED NUMBER

To change a decimal to a common fraction or mixed number, replace the decimal with the appropriate denominator (power of 10) and reduce.

examples: 1. $.25 = \frac{25}{100} = \frac{1}{4}$

2. $.375 = \frac{375}{1000} = \frac{3}{8}$

3. $.16\frac{2}{3} = \frac{16\frac{2}{3}}{100} = \frac{50}{3} \times \frac{1}{100} = \frac{1}{6}$

4. $2.08\frac{1}{3} = 2\frac{8\frac{1}{3}}{100} = 2\frac{1}{12}$

ADDING, SUBTRACTING, MULTIPLYING, AND DIVIDING DECIMALS

See pages 6, 11, 12, 14, and 15, respectively.

DECIMAL EQUIVALENTS OF COMMON FRACTIONS, USING ALIQUOT PARTS

An *aliquot* part is any number that can be divided evenly into another number. In business, we are primarily concerned with those decimal equivalents of common fractions which are aliquot parts of 1 or $1.00. They are used so frequently that they should be memorized.

When decimal equivalents that are not even ($\frac{1}{3}$, $\frac{1}{6}$, etc.) are used, they should be carried to 4 or, in some cases, 5 decimal places to assure the proper degree of accuracy, unless the fractional part is used—$33\frac{1}{3}$ instead of .33333.

Decimal equivalents of common fractions are shown below.

3rds	
$\frac{1}{3}$	$.33\frac{1}{3}$
$\frac{2}{3}$	$.66\frac{2}{3}$

4ths	
$\frac{1}{4}$	.25
$\frac{2}{4}$	.50
$\frac{3}{4}$	.75

5ths	
$\frac{1}{5}$	.20
$\frac{2}{5}$	.40
$\frac{3}{5}$	.60
$\frac{4}{5}$	.80

6ths	
$\frac{1}{6}$	$.16\frac{2}{3}$
$\frac{2}{6}$	$.33\frac{1}{3}$
$\frac{3}{6}$	.50
$\frac{4}{6}$	$.66\frac{2}{3}$
$\frac{5}{6}$	$.83\frac{1}{3}$

8ths	
$\frac{1}{8}$	.125
$\frac{2}{8}$	.25
$\frac{3}{8}$	.375
$\frac{4}{8}$	.50
$\frac{5}{8}$	.625
$\frac{6}{8}$	.75
$\frac{7}{8}$	.875

12ths	
$\frac{1}{12}$	$.08\frac{1}{3}$
$\frac{2}{12}$	$.16\frac{2}{3}$
$\frac{3}{12}$	.25
$\frac{4}{12}$	$.33\frac{1}{3}$
$\frac{5}{12}$	$.41\frac{2}{3}$
$\frac{6}{12}$	.50
$\frac{7}{12}$	$.58\frac{1}{3}$
$\frac{8}{12}$	$.66\frac{2}{3}$
$\frac{9}{12}$	.75
$\frac{10}{12}$	$.83\frac{1}{3}$
$\frac{11}{12}$	$.91\frac{2}{3}$

Tables for other decimal equivalents such as 16ths, 32nds, and 64ths are available or can be compiled. The 9ths are easy to remember since $\frac{1}{9} = .11\frac{1}{9}$, $\frac{2}{9} = .22\frac{2}{9}$, $\frac{3}{9} = .33\frac{3}{9}$ or $.33\frac{1}{3}$, etc.

If the decimal equivalents are known for the fractions being used, a great deal of time can be saved in many applications, particularly in invoice work. When values such as $.33\frac{1}{3}$ or $.16\frac{2}{3}$ are used, it is easier to divide by 3 or 6 than to multiply by the full decimal form. If $.66\frac{2}{3}$ is used, divide by 3 and multiply by 2 since this is equal to $\frac{2}{3}$.

example: 39 yds @ $$0.33\frac{1}{3}$ a yd

solution: $39 \times \frac{1}{3} = \$13.00$ since $\$0.33\frac{1}{3} = \frac{1}{3}$ of $1.00

example: 95 articles @ $12\frac{1}{2}$¢ each

solution: $95 \times \frac{1}{8} = 11.875$ *or* $11.88

example: 108 pieces @ $41\frac{2}{3}$¢ each

solution: $108 \times \frac{5}{12} = \$45.00$

example: 64 yds @ $\$0.62\frac{1}{2}$ a yd

solution: $64 \times \frac{5}{8} = \$40.00$

EXERCISE 8

Perform the following exercises orally.

I. Change to the decimal form. Record to 2 decimals and carry fraction, i.e., $\frac{2}{9} = .22\frac{2}{9}$.

1. $2\frac{1}{9}$
2. $\frac{11}{9}$
3. $\frac{1}{12}$
4. $2\frac{2}{6}$
5. $\frac{3}{16}$
6. $2\frac{5}{12}$
7. $15\frac{1}{6}$
8. $9\frac{1}{5}$
9. $6\frac{3}{8}$
10. $24\frac{3}{5}$

II. Change to the fractional form, i.e., $3.91\frac{2}{3} = 3\frac{11}{12}$.

1. $.16\frac{2}{3}$
2. $4.08\frac{1}{3}$
3. 1.875
4. .0625
5. $12.41\frac{2}{3}$
6. $.44\frac{4}{9}$
7. 3.625
8. $19.91\frac{2}{3}$
9. 4.375
10. $21.5833\frac{1}{3}$

III. Find the cost of the following purchases. All prices are per unit.

1. 36 lb @ $.08\frac{1}{3}$
2. 48 lb @ $62\frac{1}{2}$¢
3. 120 lb @ 20¢
4. 75 lb @ $33\frac{1}{3}$¢
5. 72 lb @ $37\frac{1}{2}$¢
6. 132 articles @ $8\frac{1}{3}$¢
7. 24 articles @ $91\frac{2}{3}$¢
8. 48 articles @ $16\frac{2}{3}$¢
9. 80 articles @ $.37\frac{1}{2}$¢
10. 54 articles @ $22\frac{2}{9}$¢
11. 60 qt @ $41\frac{2}{3}$¢
12. 96 qt @ $87\frac{1}{2}$¢
13. 54 qt @ $11\frac{1}{9}$¢
14. 90 qt @ 60¢
15. 24 qt @ $83\frac{1}{3}$¢
16. 120 pieces @ 75¢
17. 35 pieces @ 80¢
18. 144 pieces @ $91\frac{2}{3}$¢
19. 160 pieces @ 75¢
20. 48 pieces @ $87\frac{1}{2}$¢

EXERCISE 9 Chapter Summary

1. The following amounts represent the daily dollar sales for a department of the Hansman Specialty Shop. The office manager would like these sales grouped as follows: less than $1.00, $1.01–$5.00, $5.01–$10.00, $10.01–$20.00, and $20.01 up. Find (a) the totals for each group and (b) the average sale for all groups combined. (c) Explain to the office manager your method of proving your work. The sales are: $0.22, 1.61, 3.00, 0.45, 0.98, 8.50, 5.00, 10.20, 5.00, 10.20, 3.60, 18.00, 7.50, 21.08, 3.05, 6.75, 11.20, 8.72, 3.25, 14.90, 0.11, 3.05, 13.75, 8.12, 6.25, 11,01, 12.62, 19.00, 6.21, 17.00, 27.50, 2.16, 25.26, 16.27, 4.15, 2.19, 1.27, 0.86, 0.42, 3.13, 3.14, 0.96, 1.22, 4.10, 11.09, and 15.50.

2. (A) Complete the following partial report for the Bradford Equipment Rentals. Find (1) the total amount received in rentals for each tool or piece of equipment and (2) the amount received in rentals for each day of the week. (B) (1) If each digging blade rented for $1.75, how many were rented during the week? (2) How many pitchforks were rented during the week if the rental charges were 75¢ each? (3) What is the difference between rentals for the busiest day and the poorest day of the week?

BRADFORD EQUIPMENT RENTALS

Rental Receipts Week Ending May ____, 19____

Equipment	Mon.	Tues.	Wed.	Thurs.	Fri.	Sat.	Sun.	Total
Air compressor	$ 6.50	$	$	$21.00	$	$	$ 7.00	$
Slag hammer		2.50	7.50					
35 # Jack hammer						28.00	7.00	
Blow gun		1.00	3.00					
Digging blade	1.75			15.75		5.25		
Lawn seeder		1.00			1.00			
Power weed cutter		45.00			15.00		13.75	
Pitchfork	2.25		0.75	3.00		0.75		
Garden rake		3.00			1.50			
Total	$	$	$	$	$	$	$	$

3. The United Charities' goal was an average contribution of $7.50 each from an adult population of 30,000. The amount collected as of a certain date was $196,500. What additional average contribution was needed to meet the goal?

4. Everett and Holt made the following purchases from the Evens Haines Co. during the month: $316.20, $682.50 and $190.50. Payments made during this same period amounted to $296.00,

$500.00, and $150.00. There was a balance due at the beginning of the month amounting to $125.00. How much did Everett and Holt owe at the end of the month?

5. Mr. Eaton, Mr. Hayes, and Mr. Huff owned a business jointly and shared profits and losses in proportion to their investments which were $45,000, $15,000, and $30,000 respectively. What fractional part did each man own in the business? If profits amounted to $4506.20 for the current year, how much did each man receive as his share?

6. A man worked the following number of hours each day during a 5-day week—$6\frac{1}{2}$, $7\frac{3}{4}$, $8\frac{1}{2}$, 8, and $7\frac{2}{3}$, respectively. How many hours did he work and how much did he earn if he was paid at the rate of $8.40 an hour?

7. Mr. Whitman had a $\frac{2}{3}$ interest in a manufacturing concern. He sold $\frac{2}{5}$ of his interest for $12,000. At the same rate, what was (a) the total amount of his interest and (b) the total worth of the manufacturing concern?

8. Mr. Jones' salary is $9600. This is $\frac{1}{5}$ more than he received the previous year which, in turn, was $\frac{1}{8}$ less than his salary was 2 years ago. What was his salary (a) last year and (b) 2 years ago?

9. The following report represents rental receipts for the Bradford Equipment Rentals during the first 6 months of the current year and the same period during the preceding year. (a) Find the total receipts for each period. (b) Find the increase or decrease of this year as compared to the last year. (c) Find the average receipts per month for each period.

BRADFORD EQUIPMENT RENTALS

Month	This Year	Last Year	Increase	Decrease
January	$ 965.00	$ 926.50	$	$
February	1206.00	988.75		
March	959.50	1006.50		
April	1768.25	2100.50		
May	1856.00	1962.00		
June	2043.55	1598.75		
Total	$	$	$	$

10. Mrs. Jones bought a sample rug, 12 × 18 ft, for $249.50. (a) She wanted a hall runner cut from this rug to measure 36 in. × 18 ft. The binding cost was 25¢ a linear foot. A pad for the runner cost $1.20 per square yard. (a) How much did the binding cost? How much was the padding? (b) The remainder of the original rug measured 9 × 18 ft. This was cut into two rugs 9 × 9 ft. Each was bound at the cost of 30¢ a linear foot. What was the cost of binding both rugs? (c) What was the total cost of this purchase and service? What fractional part of the original rug was the runner and each 9 × 9 ft. rug?

11. A profit of $3120 is to be divided among A, B, and C in the ratios of $\frac{1}{4}$, $\frac{2}{3}$, and $\frac{1}{12}$. How much should each receive?

12. Mr. Bradshaw bought 250 used tires for $175 at an auction. He sold $\frac{1}{5}$ of them for $60.00, 16 at $8.50 each, 51 at $10.00 each, and junked the rest. How many tires did he throw away? What was his total profit if the cost of handling this transaction amounted to $65.00?

13. Complete the following invoice. Record all extensions correct to the nearest cent.

SHIPLEY MERCANTILE COMPANY

Sold to: J. E. Norman
Chester, New York

Invoice No. 10–563

Date: June 6, 19____

Quantity	Description	Unit Price	Amount
23 yds	Linen	$ 5.15	$
16⅔ yds	Cotton padding	.36	
100 yds	Plastic sheeting	.02½	
8⅓ yds	Wool (Baines)	9.00	
25 yds	Wool (W & F)	7.98	
89	Hooks #613	.16 cwt	
12½ yds	Gingham	.96	
		Total	$

14. A house and furnishings were valued at $52,000. The house was worth $5\frac{1}{2}$ times as much as the furnishings. Find the value of (a) the house and (b) the furnishings.

15. Find the net weight of the following carriers. (Gross weight equals the weight of the carrier plus the contents; tare weight equals the weight of the carrier; net weight equals the weight of the contents—gross weight less tare weight.) Verify the total net weight for the 5 carriers listed by the principle of crossfooting.

	Gross	Tare	Net
	13,620	462	
	27,317	462	
	32,812	462	
	35,614	462	
	29,420	462	
Total			

16. Visitors to the county museum during the first week after it had opened were as follows: Sunday, 869; Monday, 260, Tuesday, 317; Wednesday, 725; Thursday, 294; Friday, 426; and Saturday, 961. What was the average daily attendance (nearest whole number)?

17. Mr. Brown owned $\frac{3}{7}$ of a store. He sold $\frac{1}{3}$ of his share for $26,000. At the same rate, what was the value of the store?

18. A man paid $45,000 for a house and lot. If the house was worth $3\frac{1}{2}$ times as much as the lot, find (a) the value of the house and (b) the value of the lot.

19. Make the necessary extensions for the following partial inventory sheet.

HAYNES RUG COMPANY
Haynesville, Texas

Linoleum — Page Two

Stock No.	Quantity	Description	Size	Cost	Value
16	22	Vinylcrest	6 × 9 ft	$1.35 sq ft	$
1567	15	Embossed inlaid	6 × 12 ft	.29½ sq ft	
21F	3	Inlaid	8 × 10 ft	1.16 ⅓ sq ft	
32B	27	Vinyl coated	7 × 9 ft	.21 sq ft	
16D	565	Rubber tiles	9 × 9 in	.16 each	
16G	98	Rubber tiles	9 × 9 in	.18½ each	
				Total	$

20. A man earned $20,570 during the past 2 years. Find his income for each year if he earned $\frac{1}{5}$ more during the second year than in the first year.

21. The Hampton School District requested bids on the construction of a new library. One company bid $87,500 while another bid $112,000. What is the approximate difference between the two bids?

22. Sales for the People's Market dropped from $31,276.00 last week to $26,800.50 this week. What is the approximate decrease?

23. During the last four days receipts at the local theatre were: $516.25, $902.50, $628.20, and $780.50. How much, approximately, was the total of all receipts for the week?

24. Mrs. Haynes ordered carpet for her living room (56 sq yd) and linoleum for her kitchen ($27\frac{1}{2}$ sq yd). The store that was to install the floor covering charged $12.85 a square yard for the carpet and $8.75 a square yard for the linoleum. What would be a close estimate of this order?

25. A certain real estate company had sales of $36,825 for the month. If there were 18 salesmen, find the average approximate sales for each.

26. What is the approximate total of the following numbers: 36, 810, 52, 49, 310, 406?

Objectives

Since the understanding of percentage is vital to the solution of many problems in business, it is essential that all elements of it be thoroughly understood and properly applied. The aim of this chapter is to make this possible. It is devoted to developing skill in converting decimals, fractions, or whole numbers to percent, or conversely; to giving you the ability to interpret or convert information from one form to another; and finally to showing the solution of percentage problems with the proper use of three basic elements—base, rate, and percentage (amount).

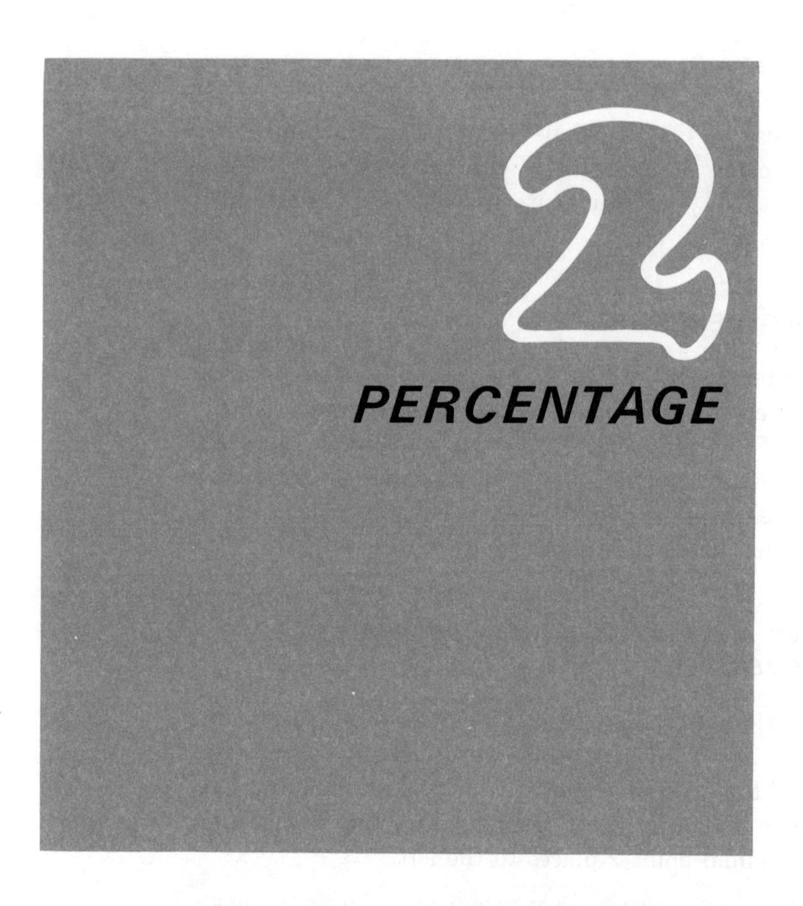

A common means of expressing relationships in business is essential. This becomes possible through the use of percentage whereby the elements of a problem can be expressed in terms of a common unit of measure. Percentage is used to express relationships of many kinds such as similarities or differences between groups, classes, fiscal periods, and increases or decreases. Percentage also is used to express rates in interest, discounts, taxes, fluctuations in prices, changes in the stock market, and many other measures too numerous to mention.

Many applications cannot be appreciated unless you are familiar with the terminology or language of the business or subject area and the forms used, many of which will be dealt with in subsequent chapters. Here we are primarily concerned with the purpose, derivation, and elements of percentage as well as some basic applications.

Any quantity may be divided into 100 equal parts, each part representing a common unit of measure which is called a percent. Thus, 2 percent represents 2 parts of 100 or

2 hundredths ($\frac{2}{100}$) of a whole thing. The fractional expression $\frac{1}{100}$ came to be replaced with the symbol %. Thus, one hundredth ($\frac{1}{100}$) of a quantity may be expressed as 1%, $\frac{2}{100}$ as 2%, $\frac{3}{100}$ as 3%, etc.

CHANGING PERCENTS

Since hundredths may also be expressed in the decimal form, any percent may be expressed in three ways: $1\% = \frac{1}{100} = .01$, $3\% = \frac{3}{100} = .03$, $20\% = \frac{20}{100} = .20$, etc. Since all percents must be converted to the decimal or fractional form in the computational work of any problem, it is important that these changes be made with ease.

CHANGING A PERCENT (%) TO A DECIMAL AND CONVERSELY

By definition, $1\% = \frac{1}{100}$ or .01 (expressed as a decimal). Conversely, $.01 = 1\%$. In other words, the % sign takes the place of the 2 decimal places which denote hundredths.

To change a percent to a decimal, drop the % sign and move the decimal point 2 places to the left.

examples: $4\% = .04$ $\quad 125\% = 1.25$ $\quad 2.5\% = .025$ $\quad \frac{1}{2}\% = .005$

To change a decimal to a percent, move the decimal point 2 places to the right and use the % sign.

examples: $.16\frac{1}{2} = 16\frac{1}{2}\%$ $\quad .035 = 3.5\%$ *or* $3\frac{1}{2}\%$ $\quad 1.25 = 125\%$ $\quad 5 = 500\%$ $\quad .0025 = .25\%$ *or* $\frac{1}{4}\%$

CHANGING A PERCENT (%) TO A FRACTION AND CONVERSELY

Since percent denotes hundredths, all percents may be thought of automatically as fractions with a denominator of 100.

To change a percent to a fraction, drop the % sign, divide the percent quantity by 100, and reduce to lowest terms.

examples: $60\% = \frac{60}{100} = \frac{3}{5}$ $\quad 4\% = \frac{4}{100} = \frac{1}{25}$

$12\% = \frac{12}{100} = \frac{3}{25}$ $\quad 17\frac{1}{2}\% = \frac{17\frac{1}{2}}{100}$ *or* $\frac{175}{1000} = \frac{7}{40}$

To change a fraction to a percent, divide the numerator by the denominator and then change the decimal form to a percent.

example: $\frac{3}{8} = .375 = 37\frac{1}{2}\%$

ALIQUOT PARTS†

Many percent values and their fractional equivalents that are used frequently should be memorized. It is often more convenient to use the fractional form than the decimal form in the solution of a problem.

example: $8\frac{1}{3}\%$ of $144 = .08333 \times 144 = 11.9995$ *or* 12

procedure: This problem is a simple one if we know that $8\frac{1}{3}\% = \frac{1}{12}$, in which case $144 \div 12 = 12$.

example: $12\frac{1}{2}\%$ of $96 = \frac{1}{8} \times 96 = 12$

procedure: This is less cumbersome than multiplying 96 by .125 (decimal equivalent of $12\frac{1}{2}\%$).

EXERCISE 10

Do this entire exercise orally. Change each of the following numbers to the percent form.

1. .36
2. $.33\frac{1}{3}$
3. 2.64
4. .05
5. .006
6. 4
7. 2.07
8. 1.25
9. $4\frac{1}{2}$
10. .26
11. 2.05
12. .08
13. 20
14. .0036
15. $2\frac{1}{4}$
16. .80
17. $.00\frac{1}{4}$
18. .9
19. $.02\frac{1}{2}$
20. $1\frac{1}{2}$

†See pages 30 and 31.

Change each of the following percents to the decimal form.

21. 3%
22. 400%
23. .6%
24. $\frac{1}{4}$%
25. 2.6%
26. 1.5%
27. 17%
28. 26%
29. 300%
30. $15\frac{1}{2}$%
31. .05%
32. $32\frac{1}{2}$%
33. 11%
34. $4\frac{1}{2}$%
35. 652%
36. $\frac{1}{2}$%
37. 13.4%
38. .67%
39. 1.7%
40. 115%

Change each of the following percents to a fraction or a mixed number.

41. 25%
42. $12\frac{1}{2}$%
43. $8\frac{1}{4}$%
44. $11\frac{1}{9}$%
45. $108\frac{1}{3}$%
46. 20%
47. 6%
48. $3\frac{1}{2}$%
49. 40%
50. 225%
51. 15%
52. $93\frac{1}{3}$%
53. 80%
54. 17%
55. .11%
56. .6%
57. $\frac{2}{5}$%
58. 48%
59. 46%
60. 475%

Change each of the following fractions or mixed numbers to the percent (%) form.

61. $\frac{3}{8}$
62. $\frac{5}{6}$
63. $\frac{1}{12}$
64. $\frac{3}{9}$
65. $\frac{2}{5}$
66. $\frac{5}{12}$
67. $\frac{1}{30}$
68. $\frac{1}{25}$
69. $\frac{1}{2}$
70. $\frac{3}{4}$
71. $1\frac{7}{8}$
72. $3\frac{3}{4}$
73. $\frac{1}{6}$
74. $\frac{2}{3}$
75. $\frac{2}{16}$
76. $\frac{4}{32}$
77. $1\frac{2}{5}$
78. $\frac{5}{9}$
79. $1\frac{1}{3}$
80. $20\frac{1}{2}$

BASIC ELEMENTS OF PERCENT PROBLEMS—THE PERCENTAGE FORMULA

There are only three basic factors or elements in any percentage problem. These are the *base* (100% or the whole amount of anything), the *rate* (percent), and the *percentage* (amount). The relationship of these factors expressed as a formula is

$$\text{Base} \times \text{Rate} = \text{Percentage} \qquad \textit{or} \qquad B \times R = P$$

All percentage problems can be solved by using this formula. The secret to successfully solving a problem is the ability to identify the factors that are known and the factor that is to be found.

FINDING THE PERCENTAGE (AMOUNT)

$$B \times R = P \qquad \text{or} \qquad P = B \times R$$

example: Find 12% of 125.

solution: $125 \times .12 = 15$ (answer)

What is wanted here is 12 parts (that is 12 hundredths) of 125 or $12/100 \times 125$. The decimal form is generally used although, as explained under aliquot parts, the fractional form may be easier to use arithmetically in some cases.

example: Mr. Smith earns $7900 a year. He saves 8% of this amount which is how much?

solution: $7900 × .08 = $632 (amount saved yearly)

example: A merchant wishes to reduce the price of a particular commodity that is listed at 75¢ by 15%. At what price should it be marked (correct to the nearest cent)?

solution: $.75 \times .15 = .1125$, $.75 - .1125 = .6375 = 64¢$

or

$.75 \times .85$ (net amount) $= .6375 = 64¢$

FINDING THE BASE OR 100%

$$B = P \div R \qquad \text{or} \qquad P \div R = B$$

example: 54 is 15% of what number?

solution: $54 \div .15 = 360$

This is the same as saying that 15% (15 hundredths) of some number equals 54.

What is that number? If 15% = 54, then

$$1\% = 54 \div .15 = 3.60$$
$$100\% = 3.60 \times 100 = 360$$

These steps can be completed at the same time, as illustrated above.

example: If the price of milk is now 22¢ a quart and represents an increase of 10% over the last month, what was the cost of milk at that time?

solution: 22¢ = 110% of last month's price.
∴ .22 ÷ 1.10 = .20 or 20¢.

proof: 10% of 20¢ = 2¢ (amount of increase)

example: If 65% or 468 of a school population were girls, how many boys were there?

solution:

$$468 \div .65 = 720 \text{ school population}$$
$$720 - 468 = 252 \text{ boys}$$

proof: If 65% were girls, then 100% − 65% or 35% = boys; 35% of 720 = 252.

FINDING THE RATE (PERCENT)

$$R = P \div B \qquad or \qquad P \div B = R$$

example: 30 is what percent of 180?

solution: $30 \div 180 = \frac{1}{6}$ *or* $.16\frac{2}{3} = 16\frac{2}{3}\%$ (answer)

In this case we wish to find what fractional part 30 is of 180, expressed as a percent. We find that it is $16\frac{2}{3}$ parts (hundredths) or $16\frac{2}{3}\%$.

example: In a class of 35 students, 28 received passing grades. What percent did this represent?

solution: $28 \div 35 = .80$ or 80%

example: What percent is 63 of 56?

solution: $63 \div 56 = 1.125 = 112.5\%$

Finding Percent (%) Increase or Decrease

Business is always interested in changes and in making comparative studies. The increase or decrease and the rate of increase or decrease in production, sales, and prices are common examples. They are important to management in determining future policies.

example: The sales for Benite Company amounted to \$38,462 last week and \$35,400 the previous week. What percent increase does this represent?

solution: Since it is always the *previous* year's, month's, week's (or any other period of time) figure that has either increased or decreased, it is the base.
Then

$$\text{rate } (\%) = \frac{\text{amount of increase or decrease}}{\text{last week}}.$$

$$\begin{aligned}\text{Increase in sales} &= \$36{,}462 - \$35{,}400 \\ &= \$1{,}062\end{aligned}$$

$$\begin{aligned}\text{Rate or percent increase} &= \$1{,}062 \div \$35{,}400 \\ &= .03 \text{ or } 3\%\end{aligned}$$

or:

\$36,462 ÷ \$35,400 = 1.03 or 103%. This means that last week's sales amounted to 103% of the previous week's sales or an increase of 3%.

example: Production in Department A of the Brown Manufacturing Company was 860 units last year and 731 this year. What rate of decrease does this represent?

solution: Percent decrease $= \dfrac{860 - 731}{860}$

$$= \frac{129}{860} = .15 \text{ or } 15\%.$$

or

$\dfrac{731}{860} = .85$ or 85 %. Percent decrease =

$$100\% - 85\% = 15\%.$$

EXERCISE 11

Find the *percentage*, given the rate and base.

1. $16\frac{1}{2}\%$ of $750
2. 23% of 96
3. $\frac{1}{4}\%$ of 9260
4. 215% of $240
5. 2.6% of 300
6. 200% of 82
7. .03% of 456
8. $3\frac{1}{2}\%$ of 120
9. $16\frac{2}{3}\%$ of 1820
10. 17% of $400

Find the *rate*, given the percentage and base. Record whole number with any remainder shown as a fraction.

1. 32 is what % of 96?
2. 90 is what % of 225?
3. What % of 64 is 1.92?
4. 300 is what % of 120?
5. 5 is what % of 150?
6. What % of 64 lb is 56 lb?
7. 2.45 is what % of 70?
8. 5 is what % of $1\frac{1}{4}$?
9. What % of 5000 is 2.5?
10. $7\frac{1}{2}$ is what % of 63?

Find the *base* (100%), given the rate and percentage. Record complete number with any remainder shown as a fraction.

1. 161 is 115 % of what number?
2. 40 is 25% of what number?
3. 35 is $33\frac{1}{3}\%$ of what number?
4. $16\frac{2}{3}\%$ of $ ______ = $264.
5. $4\frac{1}{2}\%$ of ? = $67\frac{1}{2}$.
6. If 272 = 34%, what is 100%?
7. 156 is $2\frac{1}{2}\%$ of what number?
8. $10.00 is $\frac{1}{2}\%$ of what amount?
9. $39 is 75% of what amount?
10. 125 is 210% of what number?

EXERCISE 12 Chapter Summary

1. It is estimated that the school population of Anderson County will increase by 25% during the next year. If the present enrollment is 3164, how many students will be enrolled next year?

2. Mr. Smith receives a 2% bonus on all sales he makes over $300 in any month. If his sales amounted to $850 during a certain month, what was the amount of his bonus?

3. Mrs. Jones' budget allows 12% of her income for entertainment and recreation. If she spent $624 for this purpose last year, what was her income?

4. A realtor purchased a house for $16,400. After spending $980 on improvements, he sold it for 120% of the purchase price. What was his profit?

5. If a town had a population of 9800 in 1970 and a population of 8250 in 1974, what was the percent of decrease (nearest tenth of 1%)?

6. S. J. Cain has 40% of his investments in common stock, 20% in savings and loan accounts, 15% in bonds, and the balance in real estate. The real estate investment amounts to $33,000. What is the total of his investments and how much is invested in stock, savings and loan accounts, and bonds, respectively?

7. Mrs. Haynes' salary was increased from $160 to $172 a week. What was the percent increase?

8. What number decreased by $37\frac{1}{2}$% of itself equals 946?

9. The tax on Mr. Man's home was $500. This represented an increase of $300 over the previous year. What was the rate of increase?

10. Teachers in one district received an average annual salary of $10,500 as compared to $9000 in another. How much more in terms of percent do the better paid teachers receive (nearest hundredth of 1%)?

11. After making contributions to various charities, an organization had $3562.50 left which represented 95% of its total funds. How much did the organization give to charities?

12. Mr. Ames spent 22% of his salary on rent which amounted to $198. How much did he earn each month?

13. Absenteeism for the Ajax Manufacturing Company amounted to 3% during May. This amount represented 90 persons. What was the total number of employees?

14. Sales for the Javel Rug Co. increased from $7950 to $11,200 within a 2-month period. What rate of increase did this represent (nearest tenth of 1%)?

15. Mr. Haynes bought a house for $25,500 and made a downpayment of 15%. What was the amount of the mortgage (balance due) on the house?

16. The sales for R & J Co. were $7000 more this year than last year—an increase of 16%. What did the sales total for last year? For this year?

17. The American Wholesale Co. occupied 31,500 sq ft of floor space of which the rug department used 1638 sq ft. What percent of the total floor space did this represent (nearest tenth of 1%)?

18. This report represents a list of all loans held by the Samuel Wright Loan Co. in two districts which they serve. Complete the report for the districts in the same manner as shown for the total. Notice that the percent figures for the deliquent accounts must equal the total delinquent accounts. In turn, the percent figures for delinquent accounts and current accounts must add to the total or 100%. (All percent figures should be corrected to nearest tenth of 1%).

ANALYSIS OF LOAN ACCOUNTS

Month Ending June 30, 19___

Status	Total		District A		District B	
	No.	%	No.	%	No.	%
Total loans	110	100.0	52	100.0	58	100.0
Current	86	78.2	47		40	
Delinquent:	24	21.8				
1 month	10	9.1	3		8	
2–3 months	7	6.4	—		6*	
4–6 months	4	3.6	2		2	
Over 6 months	3	2.7	—		2	

*Adjust here to balance.

19. How much is $82.50 increased by $12\frac{1}{2}\%$ of itself?

20. Mr. Smith, Mr. Jones, and Mr. Allen were business partners. Mr. Smith owned 25% of the business, Mr. Jones owned 60%, and Mr. Allen 15%. Profits were distributed according to the amount of the business each partner owned. How much did each receive if profits for the year were $8500.00?

21. The J. M. & E. Ames Company occupied floor space as follows: rug department 3600 sq ft, fabrics 2800 sq ft, linoleum 2400 sq ft, and delivery service 1600 sq ft. What percent of the total space did each department occupy (nearest whole percent)?

22. Ami Bros. commission merchants† sold produce amounting to $316.50 for the Hi-Valley Farms on which they received a commission of 15%. Freight and cartage charges were $35.20. A charge of 1% of the gross proceeds ($316.50) was made for insurance purposes. How much were the net proceeds (the amount remitted to the Hi-Valley Farms)‡

23. Don Shipman requested an agent to make the following purchase for him: 73 crates of eggs, 12 doz cartons to the crate at 37¢ a carton. The charges for making this purchase were: commission, 12%; cartage, $42.50; other expenses, $25.70. Find the total cost of this purchase.

24. H. Hurst, an apple grower, shipped 670 boxes of apples to the Central State Produce Company to be sold on a commission of $16\frac{1}{2}\%$. What was the commission if the apples were sold at $3.50 a box?

REVIEW OF CHAPTERS 1 AND 2

1. The following report represents the sales made by the West End Manufacturing Company during the past 6 months. Complete the report and find answers for the following questions: (a) What is the difference between the dollar value for the salesman with the

†A commission merchant is a merchant who buys or sells goods for others for a fee which is called a commission.

‡Net proceeds = gross proceeds ($316.50) less sum of commission, freight and cartage charges, and insurance.

highest sales as compared with the one with the lowest sales? (b) What was the average monthly sales for salesman No. 1 for the 6 months? (c) How much more did salesman No. 3 sell during the last 3 months as compared to the first 3 months?

WEST END MFG. CO.

Sales Six Months Ending June 30, 19____

Month	Salesman 1	Salesman 2	Salesman 3	Total
January	$3267	$4311	$4605	$
February	2381	3206	3172	
March	3472	4817	4299	
Subtotal	$	$	$	$
April	5600	6988	5781	
May	4865	6250	6119	
June	4698	7015	6473	
Total	$	$	$	$

2. Mr. White owned $\frac{1}{3}$ of a business, Mr. Hoyt owned $\frac{1}{2}$, Mr. Evers owned $\frac{1}{9}$, and Mr. Baines owned the remainder. If Mr. Baines' share was worth $3700, how much were the shares held by each of the other partners worth?

3. According to the United States census, the population (nearest thousand) of Niagara Falls was 102 in 1960 and 86 in 1970. What percent decrease did this represent (nearest tenth of 1%)?

4. Sales for the day in one department of a notions store were as follows: $26.50, $0.76, $12.20, $3.56, $20.09, $8.80, $5.60, $3.76, and $4.20. Find the total of all sales that were less than $10.00 and the total sales for the day.

5. Rose, Stewart, and Lynch invested $10,400, $6500, and $9100, respectively, in a business. Profits for the first year amounted to $8260. If profits and losses were paid in proportion to investments, how much did each partner receive?

6. After spending $\frac{1}{5}$ of his income for food, $\frac{1}{4}$ for rent, and $\frac{1}{8}$ for entertainment, a man had $2505.80 left. What was his total income?

7. The monthly income from a building is $475.00. Yearly operating expenses amount to $1867.00. If the owner receives a 10% annual return on his investment, how much did the building cost?

8. Express .36$\frac{1}{2}$ as a fraction, and as a percent.

9. How much is left after $\frac{1}{3}$, $\frac{1}{8}$, and $\frac{1}{5}$ have been subtracted from a quantity?

10. Sales decreased from $36,000 to $31,500 during a certain period of time. What rate of decrease did this represent (nearest tenth of 1%)?

11. What percent of an hour is 6 seconds (show whole fraction)?

12. James Ames earns $182.70 a week. This is an increase of 5% over the previous week. How much did he earn before he received the increase?

13. What was the original cost of a home that sold for 16$\frac{2}{3}$% more than it cost? The selling price was $31,752.

14. Mr. Smith left $\frac{1}{3}$ of his estate to a son and $\frac{1}{3}$ to be divided between 4 grandchildren. The balance of $7500 was left to a charity. How much did each grandchild and his son receive? What was the total amount of the estate?

15. Mrs. James received $\frac{2}{3}$ of her husband's estate. Her share was $13,700 more than all other benefactors combined. How much did she receive?

X a House & a lot are worth $44,000 if the house is worth 4½ times as much as the lot: Find
a) value of lot
B) value of house

INVENTORY TEST

This test is designed to: (1) help you identify areas in computational work in which you may need extra drill and (2) test your ability to analyze simple word problems in basic arithmetical functions and the use of percentage.

I. Copy each problem neatly and work as quickly and accurately as possible. Show fractional remainder for all division problems.

1. Add:

(a)
```
6633
 454
2117
1824
 626
```

(b)
```
  3.064
  0.0076
 12.5
230.
  7.5
```

(c) .567 + 0.0067 + 2345 + 5.43 + .073

(d) 540 + .0089 + .875 + 5.75 + .0006

(e) 67.56 + 4.5 + 0.045 + 2400 + .0764

2. Subtract:

(a)
```
750.08
 79.89
```

(b)
```
24,704
16,598
```

(c) 56.789 − 0.0897

(d) 245.087 − 67.998

(e) 500.054 − 459.060

3. .106 × 3.071

4. 367 × 125

5. .1667 × .003

6. 84 × 75

7. 21.8 × 11

8. 8675 ÷ 26

9. .216 + 100

10. 562 ÷ 75

11. 31.67 ÷ .034

12. 8250 ÷ 125

13. $\frac{1}{5} + \frac{2}{5}$

14. $11\frac{1}{3} + \frac{5}{6} + \frac{1}{7}$

15. $\frac{3}{5} - \frac{1}{4}$

16. $4\frac{5}{8} - 2\frac{7}{8}$

17. $19 - 6\frac{1}{9}$

18. $\frac{2}{3} \times \frac{4}{3} \times \frac{15}{16}$

19. $\frac{5}{6} \times \frac{12}{35}$

20. $6\frac{1}{4} \times 60$

21. $16\frac{2}{3} \times 360$

22. $12\frac{1}{2} \times 96.08$

23. $\frac{15}{16} \div \frac{5}{8}$

24. $15 \div \frac{5}{9}$

25. $\frac{5}{17} \div 25$

26. $2\frac{1}{6} \div 5\frac{1}{3}$

27. $\frac{1}{11} \div \frac{7}{9}$

28. Find $12\frac{1}{2}\%$ of 168?

29. What is 0.02% of 568?

30. $\frac{1}{4}\%$ of 6200 is how much?

31. $\frac{1}{2}$ is what % of $\frac{5}{6}$?

32. 12.32 is 22% of what number?

33. 92 increased by 5% is how much?

34. $21\frac{1}{2}$ decreased by 2% is what amount?

35. 7254 is $112\frac{1}{2}\%$ of what number?

36. What number increased by $8\frac{1}{3}\%$ equals 4881.5?

37. 26 is what % of 156?

38. $108\frac{1}{3}\%$ of 96 equals what number?

39. $37\frac{1}{2}$ is what % of 25?

40. 479.5 less 20% equals what amount?

41. 72 increased by 10% of itself equals what number?

42. $\frac{1}{4}$% of $\frac{1}{4}$ is how much?

II. Indicate all the work and the methods used in the solution of the following.

1. Divide $30,186 between 3 organizations, giving 1 organization 60% and the other 2 equal amounts.
2. The head of a photograph shop received a monthly salary of $895 plus a 2% commission on all sales. If he earned $10,954.80 last year, how much did he receive in commissions?
3. In the Havens School District, classroom supplies accounted for $37\frac{1}{2}$% of the total budget of $3,896,000. How much money was spent for these supplies?
4. This year, 144 of our employees must retire. This number represents what percent of the total working force of 1263 employees (nearest tenth of 1%)?
5. A share of common stock worth $33 earns $1.20 annually. The return equals what percent of the investment (nearest tenth of 1%)?
6. Of our employees, $\frac{1}{3}$ receive $450 monthly; $\frac{1}{4}$ receives 5% more than this amount which, in turn, represents $12\frac{1}{2}$% less than the rest of the employees. What is the total payroll if there are 24 employees?
7. A store increased its sales 16% over last year. If the increase was $36,000, what was the amount of last year's sales?
8. If 18 percent of a number is 133, what is 22% of the number?
9. Mr. Hale cut the following lengths (in feet) from a piece of lumber: $3\frac{1}{2}$, $2\frac{1}{4}$, $6\frac{1}{6}$, $5\frac{1}{3}$, and $10\frac{1}{8}$. How long was the original piece of lumber if waste is ignored?
10. Mr. Owens owned $\frac{1}{3}$ of a business; Mr. Bert owned $\frac{1}{5}$ as much as Mr. Owens, and Mr. Hays owned the remainder. If Mr. Hays' share was worth $27,600, what was the worth of the shares held by each of the other partners? What fractional part of the business did Mr. Hays own?

Objectives

Discounts are an important element in many business transactions. Therefore, it is important that everyone know what they are and how they are used. The purpose of this chapter is to develop the ability to solve problems dealing with trade and cash discounts, to use discount tables or prepare them if necessary, and to make all necessary calculations required in the solution of any problem dealing with discounts.

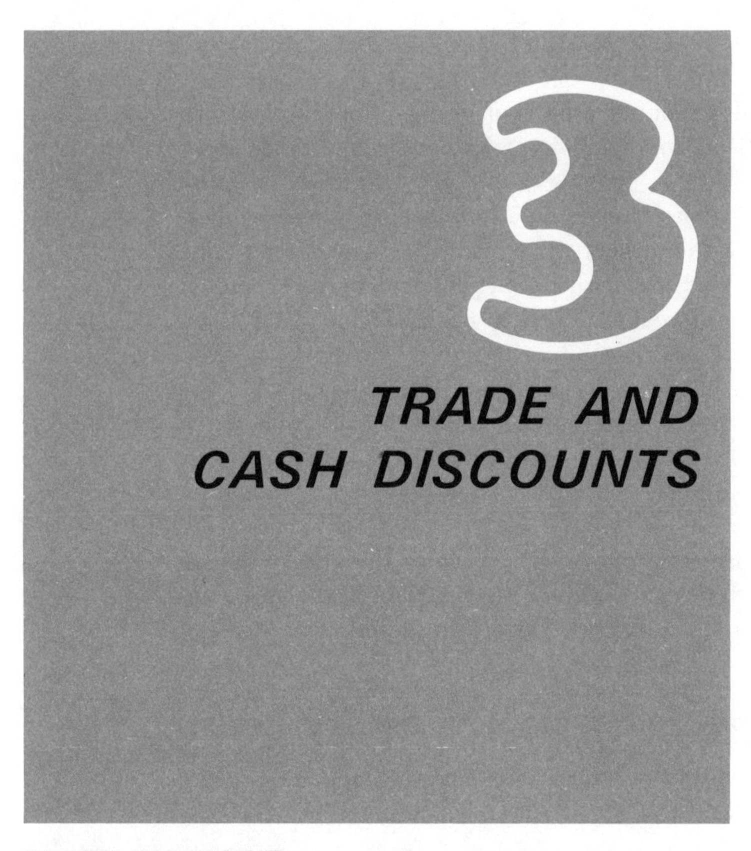

TRADE DISCOUNT

Trade discount is a deduction from the *list* or *catalog* price of an article or from the total amount of an invoice. It is expressed as a percent (%) of the list price or the total amount of an invoice. It may consist of a single discount or a series of discounts (generally referred to as a chain discount).

Trade discounts are usually associated with sales made by manufacturers or wholesale commercial dealers and not by retail stores.

USE OF TRADE DISCOUNTS

A trade discount may be used for several reasons. Customers are classified by many companies and different discount rates are applied to each group. For example, schools are often granted a more generous discount than are other customers for the same item.

Additional discounts are often granted for purchases in large quantities to secure a desirable account or to meet competition.

Many industries that handle a wide assortment of merchandise provide prospective buyers with catalogs containing detailed descriptions of their sales items. The price quoted for each item is called the list price. The cost of these catalogs would be prohibitive if they were issued to keep current with changing prices. Consequently, salesmen are furnished with inexpensive "discount" sheets that show the discounts allowed on the catalog items. In this manner, prices can be quoted to customers in keeping with fluctuating prices.

In any case, the list price quoted in the catalog is the same for all buyers. It is the discount allowed that provides for differences in the actual price paid for any item.

APPLICATION OF TRADE DISCOUNT (SINGLE DISCOUNT)

When the Amount of Discount and Net Amount are Both Required

example: A typewriter is listed at \$525.00 less 15%. Find the amount of discount and the net price.

solution:

\$525.00	list or catalog price	100%
78.75	discount (525 × .15)	15%
\$446.25	net price (amount paid)	85%

When Only the Net Amount Is Required

In the above example notice that the net price equals 85% of the list price. Consequently, the net price may be determined directly by taking 85% of the list price—\$525.00 × .85 = \$446.25.

Finding the List Price When Net Price and Discount Are Known

example: The net price of an article was \$57.00 after a 25% discount was allowed. What was the list or catalog price?

solution: Since the list price equals 100% and a discount of 25% was taken, the net price must equal 100% − 25% or 75%. If

75% = $57.00
100% = $57.00 ÷ .75 *or* $76.00 list price

proof: 25% of $76.00 = $19.00
$76.00 − $19.00 = $57.00

Finding the List Price and Net Price if the Amount and Percent of Discount Are Known

example: Mr. James bought a suit on which he received a 12% discount amounting to $14.40. What were the list price and net price of the suit?

solution: If 12% = $14.40, then

100% = $14.40 ÷ .12 *or* $120.00 list price
$120.00 − $14.40 = $105.60 net *or* selling price

proof: 12% of $120.00 = $14.40 (amount of discount which was known)

Finding the Rate of Discount when the List Price and Net Price Are Known

example: Mr. Howard bought a tractor listed at $985.00 for $861.87. What rate of discount was he allowed?

solution: Since the list price is the base or 100%, divide the discount by the list price for the rate.

$985.00 − $861.87 = $123.13 amount of discount
123.13 ÷ 985.00 = .125 *or* $12\frac{1}{2}$% rate of discount

proof: If the discount equals $12\frac{1}{2}$%, then the net price must equal $87\frac{1}{2}$%; $985.00 × .875 = $861.87.

EXERCISE 13

1. Find the amount of discount and net or billing price on a piece of furniture listed at $1500 less 15%.

2. The catalog quotation for silk yardage sold by Bourning Fabrics is $8.25 a yard less 8%. What is the net price per yard?

3. Rollins and Company offered J. P. Point and Sons a 15% discount amounting to $56.75 on a piece of garden equipment. What were the list price and the net price?

4. If the list price of an article is $280.00 and the net price is $224.00, what is the discount rate?

5. Find the net price on a suit listed at $125.00 less $15\frac{1}{2}$%.

6. Grant Bros. offered a discount of $12\frac{1}{2}$% amounting to $21.90 on their dinette tables. What were the list and selling prices on these tables?

7. If a washing machine sold for $416.00 less $5\frac{1}{2}$%, what was the amount of discount?

8. Which is the best price, a sewing machine selling for $275.00 net or one listed at $320.00 less $12\frac{1}{2}$%? How much is the difference?

TRADE DISCOUNT SERIES (CHAIN DISCOUNTS)

A trade discount may consist of a series of discounts that are deducted successively from the list price of an article or the total amount of an invoice.

example: Let us assume that the series 15%, 10%, and 5% is to be deducted from the total of an invoice amounting to $74.00. Find the net price and the amount of the discount.

solution:†		
	$74.00	total of invoice
	11.10	first discount (74 × .15)
	$62.90	net amount after first discount
	6.29	second discount (62.90 × .10)
	$56.61	net amount after second discount
	2.83	third discount (56.61 × .05)
	$53.78	final net amount (answer)

$74.00 − $53.78 = $20.22
amount of discount

The discounts in a series such as this may be taken in any order, the result will always be the same. In other words, 5–10–15% is the same as 10–5–15%.

EXERCISE 14

Find the net price paid and the amount of discount for the following invoices.

1. $375.00 less 10–10%

2. 63.75 less $33\frac{1}{3}$–10%

3. 569.00 less 25–10–5%

4. 86.20 less 10–5%

5. 514.00 less $66\frac{2}{3}$–5%

6. $126.00 less $66\frac{2}{3}$–2%

7. 15.46 less 10–10–5%

8. 18.40 less 10–5–$2\frac{1}{2}$%

9. 460.00 less 50–10%

10. 85.67 less 15–10–5%

DISCOUNT TABLES

When the same series is to be used many times, tables of *net decimal equivalents* are used. These equivalents represent the net percent left of 100% (expressed as a decimal) or the net amount left in a dollar after the series of discounts has

†Another method is to use the complements of each discount and multiply.

Discounts:	15	10	5
Complements:	85	90	95

Solution: .85 × .90 × .95 × 74 = 53.7795 Record as $53.78 net amount.

been deducted. The net amount is obtained by multiplying the dollar amount by the net decimal equivalent.

example: Find the net amount and amount of discount if the series of 15%, 10%, and 5% was deducted from $30.00.

procedure: Refer to the Discount Tables on pages 72–73.

Look for the column headed 15. Find the number in that column which is in line with 10 and 5 in the left-hand column. It is .72675. This means that after the series 15%, 10%, and 5% has been deducted, approximately 73% or 73¢ is left out of every dollar. Therefore, for $30.00, the net amount left is obtained by multiplying .72675 by 30. The result is 21.8025. Record as $21.80 *net amount.*

$30.00 − $21.80 = $8.20 *discount*

example: What is $65.00 less 25% and 5%?

solution: The *net decimal equivalent* is .7125 (check this with the table).

∴ .7125 × 65 = 46.3125. Record as $46.31 net amount.

$65.00 − $46.31 = $18.69 discount

EXERCISE 15

Find the net amount and the amount of discount for the following problems. Use the discount tables.

	Gross Amount	Discount (%)		Gross Amount	Discount (%)
1.	$ 27.00	25–5	**6.**	$600.90	40–5–5
2.	280.80	10–$2\frac{1}{2}$	**7.**	62.00	50–10
3.	300.25	$12\frac{1}{2}$–10–$2\frac{1}{2}$	**8.**	286.00	30–$7\frac{1}{2}$
4.	35.66	50–10–10	**9.**	4.95	15–10–10–10
5.	176.50	35–10–5	**10.**	109.10	$2\frac{1}{2}$–10

NET DECIMAL EQUIVALENT

Sometimes a series is used often that is not included in the discount tables. In this case, it should be computed and added to the tables.

example: Find the net decimal equivalent for the series 12%, 10%, and 2%.

solution: There are two methods which may be used to find the net decimal equivalent of any series. In the first method, discounts are deducted successively as follows.

100.00%	total
12.00%	first discount (12%)
88.00%	% left after first discount
8.80%	second discount (10% of 88.00)
79.20%	% left after second discount
1.584%	third discount (2% of 79.20)
77.616%	% left after all discounts are deducted

Since 77.616% is what remains after 12%, 10%, and 2% have been deducted, as a decimal it is .77616 or the net decimal equivalent of the series.

The second method involves the use of *complements*.† The net decimal equivalent may be obtained by multiplying the complements of the series. The complements of 12, 10, and 2 are 88, 90, and 98, respectively.

$.88 \times .90 \times .98 = .77616$ net decimal equivalent

†The *complement* of any number is the difference between it and the next highest power of 10. Thus, the complement of 7 is 3 because $7 + 3 = 10$; the complement of 18 is 82 because $18 + 82 = 100$; the complement of 125 is 875 because $125 + 875 = 1000$, etc.

PRACTICE

Compute the net decimal equivalents for the following series and compare with the tables on pages 72–73.

1. $33\frac{1}{3}$–10–10%†
2. $12\frac{1}{2}$–$7\frac{1}{2}$–10%‡
3. 50–10–5%
4. $12\frac{1}{2}$–5–$2\frac{1}{2}$%
5. $7\frac{1}{2}$–10–10–5%
6. $66\frac{2}{3}$–10–10–5%
7. $37\frac{1}{2}$–5%
8. 50–10–10–10%
9. 10–5–60%
10. 10–$2\frac{1}{2}$–15%

SINGLE DISCOUNT EQUIVALENT

The single discount equivalent of any series is the difference between 100% and the net decimal equivalent expressed as a percent.

example: The net decimal equivalent for the series 10–10–5% is .7695. Expressed as a decimal .7695 equals 76.95%. The single discount equivalent = 100% − 76.95% or 23.05%. This is the same as saying that 10–10–5% = 23.05 %.

PRACTICE

Using the table of discounts, find the single discount equivalent for the following series.

1. 30–5–$2\frac{1}{2}$%
2. $7\frac{1}{2}$–5–10%
3. 10–$12\frac{1}{2}$%
4. 35–10–10%
5. 50–10%
6. $37\frac{1}{2}$–5–5%

†This problem is simplified by deducting the 10% discount first.
‡The solution by the first method is shown below.

100.00%	
12.50	first discount (12.5%)
87.50	
6.5625	second discount (7.5%) (all digits are carried)
80.9375	
8.09375	third discount (10%) (all digits are carried)
72.84375	*or* .72844 to 5 decimals (net decimal equivalent)

Round off to 5 decimals in the final answer only.

Finding the List Price if the Net Price and Discount Are Known

example: A tire was sold for $23.90 after discounts of 10% and 5% had been taken. Find the list price.

solution: The net decimal equivalent for the series is .855 ($85\frac{1}{2}\%$). If

$$\begin{aligned} 85\tfrac{1}{2}\% &= \$23.90 \\ \text{list } (100\%) &= \$23.90 \div .855 \\ &= \$27.95 \quad \text{(answer)} \end{aligned}$$

proof: $\$27.95 \text{ less } 10\% \text{ and } 5\% = \$27.95 \times .855 = \$23.90$

Finding the List and Net Price if the Discount Series and Amount of Discount Are Known

example: A refrigerator was sold with trade discounts of 15%, 5%, and 5% which amounted to $65.20. What were the list price and the billing (net) price?

solution: The net decimal equivalent for 15–5–5% = .76713. The single discount decimal equivalent = 1.00000 − .76713 = .23287. The single discount equivalent = 23.287%. If

$$\begin{aligned} 23.287\% &= \$65.20 \\ \text{list price } (100\%) &= \$65.20 \div .23287 \\ &\quad \textit{or } \$280.00 \quad \text{(answer)} \\ \text{net price} &= \$280.00 - \$65.20 \\ &\quad \textit{or } \$214.80 \quad \text{(answer)} \end{aligned}$$

proof: $\$280.00 \text{ less } 15\text{–}5\text{–}5\% = \$280.00 \times .76713 = \$214.80$

EXERCISE 16

1. Find the net decimal equivalent and single discount equivalent for the following series.

(a) $16\frac{2}{3}$–5–5% (c) $37\frac{1}{2}$–10–10%

(b) 30–5% (d) $62\frac{1}{2}$–$33\frac{1}{3}$–10 %

2. A typewriter marked \$260 was sold at a discount of 20 %. What are the net price and the amount of discount?

3. Mr. Howard paid \$82.50 for a chair that normally sells for \$110.00. What rate of discount did he receive?

4. What was the list price of a lawn mower if it was sold for \$88.65 after a discount of 10% was allowed?

5. Which is the best buy and by how much: (a) a stove listed at \$560 less 20–10–5% or 40–10%, (b) a refrigerator with discounts of 20–5–10% or 10–20–5%

6. Discounts of 10% and 5% were allowed on ladies stockings listed at \$18.00 a dozen. What was the net price of each pair of stockings?

7. The Naylor Mercantile Company offered discounts of 10–10–$2\frac{1}{2}$% on all #627 items listed in their catalog. What do these amounts represent as a single discount?

8. A hardware dealer offered discounts of $33\frac{1}{3}$% and 10% or \$75.00 off the regular price on all camp sets. How much were the original selling price and the discounted price of these sets?

9. Find the list price and net price of an article on which a discount of \$12.50 was allowed which amounted to 20% and 10%.

10. Compare the net costs in the following: (a) \$260 less 25% and 10% or \$260 less 15%, 15%, and 5%; and (b) \$1200 less $33\frac{1}{3}$%, 10% and 5% or \$1200 less 35% and 10%.

CASH DISCOUNT

Cash discount is a deduction allowed on an invoice or bill of goods to encourage payment in accordance with the terms of the sale. The manner in which terms are expressed varies with business concerns and invoice forms. They usually appear on the upper section of the invoice.

TERMS OF SALE

2/10, n/30 means that a 2% discount is allowed if the invoice is paid within 10 days from the date of the invoice, but the total amount is due if paid on any day from the 11th to the 30th day from the date of the invoice.

2/10, 1/15, n/30 means that a 2% discount is allowed if the invoice is paid within 10 days from the date of the invoice, a 1% discount is allowed if paid from the 11th to the 15th day, and the total amount is due if paid on any day from the 16th to the 30th day from the date of the invoice.

1%–10th EOM (end of month) means that a 1% discount is allowed if the invoice is paid during the first 10 days of the next month following the date of the invoice.

2% 15th Prox. means that a 2% discount is allowed if payment is made at any time during the first 15 days of the next month following the date of the invoice.

n/30 means the invoice must be paid within 30 days from the date of the invoice and may be subject to an interest charge if paid after that time.

Postdating an invoice is common practice in some businesses. For example, a school may order text books in July for immediate shipment but request a September 1 billing. If a cash discount is allowed under such circumstances, the date of the invoice (not the date of the purchase order) is used.

R.O.G. is an abbreviation for Receipt of Goods. This means that the terms of payment are computed from the day the goods are received—not the date of the invoice. This is advantageous to the buyer when shipment requires several days. For example, an invoice for merchandise is dated July 7 with terms of 2/10 R.O.G., but the shipment is not received until August 15. In this case, the cash discount is allowed if payment is made any time up to and including August 25. When the net period is not indicated, it is generally understood to be 30 days from the receipt of the goods.

Extra dating permits a longer period of time in which the purchaser may take advantage of a cash discount. For example, 2/10–90X, 2/10–90 ex., or 2/10–90 Extra indicates that an additional 90 days or a total of 100 days from the date of the invoice is the period of time during which the 2% cash discount may be taken.

Generally, extra dating terms are granted during the off-season for certain types of merchandise such as wet weather clothing during the dry season, and heating equipment during the summer months. This is done to encourage sales at that off-season time. An example, taken from the cotton business, serves as an illustration. If a mill can use 5000 bales of cotton over several months, it is to the advantage of the broker to induce the mill owner, through attractive terms, to take all the cotton at once if the broker is anxious to move his stock. In this way the broker not only equalizes a possible off-season market but also clears his warehouse, saves storage costs, or possibly makes room for new cotton.

example: If the discount is to be granted, this bill must be paid not later than the 10th of November since the invoice was dated in October.

ALLEN PAPER COMPANY
Oakland, California

Sold to: Wohler Printing Co.
Cunningham, California

Invoice No. 00112
Date 10/25/___
Terms: 1%–10th E.O.M.

Item	Description	Unit	Price	Amount
600	Sheets #175 tagboard 28½" × 45", 336 lb	cwt	$20.75	$69.72

No mdse returnable without our permission

Taxable ( )
Nontaxable (√)*

*Cash discounts cannot be applied against sales taxes. Therefore, in cases where a sales tax is included in the invoice, such tax must be deducted from the total before the discount is computed.

solution:

$69.72	total of invoice
.70	discount (1% of $69.72)
$69.02	amount paid

example: Mr. Smith received an invoice dated January 15, terms 2/10, n/30, amounting to $156.50. He paid this obligation on January 20. How much was the discount and how much did Mr. Smith remit in payment?

solution:

\$156.50	amount of invoice
3.13	discount (2% of \$156.50)
\$153.37	amount paid

example: Mr. Green received an invoice dated June 24, terms 3/10, n/30, amounting to \$37.50 less a trade discount of 5%. How much should be remitted to pay the bill in full at the end of 10 days?† In this case, payment might not be made until July 5 (assuming it is a business day) since July 4th is a legal holiday.

solution:

\$37.50	gross
1.87	5% trade discount
\$35.63	net amount
1.07	3% cash discount
\$34.56	amount of remittance

PARTIAL PAYMENT ON ACCOUNT

Sometimes a buyer makes a partial payment on an invoice during the cash discount period. In such cases, he is entitled to a discount on that portion of the bill that has been paid.

example: Assume an invoice dated May 21 amounts to \$63.50 with terms of 2/15 Prox. On June 13 a payment of \$30 is made. Find the amount of the discount and the balance due on the invoice.

solution: Since a 2% discount is allowed, any payment made within the discount period equals 98% of each dollar paid. This means that \$30 = 98% of the sum to be credited to the account. Therefore, $30 \div .98 = 30.612$ or \$30.61 (amount to be credited).

†When the last day in which a cash discount is allowed falls on a holiday, Sunday, or nonbusiness day, the last date on which the discount is allowed is extended to the first business day following.

$63.50	amount of invoice
30.61	amount credited as payment
$32.89	balance due

$30.61 − $30.00 = $0.61 cash discount

EXERCISE 17

I. Find the last date of payment for the following problems that will allow the buyer to receive the best discount.

Invoice No.	Terms	Date of Invoice
1.	2/10, n/30	Mar. 15
2.	2/10, 1/15, n/30	Feb. 26*
3.	3/10, n/30	Jan. 13
4.	1/15, n/30	Oct. 10
5.	3/10, 1/30, n/90	July 22
6.	2/10, 1/30, n/60	March 8
7.	3/30, n/60	Nov. 11
8.	2/15, n/30	Jan. 16
9.	2% 10th E.O.M.	May 20
10.	1% 15th Prox.	June 17

*Not a leap year.

II. Find the cash discount received and the amount of payment for the following problems.

No.	Amount of Invoice	Terms	Date of Invoice	Date of Payment
1.	$ 23.16	2/10, 1/15, n/30	Jan. 5	Jan. 15
2.	89.30	1/10, n/30	Aug. 27	Sept. 1
3.	62.00	1% 10th E.O.M.	July 25	Aug. 9
4.	32.81	2/10, n/30	May 15	May 26
5.	19.75	2/10, 1/30, n/60	Oct. 10	Oct. 25
6.	135.37	3/10, n/30	Jan. 21	Feb. 1
7.	346.90	3/10, 1/30, n/60	May 30	June 10
8.	78.60	1% 15th Prox.	Dec. 19	Jan. 15

9.	503.96	2/10, 1/15, n/30	Dec. 31	Jan. 9
10.	138.00	2/15, 1/30, n/60	Oct. 30	Nov. 15

III. Find the amount of discount and balance due on the following.

Invoice No.	Amount of Invoice	Date of Invoice	Terms	Partial Payment Date	Partial Payment Amount
1.	$ 312.60	Jan. 1	3/10, n/30	Jan. 8	$150.00
2.	1079.50	May 31	1/15, n/60	June 10	550.00
3.	550.00	Oct. 10	2/10, 1/30, n/60	Oct. 25	230.00
4.	86.75	Mar. 2	5/10	Mar. 12	56.00
5.	325.00	June 15	1% 10th E.O.M.	July 9	175.00
6.	190.55	Aug. 24	2% 15th Prox.	Sept. 2	100.00
7.	95.40	Dec. 15	3/30, n/60	Jan. 10	45.00
8.	365.00	May 2	5/10, n/30	May 12	325.00
9.	116.50	July 3	2/10, 1/15, n/30	July 15	75.00
10.	38.75	Oct. 2	2/10, n/30	Oct. 8	25.00

EXERCISE 18 *Chapter Summary*

1. Complete the following invoice. If the Lane Art Co. took advantage of the cash discount, (a) what would be the latest day in which this bill could be paid and (b) how much should be remitted in payment?

Z PAPER COMPANY

Sold to: Lane Art Co.

Invoice No. K D 437
Date Jan. 15, 19___
Terms: 1% 10th E.O.M.

Quantity	Unit	No.	Description	Wt. Lb per Sheet	Price	Extension
40	Sheets	23 × 35	248 M. aqua Twiltex Cover	¼	$53.70* cwt**	$
				Less 2%		
					Net	$
				Sales Tax 4%		
					Total	$

$^{*}(40 \times \frac{1}{4}) \times \frac{53.70}{100}$

**cwt = hundred-weight

2. Find the net price on 25 pads of graph paper at 30¢ each and 50 pads of graph paper at 25¢ each if a 25% trade discount is granted.

3. What is the net selling price on an article listed at $805.70 less 20%, $16\frac{2}{3}$%, and 10%?

4. The Allan Manufacturing Co. in New York shipped merchandise amounting to $3175.00 to the Everen Stores in San Francisco. The invoice was dated June 3. The merchandise was received August 1. Terms of the sale were 2/10 R.O.G. What is the latest date that Everen Stores may pay this invoice and receive a cash discount? How much must be remitted in payment?

5. The Haight School District ordered 220 books from the Ever Ready Publishing Company on December 15 with a request for a March 1 billing. The books sold for $4.85 each less a 20% trade discount. Shipping charges amounted to $6.50. A cash discount of 1% was allowed for payment within 5 days of the billing date. If the invoice was paid on March 2, what was the amount of the payment? (Note: The cash discount cannot be applied to shipping charges.)

6. Complete the following invoice. How much was remitted in payment if this bill was paid on January 16?

ANZEL PAPER COMPANY
Olympia, Washington

Date January 6, 19___ Invoice No. 1–0148
Sold to: Duncan Publishers
Oakland, California Terms: 1% 10 days, net 30 days

Quantity	Description	Unit	Price	Amount
50	Reams 8½ × 11 – 16 # white mimeo paper	Ream	$2.62	$
		Less 50–10%		
			Total	$

7. Hale Services received the following items on which a sales tax of 3% was charged: 6 bottles correction fluid @ $0.50 each, 6 bottles correction thinner @ $0.50 each, 2 lettering guides @ $1.60 each, and 1 No. 0 stylus @ $0.50. If a discount of 15% was allowed, what was the amount of the invoice?

8. How much will it cost the buyer if the manufacturer of an item sells it for \$516.10 less $12\frac{1}{2}\%$, and $2\frac{1}{2}\%$?

9. Compute the list price on the following items for which the selling price and the discounts are known.

	Selling Price	Discounts
(a)	\$ 63.30	15%, 10%, and 5%
(b)	\$138.10	30 % and $12\frac{1}{2}$ %

10. A retail store manager bought a bill of goods at a cost of \$850.00 less 25% and 5%. If it was sold for $12\frac{1}{2}\%$ more than the net cost, what was the selling price?

11. Zane Bros. purchased a selection of floor lamps at a net price of \$875.00 which were listed by the wholesaler at \$1250.00. What was the rate of discount?

12. Which is the better discount for the buyer and by how much? (a) 40–10–$2\frac{1}{2}\%$ or 25–15–2%, (b) a single discount of 30% or 15–15%?

13. Shibley Publications received a bill for 50 reams of paper at \$2.76 a ream less 50–10% discount. A 2% cash discount was granted if payment was received within 10 days from the date of the invoice. What was the amount of the payment if advantage was taken of the cash discount?

14. The Melody Music Co. listed a group of radios for \$1650.00 less 25%. To meet competition the net price was dropped to \$1113.75. What additional discount was added to meet this drop in price?

15. Hope Co. received a bill for \$250.75 dated January 3 with terms 3/10, 1/15, n/30. The bill was paid on January 14. What was the amount remitted in payment?

16. An invoice amounting to \$625.50 less 50–10% dated March 29 with terms of 2% 10 days, net 30 days was paid on April 3. What was the amount of payment?

DISCOUNT TABLE—Net Decimal Equivalents
(Net Value of $1.00 after Discounts Have Been Deducted)

Rate (%)	5	7½	10	12½	15
Net	.95	.925	.90	.875	.85
2½	.92625	.90188	.8775	.85313	.82875
5	.9025	.87875	.855	.83125	.8075
5, 2½	.87994	.85678	.83363	.81047	.78731
5, 5	.85738	.83481	.81225	.78969	.76713
5, 5, 2½	.83594	.81394	.79194	.76995	.74795
7½	.87875	.85563	.8325	.80938	.78625
7½, 2½	.85678	.83423	.81169	.78914	.76659
7½, 5	.83481	.81284	.79088	.76891	.74694
10	.855	.8325	.81	.7875	.765
10, 2½	.83363	.81169	.78975	.76781	.74588
10, 5	.81225	.79088	.7695	.74813	.72675
10, 5, 2½	.79194	.7711	.75026	.72942	.70858
10, 7½	.79088	.77006	.74925	.72844	.70763
10, 10	.7695	.74925	.729	.70875	.6885
10, 10, 5	.73103	.71179	.69255	.67331	.65408
10, 10, 5, 2½	.71275	.69399	.67524	.65648	.63772
10, 10, 10	.69255	.67433	.6561	.63788	.61965
10, 10, 10, 10	.62330	.60689	.59049	.57409	.55769

Rate %	16⅔	20	25	30	33⅓
Net	.83333	.80	.75	.70	.66667
2½	.8125	.78	.73125	.6825	.65
5	.79167	.76	.7125	.665	.63333
5, 2½	.77187	.741	.69469	.64838	.6175
5, 5	.75208	.722	.67688	.63175	.60167
5, 5, 2½	.73328	.70395	.65995	.61596	.58663
7½	.77083	.74	.69375	.6475	.61667
7½, 2½	.75156	.7215	.67641	.63131	.60125
7½, 5	.73229	.703	.65906	.61513	.58583
10	.75	.72	.675	.63	.6
10, 2½	.73125	.702	.65813	.61425	.585
10, 5	.7125	.684	.64125	.5985	.57
10, 5, 2½	.69469	.6669	.62522	.58354	.55575
10, 7½	.69375	.666	.62438	.58275	.555
10, 10	.675	.648	.6075	.567	.54
10, 10, 5	.64125	.6156	.57713	.53865	.513
10, 10, 5, 2½	.62522	.60021	.5627	.52518	.50018
10, 10, 10	.6075	.5382	.54675	.5103	.486
10, 10, 10, 10	.54675	.52488	.49208	.45927	.4374

DISCOUNT TABLE—Net Decimal Equivalents (*continued*)
(Net Value of $1.00 after Discounts Have Been Deducted)

Rate %	35	37½	40	45
Net	.65	.625	.60	.55
2½	.63375	.60938	.585	.53625
5	.6175	.59375	.57	.5225
5, 2½	.60206	.57891	.55575	.50944
5, 5	.58663	.56406	.5415	.49638
5, 5, 2½	.57196	.54996	.52796	.48397
7½	.60125	.57813	.555	.50875
7½, 2½	.58622	.56367	.54113	.49603
7½, 5	.57119	.54922	.52725	.48331
10	.585	.5625	.54	.495
10, 2½	.57038	.54844	.5265	.48263
10, 5	.55575	.53438	.513	.47025
10, 5, 2½	.54186	.52102	.50018	.45849
10, 7½	.54113	.52031	.4995	.45788
10, 10	.5265	.50625	.486	.4455
10, 10, 5	.50018	.48094	.4617	.42323
10, 10, 5, 2½	.48767	.46891	.45016	.41264
10, 10, 10	.47385	.45563	.4374	.40095
10, 10, 10, 10	.42647	.41006	.39366	.36086

Rate %	50	60	62½	66⅔
Net	.50	.40	.375	.33333
2½	.4875	.39	.36563	.325
5	.475	.38	.35625	.31666
5, 2½	.46313	.3705	.34734	.30875
5, 5	.45125	.361	.33844	.30083
5, 5, 2½	.43997	.35198	.32998	.29331
7½	.4625	.37	.34688	.30833
7½, 2½	.45094	.36075	.3382	.30062
7½, 5	.43938	.3515	.32953	.29292
10	.45	.36	.3375	.3
10, 2½	.43875	.351	.32906	.2925
10, 5	.4275	.342	.32063	.285
10, 5, 2½	.41681	.33345	.31261	.27788
10, 7½	.41625	.333	.31219	.2775
10, 10	.405	.324	.30375	.27
10, 10, 5	.38473	.3078	.28856	.2565
10, 10, 5, 2½	.37513	.30011	.28135	.25009
10, 10, 10	.3645	.2916	.27338	.243
10, 10, 10, 10	.32805	.26244	.24604	.2187

Objectives

The objectives of this chapter are to introduce the student to problems in the field of merchandising and methods of solving them—problems dealing with markup or gross profit, markdowns, and profit and loss.

MARKUP

Merchants are in the business of buying and selling goods at a profit. The difference between the cost† and the selling price must be sufficient to cover all operating expenses with an amount remaining, called net profit, that will justify the operation of the business. This difference is referred to as *gross profit* or *markup*. How much the markup is on any article or product is based on years of experience and business practice.

The *markup* may be computed on the *cost* or the *selling* price. There is no uniform or standard rule. The method used depends on many factors such as: tradition or precedent, accounting system, method used for inventory records, type of merchandise, certain unique characteristics of a business, and size of the business.

†Cost equals the net cost (price of goods bought) plus buying expenses such as transportation and other handling charges.

In general, manufacturers, wholesalers, and jobbers use the cost price as the basis for computing markup. Furniture stores, as a rule, also figure markup on cost price. It is common practice for the larger retail stores to use the selling price as a base for computing markup. However, many retail stores (usually small ones) use the wholesale or cost price.

FINDING THE AMOUNT AND RATE OF MARKUP

By definition the *markup* is the difference between the selling price and the cost price, based on the assumption, of course, that the selling price is greater than the cost price. Then,

Markup = Selling Price − Cost Price

example: An article costs \$25.00 and sells for \$37.50. Find the markup and the rate of markup based on the (a) cost and (b) selling price.

solution: When the markup rate is based on the *cost*:

1. markup = \$37.50 − \$25.00 = \$12.50
2. *markup rate* = \$12.50 ÷ \$25.00 = .50 or 50% since the cost is the base

solution: When the markup rate is based on the *selling price*:

1. markup = \$37.50 − \$25.00 = \$12.50
2. *markup rate* = \$12.50 ÷ \$37.50 = $.33\frac{1}{3}$ *or* $33\frac{1}{3}\%$ since the selling price equals the base

EXERCISE 19

General Instructions: Record all money amounts correct to the nearest cent. Record all percent figures correct to the nearest tenth of 1% unless they are even or have a common fraction remainder such as 25% or $33\frac{1}{3}\%$, in which case record as illustrated.

I. Find the amount of markup and the rate of markup for the following examples, based on the cost price and based on the selling price (see example).

				Percent Markup	
No.	Cost Price	Selling Price	Markup	On Cost	On Selling Price
Example	$125.30	$160.80	$35.50	28.3	22.1
1.	60.00	95.00			
2.	75.50	105.70			
3.	25.50	48.00			
4.	56.00	63.00			
5.	24.95	32.00			
6.	50.00	80.00			

II. Record all processes in the solution of the following problems.

1. A retail clothier paid $15.00 for a suit and sold it for $45.00. What were the amount and rate of markup based on cost?

2. What is the rate of markup on an article selling for $80.00 that cost $44.00 if it is based on the selling price?

3. The Starr Furniture Company purchased a table for $40.00. Transportation charges were $5.00. If the table sells for $75.00 find the markup and percent markup on cost. (Note: Transportation charges are considered a part of cost.)

4. The Havens Stationary Store buys albums from the manufacturer for $11.00 less a 50% discount. The store in turn lists them at $11.00 less a 20% discount. What are the (a) rate of markup on cost and (b) the profit on 3 albums that were sold?

5. James Bros. bought 1 dozen hampers from a wholesaler for $12.50 each less 10% and 5%. If they were sold at retail for $15.00 each, what were the markup for the dozen and the percent markup based on the retail price?

6. Hoffman fixes his selling price at a set rate on the sales price. What markup rate does he use if he sells an article for $128.00 that cost him $80.00?

FINDING THE SELLING PRICE WHEN THE COST AND MARKUP RATE ARE KNOWN

example: A radio costs a dealer \$15.00. How much must he sell it for if the percent markup is (a) 15% of the cost and (b) 15% of the selling price?

solution: When the markup is based on the *cost*:

1. cost = \$15.00
2. markup = 15% of cost *or* \$15.00 × .15
   = \$2.25
3. selling price = cost + markup
   = \$15.00 + \$2.25 = \$17.25

Or:

selling price	\$17.25	115%	
cost price	15.00	100%	(base)
markup	\$ 2.25	15%	

proof: 115% of \$15.00 = \$17.25

solution: When the markup is based on the *selling* price:

1. cost = \$15.00
2. markup = 15% of selling price

The selling price in terms of dollars and cents is unknown but, since it is the base, it must equal 100%. If the selling price = 100% and the markup = 15%, then the cost price must = 100% − 15% or 85%. The cost price also = \$15.00, therefore 85% = \$15.00. Then the selling price (100%) = \$15.00 ÷ .85 = 17.647 *or* \$17.65. Or:

selling price	\$17.65	100%	(base)
cost price	15.00	85%	
markup	\$ 2.65	15%	

proof: 15% of \$17.65 = 2.6475 or \$2.65

EXERCISE 20

I. Find the selling price for the following articles (see example).

				Selling Price	
No.	Article	Cost Price	Markup Rate	Based on Cost	Based on Selling Price
Example	Table	$35.00	40%	$49.00	$58.33
1.	Highchair	9.40	25%		
2.	Desk	90.00	50%		
3.	Sofa	400.00	20%		
4.	Table	120.00	30%		
5.	Radio	45.00	15%		
6.	Picture	75.00	$33\frac{1}{3}$%		

II. Solve the following problems; show all processes.

1. L. Dakin & Co. purchased 2 dozen chairs for $360.00. Freight charges totaled $115.00. Find the selling price if a 30% gross profit (markup) on cost is to be realized. (Cost price includes freight charges.)

2. Wohler bought some purses for $4.50. At what price must they be sold if a markup of 35% is based on the selling price?

3. A hardware dealer used a markup of $16\frac{2}{3}$% on cost on some merchandise for which he paid $4.50. What was the selling price?

4. A hi-fi set cost $320.00 less 25% and 5%. What must it sell for to gain $37\frac{1}{2}$% of the selling price?

5. A dealer pays $22.80 for an article that he sells at a markup of 10% on the selling price. What is his selling price?

6. A retailer purchases suits for $35.00 on which he uses a markup of 25%. What would the selling price be if the markup is based (a) on cost? (b) on selling price?

ok
Look over
before test.

FINDING THE COST PRICE WHEN THE SELLING PRICE AND MARKUP RATE ARE KNOWN

example: A suit sold in a local store for $50.00. What was the cost if the markup of 30% was (a) based on selling price and (b) based on cost?

solution: When the markup is based on the *selling* price:

1. selling price = $50.00
2. markup = 30% of selling price
   = .30 × $50.00 = $15.00
3. cost price = selling price − markup
   = $50.00 − $15.00 = $35.00

Or:

selling price	$50.00	100%	(base)
cost price	35.00	70%	
markup	$15.00	30%	

proof: 70% of $50.00 = .70 × $50.00 = $35.00

solution: When the markup is based on the *cost*:

1. selling price = $50.00
2. markup = 30% of cost
3. cost = 100% (base)
4. selling price = cost price + markup
   = 100% + 30% = 130%

The selling price also = $50.00, therefore 130% = $50.00. Then cost (100% or base) equals $50.00 ÷ 1.30 = $38.46.

Or

selling price	$50.00	130%	
cost price	38.46	100%	(base)
markup	$11.54	30%	

proof: 30% of $38.46 = 11.5380 or $11.54

EXERCISE 21

I. Find the cost for the following items (see example).

	Given		Cost Price When Markup is Based on	
Item	Selling Price	Markup Rate	Selling Price	Cost Price
Example	$95.00	22%	$74.10	$77.87
1.	125.00	10%		
2.	126.00	$16\frac{2}{3}$%		
3.	9.26	$33\frac{1}{3}$%		
4.	54.75	25%		
5.	302.20	$12\frac{1}{2}$%		
6.	85.00	40%		

II. Solve the following problems. Indicate all steps in your solution of each problem.

1. Mr. Dixon priced his electric coffeepots to sell for $22.74. If he used a 20% markup on cost, what did he pay for them?
2. Hemp Bros. sold their TV tables at $25.00 less 10%. What was the purchase price of these tables if a markup of 40% on the net selling price was used?
3. A druggist sells a face cream for $2.50 a jar on which he realizes a 30% markup of the selling price. What is his cost price?
4. A gift shop sells a crystal vase for $18.00 at a markup of 80% on cost. What is the cost?
5. A dealer makes a gross profit of $69.60 on an electrical appliance, representing a markup of $33\frac{1}{3}$% on the selling price. What are (a) the cost price and (b) the selling price?
6. The Hammer Manufacturing Co. makes gunstocks that sell for $37.50. What is the cost if the markup on cost is 60%?

MARKDOWN

A *markdown* is a reduction in the selling price. Such a reduction is made for one or more reasons. It is a means, for example, of moving old merchandise, stimulating sales, disposing of hard-to-sell goods, or meeting competitive prices.

Markdown = Original Selling Price (previous marked price) − Actual Selling Price

% Markdown = Amount of Markdown ÷ Original Selling Price

example: A lady's suit, marked to sell for $110.00, is marked down to $85.00 as part of a promotional sale. What were the markdown and the rate of markdown?

solution: markdown = original selling price − actual selling price
= $110.00 − $85.00 = $25.00

% markdown = amount of markdown ÷ original selling price
= $25.00 ÷ $110.00
= .227 or 22.7%

example: A local store advertised a $33\frac{1}{3}$% markdown on some of its dresses that were originally marked to sell for $27.50. What was the new price?

solution: markdown = $33\frac{1}{3}$% of $27.50 = $9.17
new selling price = $27.50 − $9.17 = $18.33

EXERCISE 22

I. Find the missing factors in the following list. Show full percent for markdown rates.

No.	Original Selling Price	Actual Selling Price	Markdown	Percent Markdown
1.	$125.00	$ ______	$25.00	______
2.	37.50	______	______	$33\frac{1}{3}$%
3.	132.00	121.00	______	______

4.	______	86.25	______	25%
5.	______	68.50	______	15%
6.	______	______	58.00	$16\frac{2}{3}\%$

II. 1. What is the markdown rate on a table that normally sells for $95.00 if the price is reduced to $74.10?

2. An article costs $25.00. It was marked up 25% of cost and later reduced by $6\frac{1}{2}\%$ for a clearance sale. What were the markdown and the reduced selling price?

3. Ott's Drug Store reduced some of its Christmas cards 50% after the holidays. If the original markup was $66\frac{2}{3}\%$ of the selling price and the cost of the cards was $2.00 a box, what was the reduced selling price?

4. The Fairprice Market sold all of its jump ropes in a January sale for $0.66 each which represented a markdown of 25%. What were the amount of markdown and the original selling price?

5. Halsey's Notions sold its face cream for $2.40 which represents a markup of $33\frac{1}{3}\%$ on cost. If these creams were later reduced $8\frac{1}{3}\%$, how much would the new markup represent of cost?

6. Carlson uses a markup of 50% on cost on their shirts. The selling price on some of the shirts were reduced 20% for clearance. What is the markup on cost for these sale shirts?

PROFIT AND LOSS

As stated in the introduction to this chapter, merchants are in the business of buying and selling goods for profit. The difference between the selling price and cost price is called the markup or gross profit. The gross profit can also be thought of as the total of all operating expenses (expenses of conducting the business) and a net profit.

Operating expenses in turn may be broken down into overhead expenses (general and administrative expenses such as salaries, rent, lights, and repairs) and selling expenses (advertising, salesmen's expenses, promotion, commissions, etc.). When the gross profit is insufficient to meet the costs of operating the business, a loss occurs.

These relationships can be summarized in equation form as follows:

Gross Profit = Selling Price − Cost Price

Gross Profit = Operating Expenses + Net Profit

Net Profit or Loss = Gross Profit − Operating Expenses

The following examples illustrate these relationships.

example: An article sells for $25.00. The net profit is 7%, overhead is 18%, and the selling expense is 21%. Gross profit is based on the selling price. Find the cost.

solution:

net profit	7%	
overhead expense	18%	
selling expense	21%	
gross profit	46%	
selling price	100%	(base)
cost price	54%	
gross profit	46%	

cost price = $25.00 × .54 = $13.50

example: O'Brien Merchants wish to realize a 15% net profit on rugs that were purchased for $125.00. Buying expenses amounted to $5.00. Overhead expense is 12%; selling expense is 8%. What selling price is required, assuming profit is based on the selling price? How much are the net profit and the gross profit?

solution: gross profit = overhead + selling expense + net profit
= 12% + 8% + 15%
= 35%

cost price = selling price − gross profit
= 100% − 35%
= 65%

cost price = purchase price + buying expenses
= $125.00 + $5.00
= $130.00

If the cost (65%) = \$130.00, then the
selling price (100%) = \$130.00 ÷ .65
= \$200.00

net profit = 15% of \$200.00 = .15 × 200
= \$30.00

gross profit = selling price − cost price
= \$200.00 − \$130.00
= \$70.00

Or, arranged in report form, the solution appears as follows.

selling price	\$200.00	100%	(base)
cost price	130.00	65	
gross profit	70.00	35%	
overhead expense	24.00	12	
selling expense	16.00	8	
net profit	30.00	15	

The amount of overhead and selling expenses were not required but are shown merely to complete the problem and to show the relationship of each to the selling price.

EXERCISE 23

1. The Grand Sport Shop paid \$11.50 for sleeping bags and sold them for \$22.50. If a 20% net profit, based on cost, was realized, what was the cost of doing business (overhead and selling expenses)?

2. A gas stove sells for \$260.00. If operating expenses are 30% of sales and net profit is 8% of sales, what is the cost of the stove?

3. Mrs. Romine bought 20 suits at \$30.00 each. She sold 12 at \$47.50 each; 3 were soiled and sold for \$35.00 each. At what price must she sell the remainder to make a gross profit of 40% on total cost?

4. The A-C Furniture Co. buys coffee tables for \$27.50. (a) What is the selling price if the net profit, overhead, and selling expense (based on selling price) are 15%, 20%, and 10%, respectively? (b) What is the catalog or list price if a discount of 10% is offered?

5. Edwards Jewelers earned a net profit on sales of 22% on 12 watches which were purchased for $55.00 each. Operating expenses were $12\frac{1}{2}$% of sales. If all watches were sold, what was the total net profit?

6. A loss of $5.60 was incurred on an article, amounting to $3\frac{1}{3}$% on the selling price. What were the selling price and the cost price?

EXERCISE 24

1. Find the selling price of a dress costing $19.50 if the markup is 20% (a) based on cost, and (b) based on selling price.

2. The cost of a suit, which sold for $95.00 was $66.50. What was the percent markup based on the selling price?

3. Find the percent markup, based on cost, on an article costing $128.00 and selling for $160.00.

4. A merchant bought goods for $760.00 less 20% and 10%. He sold the goods at a markup of 15% of the cost. Find the markup and the retail price.

5. The original selling price of an article was $35.00. It was reduced to $30.25 for clearance. What was the markdown rate (nearest tenth of 1%)?

6. A store sold an article for $27.50 less 10%. If a gross profit of 30% was made, based on the selling price, what was the cost?

7. The Sonoma Music Co. purchased 15 radios on which they used a markup of 40%, based on the selling price. The purchase price was $17.50 each; buying expenses amounted to $1.50 each. (a) What was the selling price for each radio? (b) How much was the total gross profit?

8. The Fashion Shop reduced their fall coats in a postholiday sale by $8\frac{1}{3}$% of the marked price. These coats originally sold for $60.00. What was the actual selling price?

9. A grocer purchased 25 crates of berries for 70¢ each. If a crate was lost through spoilage, at what price per crate did he sell the remainder to gain a 35% profit on sales?

10. The French Appliance Shop sold an electric stove for $675.00 which included a markup of 25% on cost. The stove was purchased from the City Wholesale House of Stoves. At what price was the stove listed by the wholesale house if trade discounts of 15% and 10% were granted?

11. A loss of $12\frac{1}{2}$%, amounting to $28.50, was incurred on the sale of an article. If the loss is based on the retail, what were (a) the cost and (b) the retail prices?

12. The Remond Furniture Company makes a net profit of $6.40 or 8%, based on cost, on a table that sells for $110.00. Buying expenses amount to $15.00. What are (a) the cost exclusive of buying expenses and (b) the operating expenses?

13. Home Bros. mark an article to earn a net profit of 10% on sales. Overhead expense is 12%; selling expense is 10%. What is (a) the selling price if the cost is $74.80? What are (b) the net profit and (c) the gross profit?

14. The Fashion Shop marked up its merchandise on the selling price. Some blouses that were marked to sell for $21.00 were reduced 25% for a clearance sale. If the markup on the original selling price was 60%, what was the gross profit on each blouse sold during the clearance sale?

Objectives

The three most common methods of calculating depreciation are discussed here. Instructions are given in using each of these methods—the straight-line method, the declining-balance method, and the sum-of-the-years-digits method—in computing the annual depreciation and book value as well as preparing depreciation schedules.

5 DEPRECIATION

By definition, *depreciation* is a *decrease* in the value of property through wear, deterioration, or obsolescence. In any business enterprise it is the loss in value of fixed assets such as buildings and equipment.

Good business practice requires a means in the accounting system of providing funds for replacement of these assets when necessary. How much is set aside each year for this purpose depends on the method used, the type of asset, and the business practice. There are several methods of handling depreciation expense. Any reasonable method may be used and different methods may be used for different assets. However, the Federal Government limits the amount that can be claimed for tax purposes. The Federal Government also limits the use of certain methods which are explained in subsequent pages.

The value of any asset at the end of its useful life is called junk, scrap, or *salvage value.*† The *book value* of an asset at

†The following quote is from the Internal Revenue Service, Publication No. 17: "The determination of salvage value depends upon your policy. If it is your policy to dispose of assets which are still in good operating condition,

any time is the difference between original cost and the total of all depreciation charges to that date. Obviously, the book value equals the scrap value when the asset is due for replacement.

Three of the more common methods of computing depreciation are discussed in this chapter. They are: (1) the *straight-line* method, (2) the *declining-balance* method, and (3) the *sum-of-the-years-digits* method.

THE STRAIGHT-LINE METHOD

The straight-line method of computing depreciation is the simplest. When this method is used, the annual depreciation (amount written off) is the same throughout the useful life of the asset.

example A: Assume a typewriter that cost \$560.00 when new will have a useful life of 5 years. Scrap value is estimated to be \$63.00. What are the annual depreciation and the rate of depreciation? What is the book value at the end of 3 years?

solution: 1. depreciation
= original cost − scrap value
= \$560.00 − \$63.00 = \$497.00

annual depreciation
= \$497.00 ÷ 5 (years) = \$99.40

2. annual depreciation rate
= annual depreciation ÷ original cost
= \$99.40 ÷ \$560.00 = $17\frac{3}{4}\%$

3. book value (end of 3 years)
= original cost − (3 × annual depreciation)
= \$560.00 − (3 × \$99.40)
= \$560.00 − \$298.20 = \$261.80

---

the salvage value may represent a large part of the original cost of the asset. However, if you customarily use an asset until its inherent useful life has been substantially exhausted, salvage value may represent no more than junk value."

DEPRECIATION SCHEDULE FOR EXAMPLE A

End of Year	Annual Depreciation	Depreciation to Date	Book Value End of Year
0	—	—	$560.00
1	$99.40	$ 99.40	460.60
2	99.40	198.80	361.20
3	99.40	298.20	261.80
4	99.40	397.60	162.40
5	99.40	497.00	63.00

$497.00 + $63.00 = $560.00
total depreciation + scrap value = original cost

example B: The annual depreciation on a machine costing $1500.00 is $200.00. If the scrap value is $100.00, how many years of life were estimated?

solution: total depreciation = $1500.00 − $100.00
= $1400.00
estimated life = $1400.00 ÷ $200.00
= 7 years

EXERCISE 25

Solve the following problems by the straight-line method of depreciation

1. If a machine costing $2700.00 is worth $300.00 at the end of 15 years, find the annual depreciation charge.
2. A bus costing $12,000.00 will have a scrap value of $1170.00. The annual depreciation amounts to $902.50. How many years is it to be depreciated?
3. A machine costing $4500.00 is to be written off in 10 years. If it has a scrap value of $300.00 at the end of that time, what are (a) the annual depreciation charge and (b) the book value at the end of 5 years?
4. Furniture purchased for $1750.00 is depreciated over a 5-year period. The estimated scrap value is $300.00. Prepare a depreciation schedule for the 5 years.
5. A shop costing $19,200.00 has a life of 30 years. It has no scrap value.

(a) What is the annual depreciation charge? (b) What is the book value at the end of 10 years?

6. The yearly depreciation on tools costing $4650 was $775. What was their estimated life if there was no scrap value?

7. An automobile purchased for $3700.00 has an estimated life of 5 years at which time it will have a scrap value of $650.00. (a) What is the annual depreciation? (b) What is the book value at the end of 3 years?

8. An asset which cost $3000.00 has an estimated life of 5 years and a scrap value of $250.00. Prepare a schedule showing the depreciation to date and the book value for each year, as illustrated for Example A.

9. Mr. Holms had a new roof put on a rental house that he owned at a cost of $450.00. For income tax purposes he depreciated this expense over 10 years. How much would be written off (depreciated) at the end of 7 years?

10. Mr. Smith owned a house that he rented which he had painted at a cost of $875.00. The house was worth $35,000.00. For income tax purposes he set up a depreciation schedule of 30 years for the house and 5 years for the painting. Find the total depreciation at the end of 4 years on (a) the house and (b) the paint expense.

ACCELERATED METHODS OF DEPRECIATION

When the straight-line method of depreciation is used, the annual depreciation is the same for each year. However, sometimes a businessman may wish to charge off more to depreciation in the early life of an asset and lesser amounts as it gets older. The sum-of-the-years-digits and the declining-balance methods make this possible.

In many cases these methods are considered advantageous because a quicker recovery of the investment is realized. Also, savings in income taxes are made at a time when it may be most desirable. In addition, there tends to be less disparity between the book value and actual market value when these methods are applied to so-called style merchandise such as the automobile where the reduction in market value is often high during the first years of its life.

For tax purposes, either of these methods may be used if they meet requirements of the Internal Revenue Service. Limitations are discussed for each method under their respective headings.

THE SUM-OF-THE-YEARS-DIGITS METHOD

Internal Revenue Service Limitations: This method of depreciation may be used only if the property has a useful life of 3 or more years and was acquired new after December 13, 1953 or was constructed, reconstructed, or erected after December 13, 1953.

When this method is used, a different fraction is applied to the cost less salvage value for each year of depreciation. The denominator of the fractions used is the sum of the numbers representing the years of life of the property. For example, if the useful life of an asset is 5 years, then the sum-of-the-years-digits is $1 + 2 + 3 + 4 + 5$ or 15. The first year's depreciation will equal $\frac{5}{15}$ of the cost less salvage, the second year's depreciation will equal $\frac{4}{15}$ of the cost less salvage, etc., until the last or fifth year's depreciation will equal $\frac{1}{15}$ of the cost less salvage.

In this case it is a simple matter to add the digits 1 to 5 inclusive, but it becomes cumbersome if the life expectancy is 20 years. The sum of any series of this type may be quickly found by applying the formula

$$S = \frac{n^2 + n}{2}$$

where n represents the number of digits to be added or the life of the asset.

examples:

Life Expectancy		Sum of Digits
5 years	$S = \frac{5^2 + 5}{2} =$	15
6 years	$S = \frac{6^2 + 6}{2} =$	21
15 years	$S = \frac{15^2 + 15}{2} =$	120

The application of this method is best explained through an example.

example A: A typewriter cost $650.00. It is to be written off in 5 years with a scrap value of $50.00. Find the depreciation and book value for each year of its useful life. Prepare a depreciation schedule.

solution:
1. Since the number of years of useful life is 5, the sum-of-the-years-digits = 15 (see page 93).
2. The total amount of depreciation is $650.00 − $50.00 = $600.00
3. The depreciation is greatest during the early years. Therefore:

depreciation first year = $\frac{5}{15}$ × $600.00 = $200.00
depreciation second year = $\frac{4}{15}$ × $600.00 = $160.00
depreciation third year = $\frac{3}{15}$ × $600.00 = $120.00
depreciation fourth year = $\frac{2}{15}$ × $600.00 = $ 80.00
depreciation fifth year = $\frac{1}{15}$ × $600.00 = $ 40.00

total depreciation = $600.00

In all, $\frac{15}{15}$ or 100% will have been written off.

DEPRECIATION SCHEDULE FOR EXAMPLE A

End of Year	Annual Depreciation	Depreciation to Date	Book Value End of Year
0	—	—	$650.00
1	$200.00	$200.00	450.00
2	160.00	360.00	290.00
3	120.00	480.00	170.00
4	80.00	560.00	90.00
5	40.00	600.00	50.00*

*Scrap value.

example B: A piece of furniture costing \$725.00 has an estimated salvage value of \$25.00 at the end of 10 years. Find (a) the depreciation for the fifth year and the book value at the end of 5 years, and (b) the depreciation for the third year.

solution: (a) $S = \dfrac{10^2 + 10}{2} = 55$ (sum of digits)

total depreciation

$= \$725.00 - \$25.00 = \$700.00$

depreciation for the fifth year

$= \frac{6}{55} \times \$700.00 = \$76.36$

book value end of 5 years

$= \$725.00 -$ depreciation for first 5 years

$= \$725.00 - (40/55^{*} \times \$700.00)$

$= \$725.00 - \$509.09 = \$215.91$

*depreciation for first 5 years equals $10/55 + 9/55 + 8/55 + 7/55 + 6/55$ *or* $40/55$

(b) depreciation for third year

$= 8/55 \times \$700.00 = \$101.82$

EXERCISE 26

Solve the following problems by the sum-of-the-years-digits method of depreciation.

1. A jeep costing \$2730.00 has an estimated salvage value of \$510.00 at the end of 5 years. Find the depreciation for the (a) second year and (b) the fourth year.

2. An adding machine costing \$420.00 has an estimated life of 6 years. At the end of that time it has a resale value of 12% of its original cost. (a) How much was the depreciation for the third year? (b) What was the book value at the end of the second year?

3. Set up a depreciation schedule for a $900.00 asset whose life is 4 years. Scrap value is estimated at $75.00

4. The depreciation on an asset for the fourth year was $20.70. If the useful life of this property was 5 years, what was the original cost assuming scrap value equalled $7.50?

5. A merchant buys $8000.00 worth of store fixtures which he estimates will last 20 years. At the end of this time they will have a scrap value of $500.00. What is the amount charged to depreciation at the end of the first year?

6. Find the book value of the store fixtures in problem 5 at the end of 15 years.

7. The Alpha Manufacturing Company purchased a car for $5500 for the use of a company representative. It was to be depreciated over 5 years. Estimated salvage value was $750. Find the (a) annual depreciation by the straight-line method and (b) the depreciation for the first year by the sum-of-the-years-digits method.

8. What is the book value for the automobile in problem 7 at the end of 2 years by both methods?

9. What fractional part of an asset is depreciated during the first year for (a) 5 years, (b) 12 years, (c) 30 years? (Reduce fractions to lowest terms.)

10. Find the depreciation for the first year for an asset worth $1209.00 if it is depreciated over 5 years, 12 years, 30 years.

THE DECLINING-BALANCE METHOD

Internal Revenue Service Limitations: The maximum rate which may be used in computing depreciation by this method is twice the straight-line rate provided (1) the property has a useful life of 3 or more years and (2) was acquired new after December 31, 1953 or was constructed, reconstructed, or erected after December 31, 1953. One and one-half times the straight-line rate is the maximum rate that may be used on tangible property that does not meet the above requirements if this method results in a reasonable allowance for depreciation. This includes new or used property acquired before January 1, 1954 and used property acquired after December 31, 1953.

If this method is used, the salvage value is ignored when the annual depreciation is calculated. However, no depreciation may be allowed below a reasonable salvage value.

Since the principle involved is the same if the rate is 2 or $1\frac{1}{2}$ times the straight-line rate, the rate used for all problems in this text involving the declining-balance method is 2 times the straight-line rate.

example: An asset which cost $450.00 was to be written off in 5 years. Salvage value was estimated at $50.00. Find the depreciation and book value for each year, using the declining-balance method.

solution:

straight-line rate = 20%

2 × straight-line rate = 40%

cost of asset	$450.00	
depreciation first year	180.00	(40% of $450.00)
book value end of first year	$270.00	
depreciation second year	108.00	(40% of $270.00)
book value end of second year	$162.00	
depreciation third year	64.80	(40% of $162.00)
book value end of third year	$ 97.20	
depreciation fourth year	38.88	(40% of $ 97.20)
book value end of fourth year	$ 58.32	
depreciation fifth year	8.32*	
book value end of fifth year	$ 50.00	salvage value

*40% of $58.32 = $23.33. However, if

this figure was used, it would reduce the book value at the end of 5 years to less than the estimated salvage value. Therefore, no more than $8.32 may be charged off in the fifth year.

DEPRECIATION SCHEDULE FOR EXAMPLE

End of Year	Annual Depreciation	Depreciation to Date	Book Value End of Year
0	—	—	$450.00
1	$180.00	$180.00	270.00
2	108.00	288.00	162.00
3	64.80	352.80	97.20
4	38.88	391.68	58.32
5	8.32	400.00	50.00*

*Salvage value.

EXERCISE 27

Solve the following problems using the declining method of depreciation.

1. A piece of furniture was acquired for $765.00. It is to be written off in 10 years. The salvage value is estimated at $50.00. Find the book value at the end of 3 years.

2. A machine costing $677.50 will last 10 years and have a scrap value of $50.00. (a) Find the book value at the end of 3 years. (b) How much would the book value be if the straight-line method was used?

3. An asset costing $3000 has a life of 20 years. What is the depreciation for the third year?

4. Using a rate of 20%, prepare a depreciation schedule for the first 5 years for a machine that cost $9000.

5. What rate should be used for assets that are to be written off in 8 years? 15 years? 30 years? 12 years?

6. (a) Calculate the depreciation for the third year on an asset worth $1000 if its life is estimated to be 8 years. (b) What is the book value at the end of 3 years?

7. An asset acquired for $25,000 has a scrap value of $1500 at the end of 40 years. (a) Compute the depreciation for the first 3 years at a rate of 5%. (b) What is the book value at the end of 3 years?

8. A glass company provides for replacement of their trucks at a rate of $16\frac{2}{3}$% of cost on a declining balance. Compute the depreciation for the first 2 years on a service truck costing $5400.

9. Compute the depreciation for the following assets for the first 2 years using 2 × the straight-line rate.

	Asset	Cost	Salvage Value	Useful Life
(a)	typewriter	$360.00	$50.00	6 years
(b)	filing cabinet	125.00	35.00	10 years

10. Compute the book value at the end of the second year for the assets listed in problem 9.

ADDITIONAL FIRST-YEAR DEPRECIATION

The Internal Revenue Service permits an additional first-year allowance for depreciation on tangible personal property if it has a life of over 6 years. *This amounts to 20% of the original cost.* The allowance may be applied to any method of depreciation but must be claimed during the first year of ownership.

example: A desk calculator that cost $1205.00 is to be written off in 10 years. The salvage value is estimated to be $100.00. An additional first-year depreciation is to be taken. Prepare a depreciation schedule for the three methods discussed in this chapter.

solution:

ANNUAL DEPRECIATION

Year	Straight-Line	Declining-Balance	Sum-of-the-Years-Digits
First-yr additional depreciation (20% of original cost)	$ 241.00	$ 241.00	$ 241.00
First-yr ordinary depreciation	86.40	192.80	157.09
Total first-yr depreciation	$ 327.40	$ 433.80	$ 398.09
Second-yr depreciation	86.40	154.24	141.38
Third-yr depreciation	86.40	123.39	125.67
Fourth-yr depreciation	86.40	98.71	109.96
Fifth-yr depreciation	86.40	78.97	94.26
Sixth-yr depreciation	86.40	63.18	78.55
Seventh-yr depreciation	86.40	50.54	62.83
Eighth-yr depreciation	86.40	40.43	47.13
Ninth-yr depreciation	86.40	32.35	31.42
Tenth-yr depreciation	86.40	25.88	15.71
Total depreciation	$1105.00	$1101.49	$1105.00
Salvage value	$100.00	$103.51	$100.00

Under the straight-line method: The annual ordinary depreciation shown above is $86.40 computed as follows: (1) Deduct the salvage value and additional first-year depreciation allowance from the cost, i.e., $1205.00 – ($100.00 + $241.00) = $864.00. (2) Divide $864.00 by 10 (years of useful life), i.e., $864.00 ÷ 10 = $86.40.

Under the declining-balance method: The annual ordinary depreciation may not exceed 20% (twice the straight-line rate) for this example. Since the salvage value is ignored, the ordinary depreciation for the first year is 20% of the cost after the additional first-year depreciation is deducted, i.e., 20% of $964.00 or $192.80. The ordinary depreciation for the second year is 20% of $771.20 or $154.24, and so on.

Under the sum-of-the-years-digits method: The sum-of-the-years-digits is 55. Then the ordinary depreciation for the first year equals $\frac{10}{55}$ of \$864 (cost less salvage value and additional first-year depreciation). The second-year ordinary depreciation equals $\frac{9}{55}$ of \$864, the third year depreciation equals $\frac{8}{55}$ of \$864, and so on.

EXERCISE 28 Chapter Summary

1. (a) What is the annual depreciation, using the straight-line method, on a building worth \$65,000 if it is built to last 40 years and has an estimated salvage value of \$12,000? (b) What would be the accumulated depreciation after 10 years? (c) What would the book value equal after 25 years?

2. A manufacturing concern bought a piece of equipment for \$325. It was depreciated at the rate of \$25 annually. If the salvage value was \$25 what was its life expectancy? What was the annual rate of depreciation expressed as a percent?

3. In problem 1, if an additional first-year depreciation is taken, how much is the regular annual depreciation?

4. Property acquired for \$75,000 has an estimated life of 10 years. If the scrap value is \$3500, find (a) the depreciation for the second year and (b) the book value at the end of 8 years. Use the sum-of-the-years-digits method.

5. An asset cost \$6000 and has an estimated life of 10 years. It has no salvage value. Find the depreciation for the first year by using (a) the straight-line method, (b) the declining-balance method, and (c) the sum-of-the-years-digits method.

6. A truck cost \$6500 and has an estimated trade-in value of \$800 at the end of 5 years. Find the book value at the end of 3 years by using the declining-balance method.

7. What are the rates for the following by the declining balance method (use 2 $\times$ straight-line rate) for assets with a useful life of (a) 6 years, (b) 8 years, (c) 30 years. Express as a fraction reduced to lowest terms.

8. Set up a depreciation schedule for an asset worth $530 with a scrap value of $50 at the end of 5 years. Use the sum-of-the-years-digits method.

9. Mr. Allen owns a house that he rents for $250 a month. He is depreciating the value of the house (worth $34,500) over 30 years. (Salvage value is ignored.) Expenses for the year include a new roof costing $900 and an exterior paint job costing $675. Both of these expenses are to be "written off" in 10 years. Compute Mr. Allen's net profit or loss for the year if he had additional expenses of $850 for taxes, $31.67 for insurance, and miscellaneous expenses amounting to $21.60. (Net profit equals income less the sum of depreciation expense, repairs, and taxes.)

10. Mrs. Parker rents a home she owns which is worth $25,000 for $200 a month. Depreciation is based on 20 years or 5%. Exterior paint work cost $281.74 which is depreciated at a rate of 20% annually. Other expenses for the year are: taxes $688.64, insurance $41.00, miscellaneous repairs $19.30. Compute the net income on this investment for the year.

Objectives

This chapter explains and illustrates how property tax rates and taxes are determined and how sales taxes are applied. Ample problems are included in which these two forms of taxes are applied for practice and study.

6

PROPERTY AND SALES TAXES

A *tax* is an assessment made by law for public use and services. Taxes are derived by various means and from an increasing variety of sources. Methods of collecting taxes and the manner in which they are used also vary considerably between states, counties, municipalities, and special districts.

The chief sources of income for the Federal Government are: (1) taxes on the *income* or *earnings* of individuals, partnerships, and corporations, (2) taxes on *goods* brought into this country from another country, and (3) taxes on *luxury items* such as tobacco, gasoline, and alcoholic beverages.

State income is derived primarily from individual income taxes, general sales taxes, gasoline taxes, corporate income taxes, estate taxes, taxes on gifts, and taxes on tobacco, alcoholic beverages, licenses of various sorts, etc. Tax structure and types of taxes vary between states.

Local governments such as counties, municipalities, and special districts derive their income from many tax sources such as special fees, licenses, parking meters, sales taxes, public utilities, and special authorities (toll roads and bridges). However, the main sources of income are taxes levied against (1) real property (land and fixed improvements such as buildings) and (2) personal property (securities, cash, furniture, etc.)

PROPERTY TAXES

Although property taxes vary from state to state, the method of computation is the same. The tax rate is based on the *assessed valuation* of the property; this valuation is determined by individuals called assessors who are elected or appointed. This valuation represents a certain percent or fractional part of the current market price. Some states may have an assessment rate of 20% while others may have a rate as high as 60%.

Each state provides, within its governing framework, means for periodically adjusting assessment rates in keeping with changing fair market values. This is done through the legislature or some body such as county boards of assessment or state boards of equalization. Valuations must be at the fair market value. The assessed valuation includes both real and personal property.

FINDING THE TAX RATE

In order to determine the total funds needed, the usual procedure is to prepare a budget based on the requirements of the county, district, or municipality. After it has been accepted and all other sources of income taken into account, the tax rate is determined as follows.

$$\text{Tax Rate}^{\dagger} = \frac{\text{Total Funds Needed}}{\text{Total Assessed Valuation}}$$

$$\text{or } \frac{\text{Taxes}}{\text{Assessed Valuation}}$$

†Tax rates can and usually do vary from district to district within a state, county, or city. These variations are due to the amount of bonded indebtedness, type of improvements or services, etc.

example: Let us assume that a town needs $465,000 to meet its requirements. If the assessed valuation is $10,000,000, the tax rate equals $465,000 ÷ $10,000,000 which is .0465 or 4.65%. This means that for every dollar of assessed valuation, the tax payer is obligated to pay 4.65¢.

Tax rates may be expressed in several ways—as a percent, in mills (tenth of a cent), or in dollars and cents. The rate for the example above may be expressed as:

1. $0.0465 per dollar of assessed valuation
2. 4.65¢ per dollar of assessed valuation
3. 46.5 mills per dollar of assessed valuation
4. $4.65 per $100 of assessed valuation
5. $46.50 per $1000 of assessed valuation

Note: When there is a fractional remainder, the rate is recorded correct to the next higher digit. Therefore, .0432712 recorded to 3 decimals would equal .044, not .043.

The tax rate is actually the total of several rates applied to special purposes. For example, a total tax rate of $10.39 per $100 of assessed valuation may be made up of rates as follows: county, $2.86; city, $2.34; school, $4.84; special district, $0.35: In some cases there may be a water district, transit district, mosquito abatement district, and others.

FINDING THE TAX

The following formulas are used to find the tax:

Assessed Valuation = Assessment Rate × Market Value
Taxes = Assessed Valuation × Tax Rate

example: Mr. Riedell lives in the country and therefore has no city tax. His land is assessed at $1320, improvements at $2750. Tax rates per $100 of assessed valuation are: county, $2.95; school, $5.04; special district, $0.53: Find the amount of tax for each purpose, the total tax rate, and the total taxes to be paid.

solution: total assessed valuation
 = \$1320 + \$2750 = \$4070

county tax
 = \$40.70 × \$2.95 = \$120.065
school tax
 = \$40.70 × \$5.04 = 205.128
special district
 = \$40.70 × \$0.53 = 21.571
total taxes = \$346.764 *or* \$346.76

total tax rate
 = \$8.52 per \$100 of assessed valuation
total tax
 = \$40.70 × \$8.52 = \$346.76

example: Mrs. Cecil Jones owned a home worth \$17,500. The assessment rate was 28%. What tax did she pay if the tax rate was \$9.16667 per \$100 assessed valuation?

solution: assessed valuation = \$17,500 × .28 = \$4900
tax = \$49.00 × 9.16667 = \$449.17

EXERCISE 29

I. Change the given rate to the new rate:

1. 67 mills to percent

2. \$1.68 per \$100 to per \$1000

3. \$10.667 per \$100 to per dollar

4. 4.012¢ per dollar to dollars per \$100

5. 11.916% to dollars per \$100

6. $3\frac{1}{2}\%$ to dollars per \$1000

7. $2\frac{1}{2}$ mills to dollars per \$100

8. \$12.8125 per \$100 to nearest whole mills

9. \$3.02 per \$100 to cents per \$1.00

10. \$0.52 per \$1000 to percent

II. Find the amount of tax for the following:

No.	Assessed Value	Tax Rate
1.	$3,000	$6\frac{1}{2}\%$
2.	50,000	$11.06 per $100
3.	309,060	$0.035 per $1.00
4.	65,450	$0.04 per $1.00
5.	18,000	$15.168 per $1000
6.	17,500	$12.86 per $100

III. Find the assessed valuation and amount of tax paid for the following:

No.	Market Value	Assessment Rate	Tax Rate
1.	$20,000	26%	$12.50 per $100
2.	21,500	25%	$ 8.33 per $100
3.	35,750	20%	$10.06 per $100
4.	64,900	28%	$6\frac{1}{2}\%$
5.	125,600	25%	$8\frac{1}{3}\%$
6.	44,900	30%	$0.25 per dollar

IV. Word problems:

1. How much does Mrs. James pay in taxes if her home is assessed at $25,000 and the tax rate is $12.55 per $100?

2. Find the assessed value of property worth $16,500 if the assessment rate is 27%.

3. Taxes amounting to $5000 were paid on a piece of property in a district where the tax rate was $12\frac{1}{2}\%$. At this rate, what would be the expected market value of the property if the assessment rate was 25%?

4. The tax rate in Middletown is $6.15 per $100. The budget is established at $4,750,000. What is the total assessed valuation of property in Middletown (nearest dollar)?

5. The total budget of a certain city is $136,845,000 of which 75% must be raised by property taxes. If the tax rate is

$11.00 per $100, what is the total assessed valuation of property in the city? (Nearest dollar.)

6. H. Howen owns a piece of property worth $16,500 which is assessed at 22% of its value. He pays $0.41 per $100 for a special junior college fund, $0.02 per $100 for education of mentally retarded, $0.001 per $100 for education of institutionalized pupils (countywide), and $0.089 per $100 mandatory countywide tax in place of state equalization aid. These are additional educational levies to the school district rates. How much did Mr. Howen pay for each of these needs? What was the total tax?

7. What is the current market value of a home if current taxes amounted to $350, the tax rate is $5.00 per $100, and the assessment rate is 25%?

8. John Bradshaw purchased a home for $35,750 in a town with an assessment rate of 30%. The tax rate is $8.35 per $100. How much of his total tax is spent on education if 62% of the district taxes are spent for that purpose?

9. What is the tax rate per $100 (corrected to 2 decimals) for a community where property is assessed at $43,460,000 if the budget requirements total $6,000,000?

10. Mr. Mulliken owns a home in a city which he rents for $225 a month. The assessed value of this property is $3000 for the land, $2950 for the improvements, and $50 for personal property. (a) If the tax rate is $11.53 per $100, find the taxes paid for the year. (b) If repairs and depreciation amounted to $256 in a year, what was the net income realized?

SALES TAXES

The general sales tax has become an increasingly important and popular source of revenue for most states as well as many counties and municipalities. In addition to a tax on retail sales, many districts levy a tax on lodging and meals that are served in restaurants. Taxes levied on other services and nonretail businesses vary greatly between states and are too numerous to discuss here.

The seller acts as an agent because he collects the tax

directly from the buyer at the time of the sale. The amount collected is then remitted periodically to the city, county, or state as required by law.

Tax rates are expressed as a percent and vary between states from 2 to $6\frac{1}{2}$ percent at this time. City and county, taxes are generally much lower—$\frac{1}{2}\%$, and 1%. Although there are a few exceptions to the rule, in general a sales tax is levied only on purchases made and delivered within a tax area. Purchases delivered out of the tax area are generally tax free.

To insure a high degree of accuracy in filling out sales slips or invoices and to save time, tax schedules are used by most sales clerks. Examples on the following pages are illustrations of partial sales tax collection schedules for three states.

The *excise tax*, by definition, is a Federal tax levied on the manufacture, sale, or consumption of a commodity within a country. Since it represents part of the cost, it is passed on to the consumer or buyer as part of the purchase price. However, if this tax appears on an invoice distinct from other costs of an item, it must be added to the retail price before the sales tax is computed. There are only a few commodities on which an excise tax is so applied.

example: Mr. Haynes purchased a desk for $190.00. If the sales tax rate was 4%, how much did the desk cost?

solution:

tax = .04 × $190.00 = $7.60

total cost = $190.00 + $7.60 = $197.60

example: George Mann purchased 2 automobile tires at $15.95 each. The excise tax on each tire was $2.08. What was the total amount of the purchase if a sales tax of 3% was included?

solution: basic sales price = 2 × $15.95 = $31.90

excise tax = 2 × $2.08 = $4.16

total sales price = $31.90 + $4.16 = $36.06

sales tax = .03 × $36.06 = $1.08

total amount of purchase

= $36.06 + $1.08 = $37.14

ST-110 (1/72) NYS Sales Tax Bureau

SALES AND USE TAX BRACKET SCHEDULE FOR STATE AND LOCAL TAX PURPOSES

REVISED - EFFECTIVE JUNE 1, 1971

4% SALES AND USE TAX COLLECTION CHART

Amount of Sale	Tax to be Collected	Amount of Sale	Tax to be Collected
$0.01 to $0.12	None	$5.13 to $5.37	$.21
.13 to .33	1¢	5.38 to 5.62	.22
.34 to .58	2¢	5.63 to 5.87	.23
.59 to .83	3¢	5.88 to 6.12	.24
.84 to 1.12	4¢	6.13 to 6.37	.25
1.13 to 1.37	5¢	6.38 to 6.62	.26
1.38 to 1.62	6¢	6.63 to 6.87	.27
1.63 to 1.87	7¢	6.88 to 7.12	.28
1.88 to 2.12	8¢	7.13 to 7.37	.29
2.13 to 2.37	9¢	7.38 to 7.62	.30
2.38 to 2.62	$.10	7.63 to 7.87	.31
2.63 to 2.87	.11	7.88 to 8.12	.32
2.88 to 3.12	.12	8.13 to 8.37	.33
3.13 to 3.37	.13	8.38 to 8.62	.34
3.38 to 3.62	.14	8.63 to 8.87	.35
3.63 to 3.87	.15	8.88 to 9.12	.36
3.88 to 4.12	.16	9.13 to 9.37	.37
4.13 to 4.37	.17	9.38 to 9.62	.38
4.38 to 4.62	.18	9.63 to 9.87	.39
4.63 to 4.87	.19	9.88 to 10.00	.40
4.88 to 5.12	.20		

5% COMBINED SALES AND USE TAX COLLECTION CHART

Amount of Sale	Tax to be Collected	Amount of Sale	Tax to be Collected
$0.01 to $0.10	None	$.5.10 to $5.29	$.26
.11 to .27	1¢	5.30 to 5.49	.27
.28 to .47	2¢	5.50 to 5.69	.28
.48 to .67	3¢	5.70 to 5.89	.29
.68 to .87	4¢	5.90 to 6.09	.30
.88 to 1.09	5¢	6.10 to 6.29	.31
1.10 to 1.29	6¢	6.30 to 6.49	.32
1.30 to 1.49	7¢	6.50 to 6.69	.33
1.50 to 1.69	8¢	6.70 to 6.89	.34
1.70 to 1.89	9¢	6.90 to 7.09	.35
1.90 to 2.09	$.10	7.10 to 7.29	.36
2.10 to 2.29	.11	7.30 to 7.49	.37
2.30 to 2.49	.12	7.50 to 7.69	.38
2.50 to 2.69	.13	7.70 to 7.89	.39
2.70 to 2.89	.14	7.90 to 8.09	.40
2.90 to 3.09	.15	8.10 to 8.29	.41
3.10 to 3.29	.16	8.30 to 8.49	.42
3.30 to 3.49	.17	8.50 to 8.69	.43
3.50 to 3.69	.18	8.70 to 8.89	.44
3.70 to 3.89	.19	8.90 to 9.09	.45
3.90 to 4.09	.20	9. 10 to 9.29	.46
4.10 to 4.29	.21	9.30 to 9.49	.47
4.30 to 4.49	.22	9.50 to 9.69	.48
4.50 to 4.69	.23	9.70 to 9.89	.49
4.70 to 4.89	.24	9.90 to 10.00	.50
4.90 to 5.09	.25		

6% COMBINED SALES AND USE TAX COLLECTION CHART

Amount of Sale	Tax to be Collected	Amount of Sale	Tax to be Collected
$0.01 to $0.10	None	$5.09 to $5.24	$.31
.11 to .22	1¢	5.25 to 5.41	.32
.23 to .38	2¢	5.42 to 5.58	.33
.39 to .56	3¢	5.59 to 5.74	.34
.57 to .72	4¢	5.75 to 5.91	.35
.73 to .88	5¢	5.92 to 6.08	.36
.89 to 1.08	6¢	6.09 to 6.24	.37
1.09 to 1.24	7¢	6.25 to 6.41	.38
1.25 to 1.41	8¢	6.42 to 6.58	.39
1.42 to 1.58	9¢	6.59 to 6.74	.40
1.59 to 1.74	$.10	6.75 to 6.91	.41
1.75 to 1.91	.11	6.92 to 7.08	.42
1.92 to 2.08	.12	7.09 to 7.24	.43
2.09 to 2.24	.13	7.25 to 7.41	.44
2.25 to 2.41	.14	7.42 to 7.58	.45
2.42 to 2.58	.15	7.59 to 7.74	.46
2.59 to 2.74	.16	7.75 to 7.91	.47
2.75 to 2.91	.17	7.92 to 8.08	.48
2.92 to 3.08	.18	8.09 to 8.24	.49
3.09 to 3.24	.19	8.25 to 8.41	.50
3.25 to 3.41	.20	8.42 to 8.58	.51
3.42 to 3.58	.21	8.59 to 8.74	.52
3.59 to 3.74	.22	8.75 to 8.91	.53
3.75 to 3.91	.23	8.92 to 9.08	.54
3.92 to 4.08	.24	9.09 to 9.24	.55
4.09 to 4.24	.25	9.25 to 9.41	.56
4.25 to 4.41	.26	9.42 to 9.58	.57
4.42 to 4.58	.27	9.59 to 9.74	.58
4.59 to 4.74	.28	9.75 to 9.91	.59
4.75 to 4.91	.29	9.92 to 10.00	.60
4.92 to 5.08	.30		

7% COMBINED SALES AND USE TAX COLLECTION CHART

Amount of Sale	Tax to be Collected	Amount of Sale	Tax to be Collected
$0.01 to $0.10	None	$5.08 to $5.21	$.36
.11 to .20	1¢	5.22 to 5.35	.37
.21 to .33	2¢	5.36 to 5.49	.38
.34 to .47	3¢	5.50 to 5.64	.39
.48 to .62	4¢	5.65 to 5.78	.40
.63 to .76	5¢	5.79 to 5.92	.41
.77 to .91	6¢	5.93 to 6.07	.42
.92 to 1.07	7¢	6.08 to 6.21	.43
1.08 to 1.21	8¢	6.22 to 6.35	.44
1.22 to 1.35	9¢	6.36 to 6.49	.45
1.36 to 1.49	$.10	6.50 to 6.64	.46
1.50 to 1.64	.11	6.65 to 6.78	.47
1.65 to 1.78	.12	6.79 to 6.92	.48
1.79 to 1.92	.13	6.93 to 7.07	.49
1.93 to 2.07	.14	7.08 to 7.21	.50
2.08 to 2.21	.15	7.22 to 7.35	.51
2.22 to 2.35	.16	7.36 to 7.49	.52
2.36 to 2.49	.17	7.50 to 7.64	.53
2.50 to 2.64	.18	7.65 to 7.78	.54
2.65 to 2.78	.19	7.79 to 7.92	.55
2.79 to 2.92	.20	7.93 to 8.07	.56
2.93 to 3.07	.21	8.08 to 8.21	.57
3.08 to 3.21	.22	8.22 to 8.35	.58
3.22 to 3.35	.23	8.36 to 8.49	.59
3.36 to 3.49	.24	8.50 to 8.64	.60
3.50 to 3.64	.25	8.65 to 8.78	.61
3.65 to 3.78	.26	8.79 to 8.92	.62
3.79 to 3.92	.27	8.93 to 9.07	.63
3.93 to 4.07	.28	9.08 to 9.21	.64
4.08 to 4.21	.29	9.22 to 9.35	.65
4.22 to 4.35	.30	9.36 to 9.49	.66
4.36 to 4,49	.31	9.50 to 9.64	.67
4.50 to 4.64	.32	9.65 to 9.78	.68
4.65 to 4.78	.33	9.79 to 9.92	.69
4.79 to 4.92	.34	9.93 to 10.00	.70
4.93 to 5.07	.35		

On sales over $10.00, compute the tax by multiplying the amount of sale by the applicable tax rate and rounding the result to the nearest whole cent.

STATE OF CALIFORNIA 5% SALES TAX REIMBURSEMENT SCHEDULE

Transaction	Tax	Transaction	Tax	Transaction	Tax
.01– .10	.00	8.50– 8.69	.43	17.10–17.29	.86
.11– .27	.01	8.70– 8.89	.44	17.30–17.49	.87
.28– .47	.02	8.90– 9.09	.45	17.50–17.69	.88
.48– .68	.03	9.10– 9.29	.46	17.70–17.89	.89
.69– .89	.04	9.30– 9.49	.47	17.90–18.09	.90
.90– 1.09	.05	9.50– 9.69	.48	18.10–18.29	.91
1.10– 1.29	.06	9.70– 9.89	.49	18.30–18.49	.92
1.30– 1.49	.07	9.90–10.09	.50	18.50–18.69	.93
1.50– 1.69	.08	10.10–10.29	.51	18.70–18.89	.94
1.70– 1.89	.09	10.30–10.49	.52	18.90–19.09	.95
1.90– 2.09	.10	10.50–10.69	.53	19.10–19.29	.96
2.10– 2.29	.11	10.70–10.89	.54	19.30–19.49	.97
2.30– 2.49	.12	10.90–11.09	.55	19.50–19.69	.98
2.50– 2.69	.13	11.10–11.29	.56	19.70–19.89	.99
2.70– 2.89	.14	11.30–11.49	.57	19.90–20.09	1.00
2.90– 3.09	.15	11.50–11.69	.58	20.10–20.29	1.01
3.10– 3.29	.16	11.70–11.89	.59	20.30–20.49	1.02
3.30– 3.49	.17	11.90–12.09	.60	20.50–20.69	1.03
3.50– 3.69	.18	12.10–12.29	.61	20.70–20.89	1.04
3.70– 3.89	.19	12.30–12.49	.62	20.90–21.09	1.05
3.90– 4.09	.20	12.50–12.69	.63	21.10–21.29	1.06
4.10– 4.29	.21	12.70–12.89	.64	21.30–21.49	1.07
4.30– 4.49	.22	12.90–13.09	.65	21.50–21.69	1.08
4.50– 4.69	.23	13.10–13.29	.66	21.70–21.89	1.09
4.70– 4.89	.24	13.30–13.49	.67	21.90–22.09	1.10
4.90– 5.09	.25	13.50–13.69	.68	22.10–22.29	1.11
5.10– 5.29	.26	13.70–13.89	.69	22.30–22.49	1.12
5.30– 5.49	.27	13.90–14.09	.70	22.50–22.69	1.13
5.50– 5.69	.28	14.10–14.29	.71	22.70–22.89	1.14
5.70– 5.89	.29	14.30–14.49	.72	22.90–23.09	1.15
5.90– 6.09	.30	14.50–14.69	.73	23.10–23.29	1.16
6.10– 6.29	.31	14.70–14.89	.74	23.30–23.49	1.17
6.30– 6.49	.32	14.90–15.09	.75	23.50–23.69	1.18
6.50– 6.69	.33	15.10–15.29	.76	23.70–23.89	1.19
6.70– 6.89	.34	15.30–15.49	.77	23.90–24.09	1.20
6.90– 7.09	.35	15.50–15.69	.78	24.10–24.29	1.21
7.10– 7.29	.36	15.70–15.89	.79	24.30–24.49	1.22
7.30– 7.49	.37	15.90–16.09	.80	24.50–24.69	1.23
7.50– 7.69	.38	16.10–16.29	.81	24.70–24.89	1.24
7.70– 7.89	.39	16.30–16.49	.82	24.90–25.09	1.25
7.90– 8.09	.40	16.50–16.69	.83	25.10–25.29	1.26
8.10– 8.29	.41	16.70–16.89	.84	25.30–25.49	1.27
8.30– 8.49	.42	16.90–17.09	.85	25.50–25.69	1.28

PAUL R. LEAKE
MEMBER, STATE BOARD OF EQUALIZATION
905 Court Street, Woodland, California 95695

BT-72.3 REV. 2 (8-67)

STATE OF NEW JERSEY
DEPARTMENT OF THE TREASURY
Division of Taxation

SALES TAX COLLECTION SCHEDULE

RATE: 5% EFFECTIVE MARCH 1, 1970

Amount of Sale	Tax to be Collected	Amount of Sale	Tax to be Collected
$0.01 to $0.10	None	$6.11 to $6.25	$0.31
0.11 to 0.25	1¢	6.26 to 6.46	.32
0.26 to 0.46	2¢	6.47 to 6.67	.33
0.47 to 0.67	3¢	6.63 to 6.88	.34
0.68 to 0.88	4¢	6.89 to 7.10	.35
0.89 to 1.10	5¢	7.11 to 7.25	.36
1.11 to 1.25	6¢	7.26 to 7.46	.37
1.26 to 1.46	7¢	7.47 to 7.67	.38
1.47 to 1.67	8¢	7.63 to 7.88	.39
1.68 to 1.88	9¢	7.89 to 8.10	.40
1.89 to 2.10	$0.10	8.11 to 8.25	.41
2.11 to 2.25	.11	8.26 to 8.46	.42
2.26 to 2.46	.12	8.47 to 8.67	.43
2.47 to 2.67	.13	8.63 to 8.88	.44
2.68 to 2.88	.14	8.89 to 9.10	.45
2.89 to 3.10	.15	9.11 to 9.25	.46
3.11 to 3.25	.16	9.26 to 9.46	.47
3.26 to 3.46	.17	9.47 to 9.67	.48
3.47 to 3.67	.18	9.68 to 9.88	.49
3.68 to 3.88	.19	9.89 to 10.00	.50
3.89 to 4.10	.20	Over $ 10	.50*
4.11 to 4.25	.21	Over 20	1.00*
4.26 to 4.46	.22	Over 30	1.50*
4.47 to 4.67	.23	Over 40	2.00*
4.68 to 4.88	.24	Over 50	2.50*
4.89 to 5.10	.25	Over 60	3.00*
5.11 to 5.25	.26	Over 70	3.50*
5.26 to 5.46	.27	Over 80	4.00*
5.47 to 5.67	.28	Over 90	4.50*
5.68 to 5.88	.29	Over 100	5.00*
5.89 to 6.10	.30	Over 200	10.00*

*On amounts above $10.00, the tax shall be $0.05 on each full dollar of the amount of sale, plus the tax on each part of a dollar in excess of a full dollar in accordance with the above formula.

ST-75 (3-70) JTC 23721 PRINTED IN U.S.A.

example: J. E. Jensen ordered a lawn mower priced at \$69.75. He was allowed a discount of 5%. The sales tax equaled 3% for the state and $\frac{1}{2}$% for the city. How much did the lawn mower cost?
(Not to be confused with third example shown following. This is not a cash discount.)

solution:

$$\begin{aligned} \text{discount} &= .05 \times \$69.75 = \$3.49 \\ \text{net price} &= \$69.75 - \$3.49 = \$66.26 \\ \text{sales tax } (3\tfrac{1}{2}\%) &= .035 \times \$66.26 = \$2.32 \\ \text{total cost} &= \$66.26 + \$2.32 = \$68.58 \end{aligned}$$

example: If a coat cost \$107.10 including a sales tax of 2%, what was the list price?

solution: Since \$107.10 = 102% of the list price, the list price (100%) = \$107.10 ÷ 1.02 = \$105.00.

example: Mr. Allen purchased a pump for \$465.00 with terms of 2%, 15 days. Sales tax was 5%. If he paid this obligation in time to take advantage of the cash discount, how much did he remit in payment?

solution:

selling price	=	\$465.00	
plus sales tax	=	23.25	(5% of \$465.00)
total cost	=	\$488.25	
less cash discount	=	9.30	(2% of \$465.00)
amount paid	=	\$478.95	

Notice that a *cash* discount must be computed on the base price *before* the sales tax is added.

EXERCISE 30

1. If the sales tax in a state is 4%, what tax is collected on the following purchases: (a) \$48.20, (b) \$36.01, (c) \$3.12, (d) \$720.60, (e) \$20.16, and (f) \$0.89?

2. Mrs. O'Neil had a suit cleaned for \$2.25. What was the total charge if she paid a tax of 4.2% for the service?

3. What is the tax at 3% on the following purchase: 3 buttons at 58¢

each, 2 seam rippers at 29¢ each, 4 spools of thread at 15¢ each, 2 zippers at 57¢ each, and 2 packages of seam binding at 7¢ each?

4. Mr. Guthrie purchased 2 automobile tires, size 8.85 × 15, for $41.93 each. The excise tax on each tire was $2.50. Find the sales tax if the state rate was 4%.

5. What was the marked price for an article which cost $66.63 including a $2\frac{1}{2}$% sales tax?

6. The selling price of a coat was $45.00. If it was sold for $45.90 including the sales tax, what was the sales tax rate?

7. On July 10, J. Rowe purchased a table for $75.00 plus $6\frac{1}{2}$% sales tax with terms of 3/15. If he paid this bill on July 24, how much did he remit in payment?

8. H. Harrod made purchases amounting to $216.00. Sales taxes were 4%. If he was granted a trade discount of 15%, what was the total amount paid?

REVIEW

1. The price per share of a certain security was $27.50. It increased to $50.00 within 3 years. What were the amount and the percent increase during that time? (Record answer correct to nearest tenth of 1%.)

2. A store sold 15% of its monthly quota during 1 day. If the monthly quota was $32,500, what were the sales for the day?

3. Sales for the Homer Retail Store totaled $125,700 during 1 month. If returned merchandise amounted to $2350, what percent of sales did this represent (nearest tenth of 1%)?

4. Mr. Thomas owns rental property worth $20,000.00. What are the (a) annual depreciation and (b) the rate of depreciation if the straight-line method is used over a period of 30 years?

5. In problem 4, if taxes amounted to $375, insurance to $40, and upkeep and repairs to $415, what was Mr. Thomas' net income for the year if rentals equaled $150 monthly?

6. A mail order catalog lists a certain article at $56.75 with discounts of 15% and 10%. What is the net selling price?

7. A retail merchant received a $33\frac{1}{3}$% discount on coffee tables from the manufacturer. If the manufacturer lists them at $75, for how much must the retailer sell them to realize a 15% profit based on the cost to him?

8. The Ace Boat Supply Company sells its 40-horsepower outboard motors for $552 less 15%. If the company makes a gross profit of 20% based on the net selling price, what is the cost price?

9. The assessed valuation of property in a city equals $50,600,000. If $4,500,000 are to be raised through property taxes, what is the tax rate in dollars per $100? Per $1000? As a percent? (Carry division to 5 decimals.)

10. Property worth $35,000 is assessed at a rate of 22%. What are the taxes if the rate is 65 mills on the dollar?

11. If the sales for the Linker Co. were $95,000, what sales tax was collected at 4.2%?

12. If an item cost $300.90 including a 2% sales tax, what was the price excluding the tax?

13. An article costing $150 was sold at a 25% markup on selling price. What was the selling price?

14. Mr. Lindsay paid $364.32 in taxes on a home worth $18,000. The assessment rate was 22%. Find the tax rate per dollar (3 decimals).

15. At what price should an article costing $5.20 be marked in order to allow a 2% discount for cash and make a profit of 25% of cost?

16. A calculating machine, costing $1215, is to be written off in 10 years. If the scrap value is estimated at $175, what is the book value at the end of 4 years by using (a) the declining-balance method of depreciation and (b) the sum-of-the-years-digits method of depreciation?

17. A chair with a cost price of $85 is to be sold at a markup of 60% based on the selling price. What is the selling price?

18. A stove cost $425. Find the selling price and net profit if overhead is $12\frac{1}{2}$%, selling expense is 10%, and a net profit of 15% of selling price is desired? (Markup is based on the selling price.)

19. Compute the annual depreciation by the straight-line method for the following assets based on life expectancy as shown:

	Asset	Life Expectancy	Scrap Value
(a)	$3200	30 years	$350
(b)	750	10 years	120
(c)	4250	5 years	500

20. What is the book value at the end of 3 years for the assets listed in problem 19 using the straight-line method of depreciation?

21. The Adams Printing Shop purchases its paper supplies from the S. & R. Paper Company with terms of 1% 10 days EOM. An order amounting to $819.15 less 50% and 10% was made on April 19 and paid on May 6. How much was remitted in payment?

22. Mr. Jones purchased an automobile tire for $43.50. The excise tax was $3.05. How much did the tire cost if the sales tax was 7%?

23. Machinery worth $46,500 is to be depreciated over 30 years. What is the depreciation for the first 2 years by the sum-of-the-years-digits method?

24. A calculating machine costing $1500 is to be written off in 10 years by the straight-line method. If an additional first-year depreciation is allowed, how much is the annual depreciation? Salvage value is estimated at $250.

25. What are the rates in terms of percent for assets with a useful life of (a) 5 years, (b) 8 years, (c) 10 years, (d) 30 years, (e) 20 years? (Straight line method.)

Objectives

In this study of payroll we learn how to compute a person's gross and net earnings based on both an hourly rate or piece-rate basis. Use is made of tables to derive payroll deductions (primarily federal income tax withholding, and Social Security taxes). Instructions are also included for problems dealing with commissions, cash sheets, and the preparation of payrolls.

Prior to 1935 the preparation and maintenance of payroll records were fairly simple. However, with the passage of the Social Security Act of August 14, 1935, the Fair Labor Standards Act of 1938, and the Current Tax Payment Act of 1943 as well as state unemployment compensation laws, the problem of handling payroll records has become complex.

Many records are necessary to satisfy the requirements of these Federal and state laws. Consequently, the work of the payroll clerk has expanded greatly. Payroll records are maintained in accordance with the type and size of the business. Employees may be paid on a time or production basis.

TIME PAYMENT OR HOURLY RATE PAYMENT PAYROLL SYSTEM

When an employee is paid on a time basis, he may be paid by the hour, by the week, every 2 weeks, by the month, or by the year. The term *salary* is usually applied to compensation based on a monthly, semimonthly, or yearly basis, while the term *wages* is applied to shorter periods of time.

The manner in which time records are handled depends on the size and character of the business and on the time basis used. The record of time worked by each employee may be obtained through the use of time cards or time (sign-up) sheets. In many cases, these cards or sheets are not required for salaried personnel. However, regardless of the method used, adequate records must be maintained by the employer to satisfy legal requirements.

REGULAR EARNINGS

Regular earnings are based on a rate per hour. For example, if a man works a total of 30 hours in a week and is paid at the rate of \$1.90 an hour, his regular earnings are 30 × \$1.90 or \$57.00.

OVERTIME EARNINGS

Let us assume that *overtime*† is paid at the rate of $1\frac{1}{2}$ times the regular rate.

example A: regular rate = \$1.90
overtime rate = $1\frac{1}{2}$ × \$1.90 = \$2.85

example B: regular rate = \$2.15
overtime rate = $1\frac{1}{2}$ × \$2.15 = \$3.225
(do not round off to the nearest cent)

TOTAL OR GROSS EARNINGS

Total or *gross* earnings is the sum of regular and overtime earnings.

†In some cases, double the regular rate applies such as for work on Sundays and holidays or for night work. The manner in which overtime is computed or if it is paid at all depends on the law as applied to some employees and businesses.

example A: If Mr. Smith worked 45 hours in a week at the rate of \$1.90 per hour, what were his regular earnings, his overtime earnings, and his total or gross earnings? Overtime is paid for all hours in excess of 40.

regular earnings	$40 \times \$1.90 = \$76.00$
overtime earnings	$5 \times \$2.85 = \$14.25$
total earnings	$\$76.00 + \$14.25 = \$90.25$

example B: Mrs. Otts worked 43 hours in a week at the rate of \$2.15 an hour. If overtime is $1\frac{1}{2}$ times the regular rate, what was the total amount of her earnings for the week?

regular earnings	$40 \times \$2.15 = \$86.00$
overtime earnings	$3 \times 3.225 = \$\ 9.68$
total earnings	$\$86.00 + \$9.68 = \$95.68$

HOURLY RATES FOR EMPLOYEES PAID ON A WEEKLY OR MONTHLY BASIS

The hourly rate is often needed for personnel paid by the week or month. Many such employees are paid overtime for work exceeding a 40-hour week. Hourly rates are often needed when an employee is hired at an odd time during the week or day or to figure sick leave without pay, vacation without pay, or termination of employment.

Changing Weekly Pay Rate to Hourly Rate

Divide the weekly rate by the number of hours in the week.

example: Mr. Haynes earns \$90.00 per week. What are his hourly rate and overtime rate if overtime is $1\frac{1}{2}$ times the regular rate?

hourly rate

$$\$90.00 \div 40 \text{ (hours in week)} = \$2.25$$

overtime rate $1\frac{1}{2} \times \$2.25 = \$3.375$

Changing Monthly Rate to Hourly Rate†

Find the weekly rate and proceed as illustrated above. This may be done by two methods: (1) Multiply the monthly salary by 12 to obtain the yearly rate and then divide by 52 (number of weeks in a year) or (2) divide the monthly salary by $4\frac{1}{3}$ (average number of weeks in a month).

example: Mann Products pays its stenographers \$600.00 a month. What are the weekly, hourly, and overtime rates?

weekly rate	$\$600.00 \div 4\frac{1}{3} = \$138.46$
hourly rate	$\$138.46 \div 40 = \$3.46$**
overtime rate	$\$3.46 \times 1\frac{1}{2} = \$5.19$

EXERCISE 31

1. J. Jenkins worked 43 hours during the week at the rate of \$2.35 an hour. What was his gross earnings if he was paid $1\frac{1}{2}$ times his regular rate for any time over 40 hours?

2. Mr. Sawyer worked 39 hours regular time, 15 hours overtime at time and a half, and 9 hours at double time. If his regular rate was \$2.60 an hour, how much did he earn?

3. A man worked 7 hours on Tuesday, $5\frac{1}{2}$ hours on Saturday, and 8 hours on other days except Sunday. At $\$2.12\frac{1}{2}$ an hour, what did he earn for the week? (Use time and a half for overtime.)

4. Compute the overtime hourly rates for the following regular rates. Carry the full decimal in your answers.
(a) \$9.65, (b) \$4.20, (c) \$12.75, (d) \$25.00, (e) \$17.50, (f) \$8.34.

5. Howard Butler received \$196.00 for a 40-hour week and time and a half for overtime. Compute his earnings for a week in which he worked 48 hours.

†Sick leave and annual leave are not taken into account. In some cases the time allowed for these benefits may be deducted from the total number of weeks for the year when computing the weekly pay rate. Tables available for this purpose are generally found in the payroll offices of all large business enterprises.

‡The degree of accuracy depends on company policy; 2 decimals are used here for convenience.

6. Mrs. Hynes receives $520.00 a month as an office clerk. She is paid time and a half for overtime. If she works 15 hours overtime during the month, how much are her total earnings?

7. James Jason earned $390.00 a month as a stock clerk. During the month he took off 8 hours to which he was not entitled (leave without pay). What did he receive for the month? (Leave without pay is computed on the regular hourly rate.)

8. Find the hourly rate for the following pay schedules: (a) $450.00 a month, (b) $815.00 a month, (c) $90.00 a week, and (d) $224.00 biweekly. Record final answer to 3 decimals.

9. Convert the following hourly rates to weekly and monthly rates: (a) $4.16, (b) $2.055, (c) $1.80, (d) $3.25, (e) $5.00, and (f) $2.82½.

10. Ruth Ames earned $715.00 a month as a secretary. During 1 month she took a week off without pay. How much did she earn that month?

PAYROLL DEDUCTIONS

Payroll deductions fall into two categories, those required by law and those of a voluntary nature. The second group includes such deductions as employees association dues, credit union payments, pension plans, government bonds, and stock purchases. It is not uncommon to have twelve or more deductions listed on a paycheck voucher. The number of such deductions depends on the company and individual employee.

Deductions required by law affect all employees. To meet the requirements established for such deductions, it is important that the employer and those responsible for the preparation and maintenance of the payroll be thoroughly acquainted with the provisions of the law.

All states and the District of Columbia have unemployment insurance laws. Some states and local districts have income tax withholding laws. In some cases, benefits are paid to employees who are absent from their jobs because of disability. Because these laws are not uniform, only those deductions required by Federal law and affecting all employees are discussed and applied in this chapter.

FEDERAL INSURANCE CONTRIBUTION ACT TAX (F.I.C.A.)

This tax is commonly referred to as the *Social Security Tax*. For those who meet the requirements of the law, it provides for their retirement and other benefits as well as benefits for their dependents and survivors.

Taxes for this purpose are withheld from the employee's earnings. The employer also pays an amount equal to that withheld from each employee's earnings. The rate has increased several times since enactment of the law. As of January 1, 1973 this tax applies only to the first $10,800.00 paid to an employee in a calendar year. The maximum tax paid by any employee in the calendar year 1973 is 5.85% of $10,800.00 or $631.80.

Revised rates that apply for the years 1973 to 2011 inclusive as of January 1, 1973 are as follows:

Year	F.I.C.A. Rate
1973–1977	5.85
1978–1980	6.05
1981–1985	6.15
1986–2010	6.25
2011 and after	7.30

The Social Security Employee Tax Table on pages 136 to 140 is provided by the Internal Revenue Service for the purpose of determining the amount to be withheld on gross earnings. For earnings in excess of $310.18, apply the 5.85% rate.

example: Find the F.I.C.A. tax (taxes withheld for "Social Security") on wages amounting to (a) $250.65 and (b)$375.00.

solution:† 1973–1977 rates—

(a) $250.65 × .0585 = $14.663025 = $14.66.

†The base on which the maximum tax paid is computed for any employee has been increased five times since 1951 when it was $3600.00. It is now $10,800.00. During this period both current and projected rates have changed several times. In addition, the Federal income tax withholding tables are also subject to change. For these reasons the answers to all payroll problems involving taxes withheld are based on the rates for the year 1973, the most complete information available at the time of writing.

If the table is used (page 138), \$250.65 is "at least" \$250.52 "but less than" \$250.69. Then the amount of tax is \$14.66 appearing on the same line.

(b) $\$375.00 \times .0585 = \$21.9375 = \$21.94$ (see table to check answer).

As stated previously, the most that can be deducted for Social Security from an employee's earnings during 1973 is \$631.80 regardless of how much his total earnings may be. It is therefore very important that an employer keep accurate payroll records.

example: Mr. Brown earns \$15,000 a year or \$1250 a month. Assuming he was working as of January 1, 1973, when would he have paid all F.I.C.A. taxes for that year?

solution: F.I.C.A. taxes for January = 5.85% of \$1250

$= .0585 \times \$1250 = \$73.13$

$\$631.80 \div \$73.13 = 8$ and a fraction months. Then, by the end of August he would have paid $8 \times \$73.13$ or \$585.04. In September his tax would equal $\$631.80 - \$585.04 = \$46.76$. He would pay no tax for the remainder of the year and his "take home pay" would be \$73.13 more.

FEDERAL INCOME TAX WITHHOLDING

The Current Tax Payment Act of 1943 requires all employers to withhold a certain percent of wages paid (after exemption) to their employees† for income tax purposes. Since 1943, the rates and the amounts to be withheld have been changed frequently. Rates now vary from 14% to 36%.

There are two methods which may be used to determine the amount of income taxes to be withheld. These are called the (1) *percentage method* and (2) the *wage bracket method*. Employers may use either method; results are essentially the same.

†There are certain classifications of wage earners to whom this does not apply.

The Percentage Method

Employers who use this method must make a percent computation based on the income tax withholding table for 1 exemption for various payroll periods (weekly, biweekly, semimonthly, monthly, quarterly, semiannually, annually, and daily or miscellaneous) together with a rate table for the corresponding payroll period and the marital status of the employee. Computers are often used for this method and are especially valuable for payroll periods that are longer than a month. Because this method can be complex, many employers prefer the wage bracket method.

The Wage Bracket Method

The Federal Government provides income tax withholding tables for single and married taxpayers for the following payroll periods: weekly, biweekly, semimonthly, monthly, and daily or miscellaneous. The table to be used is determined by the marital status and payroll period of the taxpayer.

Finding the Amount of Tax

The amount of tax withheld is determined by the marital status, number of exemptions, and the amount of gross earnings. An exemption is an allowance for a person who is supported by the taxpayer. An additional exemption is allowed if a person is blind or 65 years of age or older.

The principles involved in the use of all income tax withholding tables are the same. For this reason, tables in this text are for weekly and monthly payroll periods only, for both single and married taxpayers.

example: Find the tax withheld on a weekly salary of $82.50 for a single person with 2 exemptions.

solution: Refer to the weekly payroll table on pages 144 and 145. Since $82.50 is at least $82.00 but less than $84.00, the amount to be withheld is $6.90.

example: Mr. Smith earns $875.00 a month. He has 3 children and claims 5 deductions or exemptions which include himself and his wife. Find the amount of tax withheld.

solution: See monthly payroll table on pages 146 and 147. Since $875.00 is more than $840.00 but less than $880.00, the amount of tax is $77.80 for 5 exemptions.

EXERCISE 32

1. Find the F.I.C.A. tax on the following wages: (a) $270.00, (b) $436.10, (c) $60.50, (d) $112.48, and (e) $234.21.

2. Find the amount of income taxes withheld for the following:

No.	Marital Status	No. of Exemptions	Payroll Period	Wages Earned
a.	Single	1	Week	$ 110.00
b.	Single	2	Month	228.00
c.	Single	4	Month	1550.00
d.	Married	None	Month	965.00
e.	Married	5	Week	195.00

3. Mr. Tennyson earns $165 a week. He is single and takes 1 exemption. (a) How much does his employer withhold for FICA? (b) Federal Income taxes? (c) What are his net earnings? (Net earnings = total or gross earnings *less* deductions.)

4. As of October 1, 1973 Mr. Allison earned $9300, Mr. Smith earned $12,500, and Mr. Arden earned $10,250. Compute the total F.I.C.A. tax withheld. Assume they were all working as of January 1 of the year.

5. Miss Howen's salary is $12,600 a year. What is her (a) monthly salary, (b) F.I.C.A. tax, (c) Federal Income tax, and (d) net earnings. She takes 1 exemption.

6. In problem 5, assuming Miss Howen was working as of January 1, (a) When would she have met the maximum limit of F.I.C.A. taxes for the year 1973? (b) How much would the F.I.C.A. tax equal for the last month of the year in which it was paid?

7. Complete the following partial payroll. The overtime rate ($1\frac{1}{2}$ times the regular rate) applies to all time over 40 hours during the week.

PAYROLL

For Week Ending January 23, 19___

Employee No.	Hours					Total Hours	Hourly Rate	Earnings		
	Mon.	Tue.	Wed.	Thu.	Fri.			Regular	Overtime	Total
32	8	7½	10	8	8	41½	$2.15	$86.00	$4.84	$90.84
33	7	8	8	7½	9		1.98			
34	8	8	9	8½	8		3.50			
35	10	8	8	7½	8		3.25			
36	6	9½	8	8	8		2.75			
37	9	10	10	7	7	43	2.80	112	12.60	124.60
Total							XXXX			

8. Compute the F.I.C.A. and withholding taxes and complete the following partial payroll. Assume all employees are married and that none has earned $10,800.00.

PAYROLL

For Week Ending May 11, 19___

Employee No.	Exemptions	Total Earnings	Deductions				Net Earnings
			F.I.C.A.	Tax Withheld	Other	Total	
260	2	$205.00	$11.99	$26.30	$5.20	$43.49	$161.51
217	1	321.50			3.00		
284	3	187.25			4.50		
296	2	290.84			4.50		
300	0	275.00			5.20		
301	4	167.55			5.20		
Total							

9. Mrs. J. Carroll earns $300.00 a month as a clerk for the Sabin Realty Co. She worked 4 hours overtime during the month. If she is paid time and a half for all overtime, what were her total earnings for the month? (Use a 40-hour week.)

10. Mr. Smith was employed by the Frick Iron Works at a salary of $200.00 for a 40-hour week. Compute his hourly rate and his overtime rate at time and a half.

PIECE-RATE PAYROLL SYSTEM

The *piece-rate payroll* system is one in which the employee is paid on a unit production basis. For straight piecework the amount earned equals the number of articles or units of work produced multiplied by the unit price. If an employee produces 375 articles in a day at 12¢ each, his earnings for the day equal 375 × .12 or $45.00.

There are many variations and modifications of the piece-rate plan. Sometimes special incentives or bonuses are offered for higher production. Where average standards have been established, the fast worker may be rewarded with an additional price rate for units produced in excess of the established average.

Various types of businesses use the piece-rate payroll system. For example it is used in shoe-manufacturing companies, fruit-packing plants, and plants assembling manufactured items; and for telephone solicitations for company contacts such as magazine subscriptions, insurance, and cemetery plots. Fruit pickers are paid by the unit such as the pound, box, crate, or other container.

Rates vary for the same type of work in some businesses from season to season, depending on the market or other factors unique to the business. This is particularly true in agricultural work and other food industries.

COMMISSIONS

Many businesses pay their employees wholly or in part on a commission basis. The *commission* is a *fee*, generally a percent of money received in a transaction. For example, real estate salesmen receive a certain percent of their sales.

Store sales personnel may, in addition to their salaries, receive a special commission or bonus for selling hard-to-move merchandise. As an added inducement to stimulate business, sales clerks may be paid a percent of the dollar volume of sales made above a certain amount within a specified period of time. In every instance, the commission serves as an incentive to the salesman to increase sales and as a reward for his efforts.

example: Mr. A. Hyatt sells real estate. His sales for 4 months amounted to $117,500. If he received a 3% commission, how much did he earn during that time?

solution:

commission = 3% of $117,500
= .03 × $117,500 = $3525

example: Mrs. Clark receives a salary of $95.00 a week as a sales clerk for the Comfy Shoe Store. Her sales during 1 week were $850.00 If she received a 5% commission on all sales over $300 during any week, what were her total earnings for the week?

solution:

salary	$95.00
commission (5% of $550)	27.50
total earnings	$122.50

EXERCISE 33

1. Mrs. James packs apples for the Local Apple Growers Association for which she receives 17¢ a box. How much does she earn in a week if she averages 160 boxes per day for 5 days?
2. H. Harvey assembles plastic facemasks for oxygen therapy and receives 90¢ for each unit. If he assembles 18 on Monday, 23 on Tuesday, 17 on Wednesday, 20 on Thursday, and 20 on Friday, how much does he earn for the week?
3. Find the hourly rates for problems 1 and 2 above (40-hour week).
4. Mr. Swartz received a commission of $2\frac{1}{2}$% on all sales made above $300 in any single day. If his sales for Monday through Saturday amounted to $350, $475, $500, $310, $671, and $392, respectively, how much did he earn in commissions for the week?

5. The Empire Department store gave the sales personnel a 3% commission on certain speciality items that did not sell readily. The sales for these items amounted to $936 in a particular month. If this sum was divided between 8 sales clerks, how much did each receive?

6. The Home Beauty Shop pays all operators $12 (base pay) for an 8-hour day. The operators also earn 50% of all business over $24 received during the day. If Miss Heenan does $75 worth of business during a certain day, how much does she earn?

PAYMENT OF PAYROLLS IN CASH

Many employers prefer to pay their employees in cash. When this is done, it is necessary to have the exact currency for each employee's pay envelope. A payroll cash sheet is prepared for this purpose. After the number of bills and coins required has been determined, a *change memorandum* is prepared. The bank, upon presentation of this memorandum and a check for the total amount of the payroll, will provide the denominations required.

See examples of these forms below. Notice that the amount to be paid each employee is broken down into the

CASH SHEET

Payroll Week Ending July 3, 19___

		Bills				Coins				
Employee	Net Pay	20	10	5	1	.50	.25	.10	.05	.01
113	$ 68.32	3		1	3		1		1	2
114	61.45	3			1		1	2		
117	82.60	4			2	1		1		
120	71.12	3	1		1			1		2
125	88.70	4		1	3	1		2		
126	55.81	2	1	1		1	1		1	1
Totals	$428.00	19	2	3	10	3	3	6	2	5

largest denominations possible. The currency memorandum is a summary of the number of bills and coins for the bank's use.

MEMORANDUM OF
CASH FOR PAY ROLL
FOR
A B C Company
July 3, 19

NUMBER		AMOUNT	
19	Twenties	380	00
2	Tens	20	00
3	Fives	15	00
	Twos		
10	Ones	10	00
	Silver Dollars		
3	Halves	1	50
3	Quarters		75
6	Dimes		60
2	Nickels		10
5	Pennies		05
	TOTAL	428	00

P & S FORM 96

EXERCISE 34 *Chapter Summary*

1. Complete Payroll A. Overtime is figured at time and a half.
2. Complete Payroll B.
3. Prepare a cash sheet and change memorandum for Payroll A.
4. (a) Prepare a cash sheet and change memorandum for Payroll B. (b) Compute the average hourly rate, corrected to 2 decimals, for all employees in Payroll B. Use a 40-hour week.

A. PAYROLL SUMMARY

Work Week Ending June 14, 19____ Date of Payment ____________ Sheet No. ______

Employee	Marital Status	Exempt.	Hours of Work					Total Hours	Hourly Rate	Earnings			Deductions				Net Earnings
			M	T	W	T	F			Regular	Overtime	Total Wages	F.I.C.A.	With-holding Tax	Other Deduc-tions	Total Deduc-tions	
Ames, J.	M	2	8	7	8	8½	9		$2.20						$2.50		
Bell, S.	M	3	7½	8	8	10	10		1.95						5.00		
Ellis, A.	S	1	8	8	8	8	8		3.20						5.00		
King, M.	M	3	9	8	8½	9	8		4.50						2.50		
Moore, R.	M	2	7	7	8	9	8		3.65						2.50		
Wells, H.	S	2	8	8	7½	7½	10	41	5.20	208	7.80	215.80	13.23	34.00	2.50	49.73	166.07
Total															20		

B. PAYROLL SUMMARY

Work Week Ending August 1, 19____ Date of Payment ____________ Sheet No. ______

Employee	Marital Status	Exempt.	Units Produced						Piece Rate	Gross Pay	Deductions				Net Pay
			M	T	W	T	F	Total Pieces			F.I.C.A.	Withholding Tax	Other Deductions	Total Deductions	
Bens, L.	S	0	800	850	700	920	910		.0192				$1.25		
Cain, C.	S	1	250	220	300	315	275		.061				1.25		
Dietz, F.	M	1	415	430	420	450	462		.055				1.25		
Hale, D.	M	4	300	322	360	410	390		.1025				1.25		
Hope, C.	M	3	950	990	975	998	950		.03				1.25		
Main, H.	M	2	150	165	170	167	182		.202				1.25		
Total															

SOCIAL SECURITY EMPLOYEE TAX TABLE

5.85% employee tax deductions

If Wages Are At Least	FICA TAX IS	If Wages Are At Least	FICA TAX IS	If Wages Are At Least	FICA TAX IS	If Wages Are At Least	FICA TAX IS	If Wages Are At Least	FICA TAX IS	If Wages Are At Least	FICA TAX IS	If Wages Are At Least	FICA TAX IS	If Wages Are At Least	FICA TAX IS	If Wages Are At Least	FICA TAX IS	If Wages Are At Least	FICA TAX IS
.00	.00	13.59	.80	27.27	1.60	40.95	2.40	54.62	3.20	68.30	4.00	81.97	4.80	95.65	5.60	109.32	6.40	123.00	7.20
.09	.01	13.77	.81	27.44	1.61	41.12	2.41	54.79	3.21	68.47	4.01	82.14	4.81	95.82	5.61	109.49	6.41	123.17	7.21
.26	.02	13.94	.82	27.61	1.62	41.29	2.42	54.96	3.22	68.64	4.02	82.31	4.82	95.99	5.62	109.66	6.42	123.34	7.22
.43	.03	14.11	.83	27.78	1.63	41.46	2.43	55.13	3.23	68.81	4.03	82.48	4.83	96.16	5.63	109.83	6.43	123.51	7.23
.60	.04	14.28	.84	27.95	1.64	41.63	2.44	55.30	3.24	68.98	4.04	82.65	4.84	96.33	5.64	110.00	6.44	123.68	7.24
.77	.05	14.45	.85	28.12	1.65	41.80	2.45	55.48	3.25	69.15	4.05	82.83	4.85	96.50	5.65	110.18	6.45	123.85	7.25
.95	.06	14.62	.86	28.30	1.66	41.97	2.46	55.65	3.26	69.32	4.06	83.00	4.86	96.67	5.66	110.35	6.46	124.02	7.26
1.12	.07	14.79	.87	28.47	1.67	42.14	2.47	55.82	3.27	69.49	4.07	83.17	4.87	96.84	5.67	110.52	6.47	124.19	7.27
1.29	.08	14.96	.88	28.64	1.68	42.31	2.48	55.99	3.28	69.66	4.08	83.34	4.88	97.01	5.68	110.69	6.48	124.36	7.28
1.46	.09	15.13	.89	28.81	1.69	42.48	2.49	56.16	3.29	69.83	4.09	83.51	4.89	97.18	5.69	110.86	6.49	124.53	7.29
1.63	.10	15.30	.90	28.98	1.70	42.65	2.50	56.33	3.30	70.00	4.10	83.68	4.90	97.36	5.70	111.03	6.50	124.71	7.30
1.80	.11	15.48	.91	29.15	1.71	42.83	2.51	56.50	3.31	70.18	4.11	83.85	4.91	97.53	5.71	111.20	6.51	124.88	7.31
1.97	.12	15.65	.92	29.32	1.72	43.00	2.52	56.67	3.32	70.35	4.12	84.02	4.92	97.70	5.72	111.37	6.52	125.05	7.32
2.14	.13	15.82	.93	29.49	1.73	43.17	2.53	56.84	3.33	70.52	4.13	84.19	4.93	97.87	5.73	111.54	6.53	125.22	7.33
2.31	.14	15.99	.94	29.66	1.74	43.34	2.54	57.01	3.34	70.69	4.14	84.36	4.94	98.04	5.74	111.71	6.54	125.39	7.34
2.48	.15	16.16	.95	29.83	1.75	43.51	2.55	57.18	3.35	70.86	4.15	84.53	4.95	98.21	5.75	111.89	6.55	125.56	7.35
2.65	.16	16.33	.96	30.00	1.76	43.68	2.56	57.36	3.36	71.03	4.16	84.71	4.96	98.38	5.76	112.06	6.56	125.73	7.36
2.83	.17	16.50	.97	30.18	1.77	43.85	2.57	57.53	3.37	71.20	4.17	84.88	4.97	98.55	5.77	112.23	6.57	125.90	7.37
3.00	.18	16.67	.98	30.35	1.78	44.02	2.58	57.70	3.38	71.37	4.18	85.05	4.98	98.72	5.78	112.40	6.58	126.07	7.38
3.17	.19	16.84	.99	30.52	1.79	44.19	2.59	57.87	3.39	71.54	4.19	85.22	4.99	98.89	5.79	112.57	6.59	126.24	7.39
3.34	.20	17.01	1.00	30.69	1.80	44.36	2.60	58.04	3.40	71.71	4.20	85.39	5.00	99.06	5.80	112.74	6.60	126.42	7.40
3.51	.21	17.18	1.01	30.86	1.81	44.53	2.61	58.21	3.41	71.89	4.21	85.56	5.01	99.24	5.81	112.91	6.61	126.59	7.41
3.68	.22	17.36	1.02	31.03	1.82	44.71	2.62	58.38	3.42	72.06	4.22	85.73	5.02	99.41	5.82	113.08	6.62	126.76	7.42
3.85	.23	17.53	1.03	31.20	1.83	44.88	2.63	58.55	3.43	72.23	4.23	85.90	5.03	99.58	5.83	113.25	6.63	126.93	7.43
4.02	.24	17.70	1.04	31.37	1.84	45.05	2.64	58.72	3.44	72.40	4.24	86.07	5.04	99.75	5.84	113.42	6.64	127.10	7.44
4.19	.25	17.87	1.05	31.54	1.85	45.22	2.65	58.89	3.45	72.57	4.25	86.24	5.05	99.92	5.85	113.59	6.65	127.27	7.45
4.36	.26	18.04	1.06	31.71	1.86	45.39	2.66	59.06	3.46	72.74	4.26	86.42	5.06	100.09	5.86	113.77	6.66	127.44	7.46
4.53	.27	18.21	1.07	31.89	1.87	45.56	2.67	59.24	3.47	72.91	4.27	86.59	5.07	100.26	5.87	113.94	6.67	127.61	7.47
4.71	.28	18.38	1.08	32.06	1.88	45.73	2.68	59.41	3.48	73.08	4.28	86.76	5.08	100.43	5.88	114.11	6.68	127.78	7.48
4.88	.29	18.55	1.09	32.23	1.89	45.90	2.69	59.58	3.49	73.25	4.29	86.93	5.09	100.60	5.89	114.28	6.69	127.95	7.49
5.05	.30	18.72	1.10	32.40	1.90	46.07	2.70	59.75	3.50	73.42	4.30	87.10	5.10	100.77	5.90	114.45	6.70	128.12	7.50
5.22	.31	18.89	1.11	32.57	1.91	46.24	2.71	59.92	3.51	73.59	4.31	87.27	5.11	100.95	5.91	114.62	6.71	128.30	7.51
5.39	.32	19.06	1.12	32.74	1.92	46.42	2.72	60.09	3.52	73.77	4.32	87.44	5.12	101.12	5.92	114.79	6.72	128.47	7.52
5.56	.33	19.24	1.13	32.91	1.93	46.59	2.73	60.26	3.53	73.94	4.33	87.61	5.13	101.29	5.93	114.96	6.73	128.64	7.53
5.73	.34	19.41	1.14	33.08	1.94	46.76	2.74	60.43	3.54	74.11	4.34	87.78	5.14	101.46	5.94	115.13	6.74	128.81	7.54
5.90	.35	19.58	1.15	33.25	1.95	46.93	2.75	60.60	3.55	74.28	4.35	87.95	5.15	101.63	5.95	115.30	5.75	128.98	7.55
6.07	.36	19.75	1.16	33.42	1.96	47.10	2.76	60.77	3.56	74.45	4.36	88.12	5.16	101.80	5.96	115.48	6.76	129.15	7.56
6.24	.37	19.92	1.17	33.59	1.97	47.27	2.77	60.95	3.57	74.62	4.37	88.30	5.17	101.97	5.97	115.65	6.77	129.32	7.57
6.42	.38	20.09	1.18	33.77	1.98	47.44	2.78	61.12	3.58	74.79	4.38	88.47	5.18	102.14	5.98	115.82	6.78	124.49	7.58
6.59	.39	20.26	1.19	33.94	1.99	47.61	2.79	61.29	3.59	74.96	4.39	88.64	5.19	102.31	5.99	115.99	6.79	129.66	7.59

SOCIAL SECURITY EMPLOYEE TAX TABLE

5.85% employee tax deductions

6.76	.40	20.43	1.20	34.11	2.00	47.78	2.80	61.46	3.60	75.13	4.40	88.81	5.20	102.48	6.00	116.16	6.80	129183	7.60
6.93	.41	20.60	1.21	34.28	2.01	47.95	2.81	61.63	3.61	75.30	4.41	88.98	5.21	102.65	6.01	116.33	6.81	130.00	7.61
7.10	.42	20.77	1.22	34.45	2.02	48.12	2.82	61.80	3.62	75.48	4.42	89.15	5.22	102.83	6.02	116.50	6.82	130.18	7.62
7.27	.43	20.95	1.23	34.62	2.03	48.30	2.83	61.97	3.63	75.65	4.43	89.32	5.23	103.00	6.03	116.67	6.83	130.35	7.63
7.44	.44	21.12	1.24	34.79	2.04	48.47	2.84	62.14	3.64	75.82	4.44	89.49	5.24	103.17	6.04	116.84	6.84	130.52	7.64
7.61	.45	21.29	1.25	34.96	2.05	48.64	2.85	62.31	3.65	75.99	4.45	89.66	5.25	103.34	6.05	117.01	6.85	130.69	7.65
7.78	.46	21.46	1.26	35.13	2.06	48.81	2.86	62.48	3.66	76.16	4.46	89.83	5.26	103.51	6.06	117.18	6.86	130.86	7.66
7.95	.47	21.63	1.27	35.30	2.07	48.98	2.87	62 65	3.67	76.33	4.47	90.00	5.27	103.68	6.07	117.36	6.87	131.03	7.67
8.12	.48	21.80	1.28	35.48	2.08	49.15	2.88	62.83	3.68	76.50	4.48	90.18	5.28	103.85	6.08	117.53	6.88	131.20	7.68
8.30	.49	21.97	1.29	35.65	2.09	49.32	2.89	63.00	3.69	76.67	4.49	90.35	5.29	104.02	6.09	117.70	6.89	131.37	7.69
8.47	.50	22.14	1.30	35.82	2.10	49.49	2.90	63.17	3.70	76.84	4.50	90.52	5.30	104.19	6.10	117.87	6.90	131.54	7.70
8.64	.51	22.31	1.31	35.99	2.11	49.66	2.91	63.34	3.71	77.01	4.51	90.69	5.31	104.36	6.11	118.04	6.91	131.71	7.71
8.81	.52	22.48	1.32	36.16	2.12	49.83	2.92	63.51	3.72	77.18	4.52	90.86	5.32	104.53	6.12	118.21	6.92	131.89	7.72
8.98	.53	22.65	1.33	36.33	2.13	50.00	2.93	63.68	3.73	77.36	4.53	91.03	5.33	104.71	6.13	118.38	6.93	132.06	7.73
9.15	.54	22.83	1.34	36.50	2.14	50.18	2.94	63.85	3.74	77.53	4.54	91.20	5.34	104.88	6.14	118.55	6.94	132.23	7.74
9.32	.55	23.00	1.35	36.67	2.15	50.35	2.95	64.02	3.75	77.70	4.55	91.37	5.35	105.05	6.15	118.72	6.95	132.40	7.75
9.49	.56	23.17	1.36	36.84	2.16	50.52	2.96	64.19	3.76	77.87	4.56	91.54	5.36	105.22	6.16	118.89	6.96	132.57	7.76
9.66	.57	23.34	1.37	37.01	2.17	50.69	2.97	64.36	3.77	78.04	4.57	91.71	5.37	105.39	6.17	119.06	6.97	132.74	7.77
9.83	.58	23.51	1.38	37.18	2.18	50.86	2.98	64.53	3.78	78.21	4.58	91.89	5.38	105.56	6.18	119.24	6.98	132.91	7.78
10.00	.59	23.68	1.39	37.36	2.19	51.03	2.99	64.71	3.79	78.38	4.59	92.06	5.39	105.73	6.19	119.41	6.99	133.08	7.79
10.18	.60	23.85	1.40	37.53	2.20	51.20	3.00	64.88	3.80	78.55	4.60	92.23	5.40	105.90	6.20	119.58	7.00	133.25	7.80
10.35	.61	24.02	1.41	37.70	2.21	51.37	3.01	65.05	3.81	78.72	4.61	92.40	5.41	106.07	6.21	119.75	7.01	113.42	7.81
10.52	.62	24.19	1.42	37.87	2.22	51.54	3.02	65.22	3.82	78.89	4.62	92.57	5.42	106.24	6.22	119.92	7.02	133.59	7.82
10.69	.63	24.36	1.43	38.04	2.23	51.71	3.03	65.39	3.83	79.06	4.63	92.74	5.43	106.42	6.23	120.09	7.03	133.77	7.83
10.86	.64	24.53	1.44	38.21	2.24	51.89	3.04	65.56	3.84	79.24	4.64	92.91	5.44	106.59	6.24	120.26	7.04	133.94	7.84
11.03	.65	24.71	1.45	38.38	2.25	52.06	3.05	65.73	3.85	79.41	4.65	93.08	5.45	106.76	6.25	120.43	7.05	134.11	7.85
11.20	.66	24.88	1.46	38.55	2.26	52.23	3.06	65.90	3.86	79.58	4.66	93.25	5.46	106.93	6.26	120.60	7.06	134.28	7.86
11.37	.67	25.05	1.47	38.72	2.27	52.40	3.07	66.07	3.87	79.75	4.67	93.42	5.47	107.10	6.27	120.77	7.07	134.45	7.87
11.54	.68	25.22	1.48	38.89	2.28	52.57	3.08	66.24	3.88	79.92	4.68	93.59	5.48	107.27	6.28	120.95	7.08	134.62	7.88
11.71	.69	25.39	1.49	39.06	2.29	52.74	3.09	66.42	3.89	80.09	4.69	93.77	5.49	107.44	6.29	121.12	7.09	134.79	7.89
11.89	.70	25.56	1.50	39.24	2.30	52.91	3.10	66.59	3.90	80.26	4.70	93.94	5.50	107.61	6.30	121.29	7.10	134.96	7.90
12.06	.71	25.73	1.51	39.41	2.31	53.08	3.11	66.76	3.91	80.43	4.71	94.11	5.51	107.78	6.31	121.46	7.11	135.13	7.91
12.23	.72	25.90	1.52	39.58	2.32	53.25	3.12	66.93	3.92	80.60	4.72	94.28	5.52	107.95	6.32	121.63	7.12	135.30	7.92
12.40	.73	26.07	1.53	39.75	2.33	53.42	3.13	67.10	3.93	80.77	4.73	94.45	5.53	108.12	6.33	121.80	7.13	135.48	7.93
12.57	.74	26.24	1.54	39.92	2.34	53.59	3.14	67.27	3.94	80.95	4.74	94.62	5.54	108.30	6.34	121.97	7.14	135.65	7.94
12.74	.75	26.42	1.55	40.09	2.35	53.77	3.15	67.44	3.95	81.12	4.75	94.79	5.55	108.47	6.35	122.14	7.15	135.82	7.95
12.91	.76	26.59	1.56	40.26	2.36	53.94	3.16	67.61	3.96	81.29	4.76	94.96	5.56	108.64	6.36	122.31	7.16	135.99	7.96
13.08	.77	26.76	1.57	40.43	2.37	54.11	3.17	67.78	3.97	81.46	4.77	95.13	5.57	108.81	6.37	122.48	7.17	136.16	7.97
13.25	.78	26.93	1.58	40.60	2.38	54.28	3.18	67.95	3.98	81.63	4.78	95.30	5.58	108.98	6.38	122.65	7.18	136.33	7.98
13.42	.79	27.10	1.59	40.77	2.39	54.45	3.19	68.12	3.99	81.80	4.79	95.48	5.59	109.15	6.39	122.83	7.19	136.50	7.99

continued

SOCIAL SECURITY EMPLOYEE TAX TABLE

5.85% employee tax deductions

If Wages Are At Least	FICA TAX IS	If Wages Are At Least	FICA TAX IS	If Wages Are At Least	FICA TAX IS	If Wages Are At Least	FICA TAX IS	If Wages Are At Least	FICA TAX IS	If Wages Are At Least	FICA TAX IS	If Wages Are At Least	FICA TAX IS	If Wages Are At Least	FICA TAX IS	If Wages Are At Least	FICA TAX IS	If Wages Are At Least	FICA TAX IS
136.67	8.00	150.35	8.80	164.02	9.60	177.70	10.40	191.37	11.20	205.05	12.00	218.72	12.80	232.40	13.60	246.07	14.40	259.75	15.20
136.84	8.01	150.52	8.81	164.19	9.61	177.87	10.41	191.54	11.21	205.22	12.01	218.89	12.81	232.57	13.61	246.24	14.41	259.92	15.21
137.01	8.02	150.69	8.82	164.36	9.62	178.04	10.42	191.71	11.22	205.39	12.02	219.06	12.82	232.74	13.62	246.42	14.42	260.09	15.22
137.18	8.03	150.86	8.83	164.53	9.63	178.21	10.43	191.89	11.23	205.56	12.03	219.24	12.83	232.91	13.63	246.59	14.43	260.26	15.23
137.36	8.04	151.03	8.84	164.71	9.64	178.38	10.44	192.06	11.24	205.73	12.04	219.41	12.84	233.08	13.64	246.76	14.44	260.43	15.24
137.53	8.05	151.20	8.85	164.88	9.65	178.55	10.45	192.23	11.25	205.90	12.05	219.58	12.85	233.25	13.65	246.93	14.45	260.60	15.25
137.70	8.06	151.37	8.86	165.05	9.66	178.72	10.46	192.40	11.26	206.07	12.06	219.75	12.86	233.42	13.66	247.10	14.46	260.77	15.26
137.87	8.07	151.54	8.87	165.22	9.67	178.89	10.47	192.57	11.27	206.24	12.07	219.92	12.87	233.59	13.67	247.27	14.47	260.95	15.27
138.04	8.08	151.71	8.88	165.39	9.68	179.06	10.48	192.74	11.28	206.42	12.08	220.09	12.88	233.77	13.68	247.44	14.48	261.12	15.28
138.21	8.09	151.89	8.89	165.56	9.69	179.24	10.49	192.91	11.29	206.59	12.09	220.26	12.89	233.94	13.69	247.61	14.49	261.29	15.29
138.38	8.10	152.06	8.90	165.73	9.70	179.41	10.50	193.08	11.30	206.76	12.10	220.43	12.90	234.11	13.70	247.78	14.50	261.46	15.30
138.55	8.11	152.23	8.91	165.90	9.71	179.58	10.51	193.25	11.31	206.93	12.11	220.60	12.91	234.28	13.71	247.95	14.51	261.63	15.31
138.72	8.12	152.40	8.92	166.07	9.72	179.75	10.52	193.42	11.32	207.10	12.12	220.77	12.92	234.45	13.72	248.12	14.52	261.80	15.32
138.89	8.13	152.57	8.93	166.24	9.73	179.92	10.53	193.59	11.33	207.27	12.13	220.95	12.93	234.62	13.73	248.30	14.53	261.97	15.33
139.06	8.14	152.74	8.94	166.42	9.74	180.09	10.54	193.77	11.34	207.44	12.14	221.12	12.94	234.79	13.74	248.47	14.54	262.14	15.34
139.24	8.15	152.91	8.95	166.59	9.75	180.26	10.55	193.94	11.35	207.61	12.15	221.29	12.95	234.96	13.75	248.64	14.55	262.31	15.35
139.41	8.16	153.08	8.96	166.76	9.76	180.43	10.56	194.11	11.36	207.78	12.16	221.46	12.96	235.13	13.76	248.81	14.56	262.48	15.36
139.58	8.17	153.25	8.97	166.93	9.77	180.60	10.57	194.28	11.37	207.95	12.17	221.63	19.97	235.30	13.77	248.98	14.57	262.65	15.37
139.75	8.18	153.42	8.98	167.10	9.78	180.77	10.58	194.45	11.38	208.12	12.18	221.80	12.98	235.48	13.78	249.15	14.58	262.83	15.38
139.92	8.19	153.59	8.99	167.27	9.79	180.95	10.59	194.62	11.39	208.30	12.19	221.97	12.99	235.65	13.79	249.32	14.59	263.00	15.39
140.09	8.20	153.77	9.00	167.44	9.80	181.12	10.60	194.79	11.40	208.47	12.20	222.14	13.00	235.82	13.80	249.49	14.60	263.17	15.40
140.26	8.21	153.94	9.01	167.61	9.81	181.29	10.61	194.96	11.41	208.64	12.21	222.31	13.01	235.99	13.81	249.66	14.61	263.34	15.41
140.43	8.22	154.11	9.02	167.78	9.82	181.46	10.62	195.13	11.42	208.81	12.22	222.48	13.02	236.16	13.82	249.83	14.62	263.51	15.42
140.60	8.23	154.28	9.03	167.95	9.83	181.63	10.63	195.30	11.43	208.98	12.23	222.65	13.03	236.33	13.83	250.00	14.63	263.68	15.43
140.77	8.24	154.45	9.04	168.12	9.84	181.80	10.64	195.48	11.44	209.15	12.24	222.83	13.04	236.50	13.84	250.18	14.64	263.85	15.44
140.95	8.25	154.62	9.05	168.30	9.85	181.97	10.65	195.65	11.45	209.32	12.25	223.00	13.05	236.67	13.85	250.35	14.65	264.02	15.45
141.12	8.26	154.79	9.06	168.47	9.86	182.14	10.66	195.82	11.46	209.49	12.26	223.17	13.06	236.84	13.86	250.52	14.66	264.19	15.46
141.29	8.27	154.96	9.07	168.64	9.87	182.31	10.67	195.99	11.47	209.66	12.27	223.34	13.07	237.01	13.87	250.69	14.67	264.36	15.47
141.46	8.28	155.13	9.08	168.81	9.88	182.48	10.68	196.16	11.48	209.83	12.28	223.51	13.08	237.18	13.88	250.86	14.68	264.53	15.48
141.63	8.29	155.30	9.09	168.98	9.89	182.65	10.69	196.33	11.49	210.00	12.29	223.68	13.09	236.36	13.89	251.03	14.69	264.71	15.49
141.80	8.30	155.48	9.10	169.15	9.90	182.83	10.70	196.50	11.50	210.18	12.30	223.85	13.10	237.53	13.90	251.20	14.70	264.88	15.50
141.97	8.31	155.65	9.11	169.32	9.91	183.00	10.71	196.67	11.51	210.35	12.31	224.02	13.11	237.70	13.91	251.37	14.71	265.05	15.51
142.14	8.32	155.82	9.12	169.49	9.92	183.17	10.72	196.84	11.52	210.52	12.32	224.19	13.12	237.87	13.92	251.54	14.72	265.22	15.52
142.31	8.33	155.99	9.13	169.66	9.93	183.34	10.73	197.01	11.53	210.69	12.33	224.36	13.13	238.04	13.93	251.71	14.73	265.39	15.53
142.48	8.34	156.16.9.14		169.83	9.94	183.51	10.74	197.18	11.54	210.86	12.34	224.53	13.14	238.21	13.94	251.89	14.74	265.56	15.54
142.65	8.35	156.33	9.15	170.00	9.95	183.68	10.75	197.36	11.55	211.03	12.35	224.71	13.15	238.38	13.95	252.06	14.75	265.73	15.55
142.83	8.36	156.50	9.16	170.18	9.96	183.85	10.76	197.53	11.56	211.20	12.36	224.88	13.16	238.55	13.96	252.23	14.76	265.90	15.56
143.00	8.37	156.67	9.17	170.35	9.97	184.02	10.77	197.70	11.57	211.37	12.37	225.05	13.17	238.72	13.97	252.40	14.77	266.07	15.57
143.17	8.38	156.84	9.18	170.52	9.98	184.19	10.78	197.87	11.58	211.54	12.38	225.22	13.18	238.89	13.98	252.57	14.78	266.24	15.58
143.34	8.39	157.01	9.19	170.69	9.99	184.36	10.79	198.04	11.59	211.71	12.39	225.39	13.19	239.06	13.99	252.74	14.79	266.42	15.59

SOCIAL SECURITY EMPLOYEE TAX TABLE

5.85% employee tax deductions

143.51	8.40	157.18	9.20	170.86	10.00	184.53	10.80	198.21	11.60	211.89	12.40	225.56	13.20	239.24	14.00	252.91	14.80	266.59	15.60
143.68	8.41	157.36	9.21	171.03	10.01	184.71	10.81	198.38	11.61	212.06	12.41	225.73	13.21	239.41	14.01	253.08	14.81	266.76	15.61
143.85	8.42	157.53	9.22	171.20	10.02	184.88	10.82	198.55	11.62	212.23	12.42	255.90	13.22	239.58	14.02	253.25	14.82	266.93	15.62
144.02	8.43	157.70	9.23	171.37	10.03	185.05	10.83	198.72	11.63	212.40	12.43	226.07	13.23	239.75	14.03	253.42	14.83	267.10	15.63
144.19	8.44	157.87	9.24	171.54	10.04	185.22	10.84	198.89	11.64	212.57	12.44	226.24	13.24	239.92	14.04	253.59	14.84	267.27	15.64
144.36	8.45	158.04	9.25	171.71	10.05	185.39	10.85	199.06	11.65	212.74	12.45	226.42	13.25	240.09	14.05	253.77	14.85	267.44	15.65
144.53	8.46	158.21	9.26	171.89	10.06	185.56	10.86	199.24	11.66	212.91	12.46	226.59	13.26	240.26	14.06	253.94	14.86	267.61	15.66
144.71	8.47	158.38	9.27	172.06	10.07	185.73	10.87	199.41	11.67	213.08	12.47	226.76	13.27	240.43	14.07	254.11	14.87	267.78	15.67
144.88	8.48	158.55	9.28	172.23	10.08	185.90	10.88	199.58	11.68	213.25	12.48	226.93	13.28	240.60	14.08	254.28	14.88	267.95	15.68
145.05	8.49	158.72	9.29	172.40	10.09	186.07	10.89	199.75	11.69	213.42	12.49	227.10	13.29	240.77	14.09	254.45	14.89	268.12	15.69
145.22	8.50	158.89	9.30	172.57	10.10	186.24	10.90	199.92	11.70	213.59	12.50	227.27	13.30	240.95	14.10	254.62	14.90	268.30	15.70
145.39	8.51	159.06	9.31	172.74	10.11	186.42	10.91	200.09	11.71	213.77	12.51	227.44	13.31	241.12	14.11	254.79	14.91	268.47	15.71
145.56	8.52	159.24	9.32	172.91	10.12	186.59	10.92	200.26	11.72	213.94	12.52	227.61	13.32	241.29	14.12	254.96	14.92	268.64	15.72
145.73	8.53	159.41	9.33	173.08	10.13	186.76	10.93	200.43	11.73	214.11	12.53	227.78	13.33	241.46	14.13	255.13	14.93	268.81	15.73
145.90	8.54	159.58	9.34	173.25	10.14	186.93	10.94	200.60	11.74	214.28	12.54	227.95	13.34	241.63	14.14	255.30	14.94	268.98	15.74
146.07	8.55	159.75	9.35	173.42	10.15	187.10	10.95	200.77	11.75	214.45	12.55	228.12	13.35	241.80	14.15	255.48	14.95	269.15	15.75
146.24	8.56	159.92	9.36	173.59	10.16	187.27	10.96	200.95	11.76	214.62	12.56	228.30	13.36	241.97	14.16	255.65	14.96	269.32	15.76
146.42	8.57	160.09	9.37	173.77	10.17	187.44	10.97	201.12	11.77	214.79	12.57	228.47	13.37	242.14	14.17	255.82	14.97	269.49	15.77
146.59	8.58	160.26	9.38	173.94	10.18	187.61	10.98	201.29	11.78	214.96	12.58	228.64	13.38	242.31	14.18	255.99	14.98	269.66	15.78
146.76	8.59	160.43	9.39	174.11	10.19	187.78	10.99	201.46	11.79	215.13	12.59	228.81	13.39	242.48	14.19	256.16	14.99	269.83	15.79
146.93	8.60	160.60	9.40	174.28	10.20	187.95	11.00	201.63	11.80	215.30	12.60	228.98	13.40	242.65	14.20	256.33	15.00	270.00	15.80
147.10	8.61	160.77	9.41	174.45	10.21	188.12	11.01	201.80	11.81	215.48	12.61	229.15	13.41	242.83	14.21	256.50	15.01	270.18	15.81
147.27	8.62	160.95	9.42	174.62	10.22	188.30	11.02	201.97	11.82	215.65	12.62	229.32	13.42	243.00	14.22	256.67	15.02	270.35	15.82
147.44	8.63	161.12	9.43	174.79	10.23	188.47	11.03	202.14	11.83	215.82	12.63	229.49	13.43	243.17	14.23	256.84	15.03	270.52	15.83
147.61	8.64	161.29	9.44	174.96	10.24	188.64	11.04	202.31	11.84	215.99	12.64	229.66	13.44	243.34	14.24	257.01	15.04	270.69	15.84
147.78	8.65	161.46	9.45	175.13	10.25	188.81	11.05	202.48	11.85	216.16	12.65	229.83	13.45	243.51	14.25	257.18	15.05	270.86	15.85
147.95	8.66	161.63	9.46	175.30	10.26	188.98	11.06	202.65	11.86	216.33	12.66	230.00	13.46	243.68	14.26	257.36	15.06	271.03	15.86
148.12	8.67	161.80	9.47	175.48	10.27	189.15	11.07	202.83	11.87	216.50	12.67	230.18	13.47	243.85	14.27	257.53	15.07	271.20	15.87
148.30	8.68	161.97	9.48	175.65	10.28	189.32	11.08	203.00	11.88	216.67	12.68	230.35	13.48	244.02	14.28	257.70	15.08	271.37	15.88
148.47	8.69	162.14	9.49	175.82	10.29	189.49	11.09	203.17	11.89	216.84	12.69	230.52	13.49	244.19	14.29	257.87	15.09	271.54	15.89
148.64	8.70	162.31	9.50	175.99	10.30	189.66	11.10	203.34	11.90	217.01	12.70	230.69	13.50	244.36	14.30	258.04	15.10	271.71	15.90
148.81	8.71	162.48	9.51	176.16	10.31	189.83	11.11	203.51	11.91	217.18	12.71	230.86	13.51	244.53	14.31	258.21	15.11	271.89	15.91
148.98	8.72	162.65	9.52	176.33	10.32	190.00	11.12	203.68	11.92	217.36	12.72	231.03	13.52	244.71	14.32	258.38	15.12	272.06	15.92
149.15	8.73	162.83	9.53	176.50	10.33	190.18	11.13	203.85	11.93	217.53	12.73	231.20	13.53	244.88	14.33	258.55	15.13	272.23	15.93
149.32	8.74	163.00	9.54	176.67	10.34	190.35	11.14	204.02	11.94	217.70	12.74	231.37	13.54	245.05	14.34	258.72	15.14	272.40	15.94
149.49	8.75	163.17	9.55	176.84	10.35	190.52	11.15	204.19	11.95	217.87	12.75	231.54	13.55	245.22	14.35	258.89	15.15	272.57	15.95
149.66	8.76	163.34	9.56	177.01	10.36	190.69	11.16	204.36	11.96	218.04	12.76	231.71	13.56	245.39	14.36	259.06	15.16	272.74	15.96
149.83	8.77	163.51	9.57	177.18	10.37	190.86	11.17	204.53	11.97	218.21	12.77	231.89	13.57	245.56	14.37	259.24	15.17	272.91	15.97
150.00	8.78	163.68	9.58	177.36	10.38	191.03	11.18	204.71	11.98	218.38	12.78	232.06	13.58	245.73	14.38	259.41	15.18	273.08	15.98
150.18	8.79	163.85	9.59	177.53	10.39	191.20	11.19	204.88	11.99	218.55	12.79	232.23	13.59	245.90	14.39	259.58	15.19	273.25	15.99

continued

SOCIAL SECURITY EMPLOYEE TAX TABLE

5.85% employee tax deductions

If Wages Are At Least	FICA TAX IS	If Wages Are At Least	FICA TAX IS	If Wages Are At Least	FICA TAX IS	If Wages Are At Least	FICA TAX IS	If Wages Are At Least	FICA TAX IS	If Wages Are At Least	FICA TAX IS	If Wages Are At Least	FICA TAX IS	If Wages Are At Least	FICA TAX IS	If Wages Are At Least	FICA TAX IS
273.42	16.00	277.70	16.25	281.97	16.50	286.24	16.75	290.52	17.00	294.79	17.25	299.06	17.50	303.34	17.75	307.61	18.00
273.59	16.01	277.87	16.26	282.14	16.51	286.42	16.76	290.69	17.01	294.96	17.26	299.24	17.51	303.51	17.76	307.78	18.01
273.77	16.02	278.04	16.27	282.31	16.52	286.59	16.77	290.86	17.02	295.13	17.27	299.41	17.52	303.68	17.77	307.95	18.02
273.94	16.03	278.21	16.28	282.48	16.53	286.76	16.78	291.03	17.03	295.30	17.28	299.58	17.53	303.85	17.78	308.12	18.03
274.11	16.04	278.38	16.29	282.65	16.54	286.93	16.79	291.20	17.04	295.48	17.29	299.75	17.54	304.02	17.79	308.30	18.04
274.28	16.05	278.55	16.30	282.83	16.55	287.10	16.80	291.37	17.05	295.65	17.30	299.92	17.55	304.19	17.80	308.47	18.05
274.45	16.06	278.72	16.31	283.00	16.56	287.27	16.81	291.54	17.06	295.82	17.31	300.09	17.56	304.36	17.81	208.64	18.06
274.62	16.07	278.89	16.32	283.17	16.57	287.44	16.82	291.71	17.07	295.99	17.32	300.26	17.57	304.53	17.82	308.81	18.07
274.79	16.08	279.06	16.33	283.34	16.58	287.61	16.83	271.89	17.08	296.16	17.33	300.43	17.58	304.71	17.83	308.98	18.08
274.96	16.09	279.24	16.34	283.51	16.59	287.78	16.84	292.06	17.09	296.33	17.34	300.60	17.59	304.88	17.84	309.15	18.09
275.13	16.10	279.41	16.35	283.68	16.60	287.95	16.85	292.23	17.10	296.50	17.35	300.77	17.60	305.05	17.85	309.32	18.10
275.30	16.11	279.58	16.36	283.85	16.61	288.12	16.86	292.40	17.11	296.67	17.36	300.95	17.61	305.22	17.86	309.49	18.11
275.48	16.12	279.75	16.37	284.02	16.62	288.30	16.87	292.57	17.12	296.84	17.37	301.12	17.62	305.39	17.87	309.66	18.12
275.65	16.13	279.92	16.38	284.19	16.63	288.47	16.88	292.74	17.13	297.01	17.38	301.29	17.63	305.56	17.88	309.83	18.13
275.82	16.14	280.09	16.39	284.36	16.64	288.64	16.89	292.91	17.14	297.18	17.39	301.46	17.64	305.73	17.89	310.00	18.14
275.99	16.15	280.26	16.40	284.53	16.65	288.81	16.90	293.08	17.15	297.36	17.40	301.63	17.65	305.90	17.90	310.18	18.15
276.16	16.16	280.43	16.41	284.71	16.66	288.98	16.91	293.25	17.16	297.53	17.41	301.80	17.66	306.07	17.91	310.35	18.16
276.33	16.17	280.60	16.42	284.88	16.67	289.15	16.92	293.42	17.17	297.70	17.42	301.97	17.67	306.24	17.92	310.52	18.17
276.50	16.18	280.77	16.43	285.05	16.68	289.32	16.93	293.59	17.18	297.87	17.43	302.14	17.68	306.42	17.93	310.69	18.18
276.67	16.19	280.95	16.44	285.22	16.69	289.49	16.94	293.77	17.19	298.04	17.44	302.31	17.69	306.59	17.94	310.86	18.19
276.84	16.20	281.12	16.45	285.39	16.70	289.66	16.95	293.94	17.20	298.21	17.45	302.48	17.70	306.76	17.95		
277.01	16.21	281.29	16.46	285.56	16.71	289.83	16.96	294.11	17.21	298.38	17.46	302.65	17.71	306.93	17.96		
277.18	16.22	281.46	16.47	285.73	16.72	290.00	16.97	294.28	17.22	298.55	17.47	302.83	17.72	307.10	17.97		
277.36	16.23	281.63	16.48	285.90	16.73	290.18	16.98	294.45	17.23	298.72	17.48	303.00	17.73	307.27	17.98		
277.53	16.24	281.80	16.49	286.07	16.74	290.35	16.99	294.62	17.24	298.89	17.49	303.17	17.74	307.44	17.99		

Over $310.86, apply the 5.85% rate.

FEDERAL INCOME TAX WITHHOLDING TABLE

Weekly Payroll for *Married* Wage Earners

And the Wages are		And the Number of Withholding Exemptions Claimed is—						
At Least	But Less Than	0	1	2	3	4	5	6 Or More
		The Amount of Income Tax to be Withheld Shall be—						
$ 0	$11	$ 0	$ 0	$ 0	$ 0	$ 0	$ 0	$ 0
11	12	.10	0	0	0	0	0	0
12	13	.30	0	0	0	0	0	0
13	14	.40	0	0	0	0	0	0
14	15	.50	0	0	0	0	0	0
15	16	.70	0	0	0	0	0	0
16	17	.80	0	0	0	0	0	0
17	18	1.00	0	0	0	0	0	0
18	19	1.10	0	0	0	0	0	0
19	20	1.20	0	0	0	0	0	0
20	21	1.40	0	0	0	0	0	0
21	22	1.50	0	0	0	0	0	0
22	23	1.70	0	0	0	0	0	0
23	24	1.80	0	0	0	0	0	0
24	25	1.90	0	0	0	0	0	0
25	26	2.10	.10	0	0	0	0	0
26	27	2.20	.20	0	0	0	0	0
27	28	2.40	.40	0	0	0	0	0
28	29	2.50	.50	0	0	0	0	0
29	30	2.60	.60	0	0	0	0	0
30	31	2.80	.80	0	0	0	0	0
31	32	2.90	.90	0	0	0	0	0
32	33	3.10	1.10	0	0	0	0	0
33	34	3.20	1.20	0	0	0	0	0
34	35	3.30	1.30	0	0	0	0	0
35	36	3.50	1.50	0	0	0	0	0
36	37	3.60	1.60	0	0	0	0	0
37	38	3.80	1.80	0	0	0	0	0
38	39	3.90	1.90	0	0	0	0	0

And the Wages are		And the Number of Withholding Exemptions Claimed is—										
At Least	But Less Than	0	1	2	3	4	5	6	7	8	9	10 Or More
		The Amount of Income Tax to be Withheld Shall be—										
$90	$92	$12.30	$10.00	$ 7.70	$ 5.40	$ 3.20	$ 1.20	$ 0	$ 0	$ 0	$ 0	$ 0
92	94	12.60	10.30	8.00	5.70	3.50	1.40	0	0	0	0	0
94	96	12.90	10.60	8.30	6.00	3.70	1.70	0	0	0	0	0
96	98	13.30	10.90	8.60	6.30	4.00	2.00	0	0	0	0	0
98	100	13.60	11.30	9.00	6.60	4.30	2.30	.30	0	0	0	0
100	105	14.10	11.80	9.50	7.20	4.90	2.80	.80	0	0	0	0
105	110	14.90	12.60	10.30	8.00	5.70	3.50	1.50	0	0	0	0
110	115	15.70	13.40	11.10	8.80	6.50	4.20	2.20	.10	0	0	0
115	120	16.50	14.20	11.90	9.60	7.30	5.00	2.90	.80	0	0	0
120	125	17.30	15.00	12.70	10.40	8.10	5.80	3.60	1.50	0	0	0
125	130	18.10	15.80	13.50	11.20	8.90	6.60	4.30	2.20	.20	0	0
130	135	18.90	16.60	14.30	12.00	9.70	7.40	5.10	2.90	.90	0	0
135	140	19.70	17.40	15.10	12.80	10.50	8.20	5.90	3.60	1.60	0	0
140	145	20.50	18.20	15.90	13.60	11.30	9.00	6.70	4.40	2.30	.30	0
145	150	21.30	19.00	16.70	14.40	12.10	9.80	7.50	5.20	3.00	1.00	0
150	160	22.50	20.20	17.90	15.60	13.30	11.00	8.70	6.40	4.10	2.00	0
160	170	24.10	21.80	19.50	17.20	14.90	12.60	10.30	8.00	5.70	3.40	1.40
170	180	26.00	23.40	21.10	18.80	16.50	14.20	11.90	9.60	7.30	5.00	2.80
180	190	28.00	25.20	22.70	20.40	18.10	15.80	13.50	11.20	8.90	6.60	4.30
190	200	30.00	27.20	24.30	22.00	19.70	17.40	15.10	12.80	10.50	8.20	5.90
200	210	32.00	29.20	26.30	23.60	21.30	19.00	16.70	14.40	12.10	9.80	7.50
210	220	34.40	31.20	28.30	25.40	22.90	20.60	18.30	16.00	13.70	11.40	9.10
220	230	36.80	33.30	30.30	27.40	24.50	22.20	19.90	17.60	15.30	13.00	10.70
230	240	39.20	35.70	32.30	29.40	26.50	23.80	21.50	19.20	16.90	14.60	12.30
240	250	41.60	38.10	34.60	31.40	28.50	25.60	23.10	20.80	18.50	16.20	13.90
250	260	44.00	40.50	37.00	33.60	30.50	27.60	24.70	22.40	20.10	17.80	15.50
260	270	46.40	42.90	39.40	36.00	32.50	29.60	26.70	24.00	21.70	19.40	17.10
270	280	48.80	45.30	41.80	38.40	34.90	31.60	28.70	25.80	23.30	21.00	18.70
280	290	51.20	47.70	44.20	40.80	37.30	33.90	30.70	27.80	25.00	22.60	20.30

39	40	4.10	2.00	0	0	0	0	0
40	41	4.20	2.20	.20	0	0	0	0
41	42	4.40	2.30	.30	0	0	0	0
42	43	4.50	2.50	.40	0	0	0	0
43	44	4.70	2.60	.60	0	0	0	0
44	45	4.90	2.70	.70	0	0	0	0
45	46	5.00	2.90	.90	0	0	0	0
46	47	5.20	3.00	1.00	0	0	0	0
47	48	5.30	3.20	1.10	0	0	0	0
48	49	5.50	3.30	1.30	0	0	0	0
49	50	5.70	3.40	1.40	0	0	0	0
50	51	5.80	3.60	1.60	0	0	0	0
51	52	6.00	3.70	1.70	0	0	0	0
52	53	6.10	3.90	1.80	0	0	0	0
53	54	6.30	4.00	2.00	0	0	0	0
54	55	6.50	4.10	2.10	.10	0	0	0
55	56	6.60	4.30	2.30	.20	0	0	0
56	57	6.80	4.50	2.40	.40	0	0	0
57	58	6.90	4.60	2.50	.50	0	0	0
58	59	7.10	4.80	2.70	.70	0	0	0
59	60	7.30	4.90	2.80	.80	0	0	0
60	62	7.50	5.20	3.00	1.00	0	0	0
62	64	7.80	5.50	3.30	1.30	0	0	0
64	66	8.10	5.80	3.60	1.60	0	0	0
66	68	8.50	6.10	3.90	1.80	0	0	0
68	70	8.80	6.50	4.20	2.10	.10	0	0
70	72	9.10	6.00	4.50	2.40	.40	0	0
72	74	9.40	7.10	4.80	2.70	.70	0	0
74	76	9.70	7.40	5.10	3.00	.90	0	0
76	78	10.10	7.70	5.40	3.20	1.20	0	0
78	80	10.40	8.10	5.80	3.50	1.50	0	0
80	82	10.70	8.40	6.10	3.80	1.80	0	0
82	84	11.00	8.70	6.40	4.10	2.10	0	0
84	86	11.30	9.00	6.70	4.40	2.30	.30	0
86	88	11.70	9.30	7.00	4.70	2.60	.60	0
88	90	12.00	9.70	7.40	5.00	2.90	.90	0

290	300	53.60	50.10	46.60	43.20	39.70	36.30	32.80	29.80	27.00	24.20	21.90
300	310	56.00	52.50	49.00	45.60	42.10	38.70	35.20	31.80	29.00	26.10	23.50
310	320	58.40	54.90	51.40	48.00	44.50	41.10	37.60	34.10	31.00	28.10	25.20
320	330	60.80	57.30	53.80	50.40	46.90	43.50	40.00	36.50	33.10	30.10	27.20
330	340	63.60	59.70	56.20	52.80	49.30	45.90	42.40	38.90	35.50	32.10	29.20
340	350	66.40	62.40	58.60	55.20	51.70	48.30	44.80	41.30	37.90	34.40	31.20
350	360	69.20	65.20	61.10	57.60	54.10	50.70	47.20	43.70	40.30	36.80	33.40
360	370	72.00	68.00	63.90	60.00	56.50	53.10	49.60	46.10	42.70	39.20	35.80
370	380	74.80	70.80	66.70	62.70	58.90	55.50	52.00	48.50	45.10	41.60	38.20
380	390	77.60	73.60	69.50	65.50	61.50	57.90	54.40	50.90	47.50	44.00	40.60
390	400	80.40	76.40	72.30	68.30	64.30	60.30	56.80	53.30	49.90	46.40	43.00
400	410	83.20	79.20	75.10	71.10	67.10	63.00	59.20	55.70	52.30	48.80	45.40
410	420	86.30	82.00	77.90	73.90	69.90	65.80	61.80	58.10	54.70	51.20	47.80
420	430	89.50	84.80	80.70	76.70	72.70	68.60	64.60	60.50	57.10	53.60	50.20
430	440	92.70	88.00	83.50	79.50	75.50	71.40	67.40	63.30	59.50	56.00	52.60
440	450	95.90	91.20	86.60	82.30	78.30	74.20	70.20	66.10	62.10	58.40	55.00
450	460	99.10	94.40	89.80	85.20	81.10	77.00	73.00	68.90	64.90	60.90	57.40
460	470	102.30	97.60	93.00	88.40	83.90	79.80	75.80	71.70	67.70	63.70	59.80
470	480	105.50	100.80	96.20	91.60	87.00	82.60	78.60	74.50	70.50	66.50	62.40
480	490	108.70	104.00	99.40	94.80	90.20	85.60	81.40	77.30	73.30	69.30	65.20
490	500	112.20	107.20	102.60	98.00	93.40	88.80	84.20	80.10	76.10	72.10	68.00
500	510	115.80	110.60	105.80	101.20	96.60	92.00	87.40	82.90	78.90	74.90	70.80
510	520	119.40	114.20	109.10	104.40	99.80	95.20	90.60	86.00	81.70	77.70	73.60
520	530	123.00	117.80	112.70	107.60	103.00	98.40	93.80	89.20	84.50	80.50	76.40
530	540	126.60	121.40	116.30	111.10	106.20	101.60	97.00	92.40	87.70	83.30	79.20
540	550	130.20	125.00	119.90	114.70	109.50	104.80	100.20	95.60	90.90	86.30	82.00
550	560	133.80	128.60	123.50	118.30	113.10	108.00	103.40	98.80	94.10	89.50	84.90
560	570	137.40	132.20	127.10	121.90	116.70	111.50	106.60	102.00	97.30	92.70	88.10
570	580	141.00	135.80	130.70	125.50	120.30	115.10	109.90	105.20	100.50	95.90	91.30
580	590	144.60	139.40	134.30	129.10	123.90	118.70	113.50	108.40	103.70	99.10	94.50
590	600	148.20	143.00	137.90	132.70	127.50	122.30	117.10	111.90	106.90	102.30	97.70
600	610	151.80	146.60	141.50	136.30	131.10	125.90	120.70	115.50	110.30	105.50	100.90
610	620	155.40	150.20	145.10	139.90	134.70	129.50	124.30	119.10	113.90	108.70	104.10
620	630	159.00	153.80	148.70	143.50	138.30	133.10	127.90	122.70	117.50	112.30	107.30
630	640	162.60	157.40	152.30	147.10	141.90	136.70	131.50	126.30	121.10	115.90	110.70
$640 and over		36 percent of the excess over $640 plus										
		164.40	159.20	154.10	148.90	143.70	138.50	133.30	128.10	122.90	117.70	112.50

FEDERAL INCOME TAX WITHHOLDING TABLE — *Weekly* Payroll for *Single* Wage Earners

And the Wages are		And the Number of Withholding Exemptions Claimed is—						
At Least	But Less Than	0	1	2	3	4	5	6 Or More
		The Amount of Income Tax to be Withheld Shall be—						
$ 0	$11	$ 0	$ 0	$ 0	$ 0	$ 0	$ 0	$ 0
11	12	.10	0	0	0	0	0	0
12	13	.30	0	0	0	0	0	0
13	14	.40	0	0	0	0	0	0
14	15	.50	0	0	0	0	0	0
15	16	.70	0	0	0	0	0	0
16	17	.80	0	0	0	0	0	0
17	18	1.00	0	0	0	0	0	0
18	19	1.10	0	0	0	0	0	0
19	20	1.20	0	0	0	0	0	0
20	21	1.40	0	0	0	0	0	0
21	22	1.50	0	0	0	0	0	0
22	23	1.70	0	0	0	0	0	0
23	24	1.80	0	0	0	0	0	0
24	25	1.90	0	0	0	0	0	0
25	26	2.10	.10	0	0	0	0	0
26	27	2.20	.20	0	0	0	0	0
27	28	2.40	.40	0	0	0	0	0
28	29	2.50	.50	0	0	0	0	0
29	30	2.60	.60	0	0	0	0	0
30	31	2.80	.80	0	0	0	0	0
31	32	2.90	.90	0	0	0	0	0
32	33	3.10	1.10	0	0	0	0	0
33	34	3.20	1.20	0	0	0	0	0
34	35	3.30	1.30	0	0	0	0	0

And the Wages are		And the Number of Withholding Exemptions Claimed is—										
At Least	But Less Than	0	1	2	3	4	5	6	7	8	9	10 Or More
		The Amount of Income Tax to be Withheld Shall be—										
$80	$82	$12.00	$ 9.10	$ 6.50	$ 3.90	$ 1.80	$ 0	$ 0	$ 0	$ 0	$ 0	$ 0
82	84	12.40	9.50	6.90	4.30	2.10	0	0	0	0	0	0
84	86	12.80	9.80	7.20	4.60	2.30	.30	0	0	0	0	0
86	88	13.20	10.20	7.60	5.00	2.60	.60	0	0	0	0	0
88	90	13.60	10.60	8.00	5.40	2.90	.90	0	0	0	0	0
90	92	14.10	11.00	8.30	5.70	3.20	1.20	0	0	0	0	0
92	94	14.50	11.40	8.70	6.10	3.50	1.40	0	0	0	0	0
94	96	14.90	11.90	9.00	6.40	3.90	1.70	0	0	0	0	0
96	98	15.30	12.30	9.40	6.80	4.20	2.00	0	0	0	0	0
98	100	15.70	12.70	9.80	7.20	4.60	2.30	.30	0	0	0	0
100	105	16.50	13.40	10.40	7.80	5.20	2.80	.80	0	0	0	0
105	110	17.50	14.50	11.50	8.70	6.10	3.50	1.50	0	0	0	0
110	115	18.60	15.50	12.50	9.60	7.00	4.40	2.20	.10	0	0	0
115	120	19.60	16.60	13.60	10.50	7.90	5.30	2.90	.80	0	0	0
120	125	20.70	17.60	14.60	11.60	8.80	6.20	3.60	1.50	0	0	0
125	130	21.70	18.70	15.70	12.60	9.70	7.10	4.50	2.20	.20	0	0
130	135	22.80	19.70	16.70	13.70	10.70	8.00	5.40	2.90	.90	0	0
135	140	23.80	20.80	17.80	14.70	11.70	8.90	6.30	3.70	1.60	0	0
140	145	24.90	21.80	18.80	15.80	12.80	9.80	7.20	4.60	2.30	.30	0
145	150	25.90	22.90	19.90	16.80	13.80	10.80	8.10	5.50	3.00	1.00	0
150	160	27.50	24.50	21.40	18.40	15.40	12.30	9.50	6.90	4.30	2.00	0
160	170	29.60	26.60	23.50	20.50	17.50	14.40	11.40	8.70	6.10	3.50	1.40
170	180	31.70	28.70	25.60	22.60	19.60	16.50	13.50	10.50	7.90	5.30	2.80
180	190	33.80	30.80	27.70	24.70	21.70	18.60	15.60	12.60	9.70	7.10	4.50
190	200	35.90	32.90	29.80	26.80	23.80	20.70	17.70	14.70	11.70	8.90	6.30

35	36	3.50	1.50	0	0	0	0	0
36	37	3.70	1.60	0	0	0	0	0
37	38	3.90	1.80	0	0	0	0	0
38	39	4.10	1.90	0	0	0	0	0
39	40	4.20	2.00	0	0	0	0	0
40	41	4.40	2.20	.20	0	0	0	0
41	42	4.60	2.30	.30	0	0	0	0
42	43	4.80	2.50	.40	0	0	0	0
43	44	5.00	2.60	.60	0	0	0	0
44	45	5.10	2.70	.70	0	0	0	0
45	46	5.30	2.90	.90	0	0	0	0
46	47	5.50	3.00	1.00	0	0	0	0
47	48	5.70	3.20	1.10	0	0	0	0
48	49	5.90	3.30	1.30	0	0	0	0
49	50	6.00	3.40	1.40	0	0	0	0
50	51	6.20	3.60	1.60	0	0	0	0
51	52	6.40	3.80	1.70	0	0	0	0
52	53	6.60	4.00	1.80	0	0	0	0
53	54	6.80	4.20	2.00	0	0	0	0
54	55	6.90	4.30	2.10	.10	0	0	0
55	56	7.10	4.50	2.30	.20	0	0	0
56	57	7.30	4.70	2.40	.40	0	0	0
57	58	7.50	4.90	2.50	.50	0	0	0
58	59	7.70	5.10	2.70	.70	0	0	0
59	60	7.80	5.20	2.80	.80	0	0	0
60	62	8.10	5.50	3.00	1.00	0	0	0
62	64	8.50	5.90	3.30	1.30	0	0	0
64	66	8.80	6.20	3.60	1.60	0	0	0
66	68	9.20	6.60	4.00	1.80	0	0	0
68	70	9.60	7.00	4.40	2.10	.10	0	0
70	72	9.90	7.30	4.70	2.40	.40	0	0
72	74	10.30	7.70	5.10	2.70	.70	0	0
74	76	10.70	8.00	5.40	3.00	.90	0	0
76	78	11.10	8.40	5.80	3.20	1.20	0	0
78	80	11.50	8.80	6.20	3.60	1.50	0	0

200	210	38.10	35.00	31.90	28.90	25.90	22.80	19.80	16.80	13.80	10.70	8.10
210	220	40.40	37.10	34.00	31.00	28.00	24.90	21.90	18.90	15.90	12.80	9.90
220	230	42.70	39.30	36.10	33.10	30.10	27.00	24.00	21.00	18.00	14.90	11.90
230	240	45.10	41.60	38.30	35.20	32.20	29.10	26.10	23.10	20.10	17.00	14.00
240	250	47.80	43.90	40.60	37.30	34.30	31.20	28.20	25.20	22.20	19.10	16.10
250	260	50.50	46.60	42.90	39.60	36.40	33.30	30.30	27.30	24.30	21.20	18.20
260	270	53.20	49.30	45.40	41.90	38.60	35.40	32.40	29.40	26.40	23.30	20.30
270	280	56.20	52.00	48.10	44.20	40.90	37.60	34.50	31.50	28.50	25.40	22.40
280	290	59.30	54.80	50.80	46.90	43.20	39.90	36.60	33.60	30.60	27.50	24.50
290	300	62.40	57.90	53.50	49.60	45.70	42.20	38.90	35.70	32.70	29.60	26.60
300	310	65.50	61.00	56.50	52.30	48.40	44.60	41.20	37.80	34.80	31.70	28.70
310	320	68.60	64.10	59.60	55.10	51.10	47.30	43.50	40.10	36.90	33.80	30.80
320	330	71.70	67.20	62.70	58.20	53.80	50.00	46.10	42.40	39.10	35.90	32.90
330	340	74.80	70.30	65.80	61.30	56.90	52.70	48.80	44.90	41.40	38.10	35.00
340	350	78.30	73.40	68.90	64.40	60.00	55.50	51.50	47.60	43.70	40.40	37.10
350	360	81.80	76.80	72.00	67.50	63.10	58.60	54.20	50.30	46.40	42.70	39.40
360	370	85.30	80.30	75.30	70.60	66.20	61.70	57.20	53.00	49.10	45.20	41.70
370	380	88.80	83.80	78.80	73.70	69.30	64.80	60.30	55.90	51.80	47.90	44.00
380	390	92.30	87.30	82.30	77.20	72.40	67.90	63.40	59.00	54.50	50.60	46.70
390	400	95.80	90.80	85.80	80.70	75.70	71.00	66.50	62.10	57.60	53.30	49.40
400	410	99.30	94.30	89.30	84.20	79.20	74.10	69.60	65.20	60.70	56.20	52.10
410	420	102.80	97.80	92.80	87.70	82.70	77.60	72.70	68.30	63.80	59.30	54.80
420	430	106.30	101.30	96.30	91.20	86.20	81.10	76.10	71.40	66.90	62.40	57.90
430	440	109.80	104.80	99.80	94.70	89.70	84.60	79.60	74.50	70.00	65.50	61.00
440	450	113.30	108.30	103.30	98.20	93.20	88.10	83.10	78.00	73.10	68.60	64.10
450	460	116.80	111.80	106.80	101.70	96.70	91.60	86.60	81.50	76.50	71.70	67.20
460	470	120.30	115.30	110.30	105.20	100.20	95.10	90.10	85.00	80.00	74.90	70.30
470	480	123.80	118.80	113.80	108.70	103.70	98.60	93.60	88.50	83.50	78.40	73.40
480	490	127.30	122.30	117.30	112.20	107.20	102.10	97.10	92.00	87.00	81.90	76.90
$490 and over		35 percent of the excess over $490 plus										
		129.10	124.00	119.00	114.00	108.90	103.90	98.80	93.80	88.70	83.70	78.60

FEDERAL INCOME TAX WITHHOLDING TABLE *Monthly* Payroll for *Single* Wage Earners

And the Wages are		And the Number of Withholding Exemptions Claimed is—						
At Least	But Less Than	0	1	2	3	4	5	6 Or More
		The Amount of Income Tax to be Withheld Shall be—						
$ 0	$48	$ 0	$ 0	$ 0	$ 0	$ 0	$ 0	$ 0
48	52	.60	0	0	0	0	0	0
52	56	1.10	0	0	0	0	0	0
56	60	1.70	0	0	0	0	0	0
60	64	2.30	0	0	0	0	0	0
64	68	2.80	0	0	0	0	0	0
68	72	3.40	0	0	0	0	0	0
72	76	3.90	0	0	0	0	0	0
76	80	4.50	0	0	0	0	0	0
80	84	5.10	0	0	0	0	0	0
84	88	5.60	0	0	0	0	0	0
88	92	6.20	0	0	0	0	0	0
92	96	6.70	0	0	0	0	0	0
96	100	7.30	0	0	0	0	0	0
100	104	7.90	0	0	0	0	0	0
104	108	8.40	0	0	0	0	0	0
108	112	9.00	.20	0	0	0	0	0
112	116	9.50	.80	0	0	0	0	0
116	120	10.10	1.40	0	0	0	0	0
120	124	10.70	1.90	0	0	0	0	0
124	128	11.20	2.50	0	0	0	0	0
128	132	11.80	3.00	0	0	0	0	0
132	136	12.30	3.60	0	0	0	0	0
136	140	12.90	4.20	0	0	0	0	0
140	144	13.50	4.70	0	0	0	0	0

And the Wages are		And the Number of Withholding Exemptions Claimed is—										
At Least	But Less Than	0	1	2	3	4	5	6	7	8	9	10 Or More
		The Amount of Income Tax to be Withheld Shall be—										
$ 328	$ 336	$47.80	$36.10	$24.80	$13.80	$ 5.10	$ 0	$ 0	$ 0	$ 0	$ 0	$ 0
336	344	49.50	37.50	26.30	15.00	6.20	0	0	0	0	0	0
344	352	51.20	39.00	27.70	16.50	7.30	0	0	0	0	0	0
352	360	52.80	40.40	29.20	17.90	8.40	0	0	0	0	0	0
360	368	54.50	41.90	30.60	19.40	9.50	.80	0	0	0	0	0
368	376	56.20	43.30	32.00	20.80	10.70	1.90	0	0	0	0	0
376	384	57.90	44.80	33.50	22.20	11.80	3.00	0	0	0	0	0
384	392	59.60	46.40	34.90	23.70	12.90	4.20	0	0	0	0	0
392	400	61.20	48.10	36.40	25.10	14.00	5.30	0	0	0	0	0
400	420	64.20	51.10	38.90	27.60	16.40	7.20	0	0	0	0	0
420	440	68.40	55.30	42.50	31.20	20.00	10.00	1.30	0	0	0	0
440	460	72.60	59.50	46.30	34.80	23.60	12.80	4.10	0	0	0	0
460	480	76.80	63.70	50.50	38.40	27.20	15.90	6.90	0	0	0	0
480	500	81.00	67.90	54.70	42.00	30.80	19.50	9.70	.90	0	0	0
500	520	85.20	72.10	58.90	45.80	34.40	23.10	12.50	3.70	0	0	0
520	540	89.40	76.30	63.10	50.00	38.00	26.70	15.50	6.50	0	0	0
540	560	93.60	80.50	67.30	54.20	41.60	30.30	19.10	9.30	.60	0	0
560	580	97.80	84.70	71.50	58.40	45.30	33.90	22.70	12.10	3.40	0	0
580	600	102.00	88.90	75.70	62.60	49.50	37.50	26.30	15.00	6.20	0	0
600	640	108.30	95.20	82.00	68.90	55.80	42.90	31.70	20.40	10.40	1.60	0
640	680	116.70	103.60	90.40	77.30	64.20	51.10	38.90	27.60	16.40	7.20	0
680	720	125.10	112.00	98.80	85.70	72.60	59.50	46.30	34.80	23.60	12.80	4.10
720	760	133.50	120.40	107.20	94.10	81.00	67.90	54.70	42.00	30.80	19.50	9.70
760	800	141.90	128.80	115.60	102.50	89.40	76.30	63.10	50.00	38.00	26.70	15.50
800	840	150.30	137.20	124.00	110.90	97.80	84.70	71.50	58.40	45.30	33.90	22.70
840	880	158.70	145.60	132.40	119.30	106.20	93.10	79.90	66.80	53.70	41.10	29.90

144	148	14.00	5.30	0	0	0	0	0
148	152	14.60	5.80	0	0	0	0	0
152	156	15.30	6.40	0	0	0	0	0
156	160	16.00	7.00	0	0	0	0	0
160	164	16.70	7.50	0	0	0	0	0
164	168	17.50	8.10	0	0	0	0	0
168	172	18.20	8.60	0	0	0	0	0
172	176	18.90	9.20	.40	0	0	0	0
176	180	19.60	9.80	1.00	0	0	0	0
180	184	20.30	10.30	1.60	0	0	0	0
184	188	21.10	10.90	2.10	0	0	0	0
188	192	21.80	11.40	2.70	0	0	0	0
192	196	22.50	12.00	3.20	0	0	0	0
196	200	23.20	12.60	3.80	0	0	0	0
200	204	23.90	13.10	4.40	0	0	0	0
204	208	24.70	13.70	4.90	0	0	0	0
208	212	25.40	14.20	5.50	0	0	0	0
212	216	26.10	14.90	6.00	0	0	0	0
216	220	26.80	15.60	6.60	0	0	0	0
220	224	27.50	16.30	7.20	0	0	0	0
224	228	28.30	17.00	7.70	0	0	0	0
228	232	29.00	17.70	8.30	0	0	0	0
232	236	29.70	18.50	8.80	.10	0	0	0
236	240	30.40	19.20	9.40	.70	0	0	0
240	248	31.50	20.30	10.20	1.50	0	0	0
248	256	32.90	21.70	11.40	2.60	0	0	0
256	264	34.40	23.10	12.50	3.70	0	0	0
264	272	35.80	24.60	13.60	4.90	0	0	0
272	280	37.30	26.00	14.80	6.00	0	0	0
280	288	38.70	27.50	16.20	7.10	0	0	0
288	296	40.10	28.90	17.60	8.20	0	0	0
296	304	41.60	30.30	19.10	9.30	.60	0	0
304	312	43.00	31.80	20.50	10.50	1.70	0	0
312	320	44.50	33.20	22.00	11.60	2.80	0	0
320	328	46.10	34.70	23.40	12.70	3.90	0	0

880	920	167.60	154.00	140.80	127.70	114.60	101.50	88.30	75.20	62.10	49.00	37.10
920	960	176.80	162.40	149.20	136.10	123.00	109.90	96.70	83.60	70.50	57.40	44.30
960	1,000	186.00	171.60	157.60	144.50	131.40	118.30	105.10	92.00	78.90	65.80	52.60
1,000	1,040	196.00	180.80	166.40	152.90	139.80	126.70	113.50	100.40	87.30	74.20	61.00
1,040	1,080	206.80	190.00	175.60	161.30	148.20	135.10	121.90	108.80	95.70	82.60	69.40
1,080	1,120	217.60	200.70	184.80	170.50	156.60	143.50	130.30	117.20	104.10	91.00	77.80
1,120	1,160	228.40	211.50	194.60	179.70	165.30	151.90	138.70	125.60	112.50	99.40	86.20
1,160	1,200	239.70	222.30	205.40	188.90	174.50	160.30	147.10	134.00	120.90	107.80	94.60
1,200	1,240	252.10	233.10	216.20	199.40	183.70	169.30	155.50	142.40	129.30	116.20	103.00
1,240	1,280	264.50	245.10	227.00	210.20	193.30	178.50	164.10	150.80	137.70	124.60	111.40
1,280	1,320	276.90	257.50	238.20	221.00	204.10	187.70	173.30	159.20	146.10	133.00	119.80
1,320	1,360	289.30	269.90	250.60	231.80	214.90	198.00	182.50	168.20	154.50	141.40	128.20
1,360	1,400	301.70	282.30	263.00	243.60	225.70	208.80	191.90	177.40	163.00	149.80	136.60
1,400	1,440	314.10	294.70	275.40	256.00	236.60	219.60	202.70	186.60	172.20	158.20	145.00
1,440	1,480	327.30	307.10	287.80	268.40	249.00	230.40	213.50	196.70	181.40	167.00	153.40
1,480	1,520	341.30	319.50	300.20	280.80	261.40	242.00	224.30	207.50	190.60	176.20	161.80
1,520	1,560	355.30	333.40	312.60	293.20	273.60	254.40	235.10	218.30	201.40	185.40	171.00
1,560	1,600	369.30	347.40	325.50	305.60	286.20	266.80	247.50	229.10	212.20	195.30	180.20
1,600	1,640	383.30	361.40	339.50	318.00	298.60	279.20	259.90	240.50	223.00	206.10	189.40
1,640	1,680	397.30	375.40	353.50	331.60	311.00	291.60	272.30	252.90	233.80	216.90	200.00
1,680	1,720	411.30	389.40	367.50	345.60	323.80	304.00	284.70	265.30	245.90	227.70	210.80
1,720	1,760	425.30	403.40	381.50	359.60	337.80	316.40	297.10	277.70	258.30	238.90	221.60
1,760	1,800	439.30	417.40	395.50	373.60	351.80	329.90	309.50	290.10	270.70	251.30	232.40
1,800	1,840	453.30	431.40	409.50	387.60	365.80	343.90	322.00	302.50	283.10	263.70	244.40
1,840	1,880	467.30	445.40	423.50	401.60	379.80	357.90	336.00	314.90	295.50	276.10	256.80
1,880	1,920	481.30	459.40	437.50	415.60	393.80	371.90	350.00	328.10	307.90	288.50	269.20
1,920	1,960	495.30	473.40	451.50	429.60	407.80	385.90	364.00	342.10	320.30	300.90	281.60
1,960	2,000	509.30	487.40	465.50	443.60	421.80	399.90	378.00	356.10	334.30	313.30	294.00
2,000	2,040	523.30	501.40	479.50	457.60	435.80	413.90	392.00	370.10	348.30	326.40	306.40
2,040	2,080	537.30	515.40	493.50	471.60	449.80	427.90	406.00	384.10	362.30	340.40	318.80
2,080	2,120	551.30	529.40	507.50	485.60	463.80	441.90	420.00	398.10	376.30	354.40	332.50
$2,120 and over		35 percent of the excess over $2,120 plus										
		558.30	536.40	514.50	492.60	470.80	448.90	427.00	405.10	383.30	361.40	339.50

FEDERAL INCOME TAX WITHHOLDING TABLE — *Monthly* Payroll for *Married* Wage Earners

And the Wages are		And the Number of Withholding Exemptions Claimed is—						
At Least	But Less Than	0	1	2	3	4	5	6 Or More
		The Amount of Income Tax to be Withheld Shall be—						
$ 0	$48	$ 0	$ 0	$ 0	$ 0	$ 0	$ 0	$ 0
48	52	.60	0	0	0	0	0	0
52	56	1.10	0	0	0	0	0	0
56	60	1.70	0	0	0	0	0	0
60	64	2.30	0	0	0	0	0	0
64	68	2.80	0	0	0	0	0	0
68	72	3.40	0	0	0	0	0	0
72	76	3.90	0	0	0	0	0	0
76	80	4.50	0	0	0	0	0	0
80	84	5.10	0	0	0	0	0	0
84	88	5.60	0	0	0	0	0	0
88	92	6.20	0	0	0	0	0	0
92	96	6.70	0	0	0	0	0	0
96	100	7.30	0	0	0	0	0	0
100	104	7.90	0	0	0	0	0	0
104	108	8.40	0	0	0	0	0	0
108	112	9.00	.20	0	0	0	0	0
112	116	9.50	.80	0	0	0	0	0
116	120	10.10	1.40	0	0	0	0	0
120	124	10.70	1.90	0	0	0	0	0
124	128	11.20	2.50	0	0	0	0	0
128	132	11.60	3.00	0	0	0	0	0
132	136	12.30	3.60	0	0	0	0	0
136	140	12.90	4.20	0	0	0	0	0
140	144	13.50	4.70	0	0	0	0	0
144	148	14.00	5.30	0	0	0	0	0
148	152	14.60	5.80	0	0	0	0	0
152	156	15.10	6.40	0	0	0	0	0
156	160	15.70	7.00	0	0	0	0	0

And the Wages are		And the Number of Withholding Exemptions Claimed is—										
At Least	But Less Than	0	1	2	3	4	5	6	7	8	9	10 Or More
		The Amount of Income Tax to be Withheld Shall be—										
$ 368	$ 376	$49.70	$39.70	$29.70	$19.70	$10.70	$ 1.90	$ 0	$ 0	$ 0	$ 0	$ 0
376	384	51.00	41.00	31.00	21.00	11.80	3.00	0	0	0	0	0
384	392	52.20	42.20	32.20	22.20	12.90	4.20	0	0	0	0	0
392	400	53.50	43.50	33.50	23.50	14.00	5.30	0	0	0	0	0
400	420	55.80	45.80	35.80	25.80	16.00	7.20	0	0	0	0	0
420	440	59.00	49.00	39.00	29.00	19.00	10.00	1.30	0	0	0	0
440	460	62.20	52.20	42.20	32.20	22.20	12.80	4.10	0	0	0	0
460	480	65.40	55.40	45.40	35.40	25.40	15.60	6.90	0	0	0	0
480	500	68.60	58.60	48.60	38.60	28.60	18.60	9.70	.90	0	0	0
500	520	71.80	61.80	51.80	41.80	31.80	21.80	12.50	3.70	0	0	0
520	540	75.00	65.00	55.00	45.00	35.00	25.00	15.30	6.50	0	0	0
540	560	78.20	68.20	58.20	48.20	38.20	28.20	18.20	9.30	.60	0	0
560	580	81.40	71.40	61.40	51.40	41.40	31.40	21.40	12.10	3.40	0	0
580	600	84.60	74.60	64.60	54.60	44.60	34.60	24.60	14.90	6.20	0	0
600	640	89.40	79.40	69.40	59.40	49.40	39.40	29.40	19.40	10.40	1.60	0
640	680	95.80	85.80	75.80	65.80	55.80	45.80	35.80	25.80	16.00	7.20	0
680	720	102.20	92.20	82.20	72.20	62.20	52.20	42.20	32.20	22.20	12.80	4.10
720	760	109.20	98.60	88.60	78.60	68.60	58.60	48.60	38.60	28.60	18.60	9.70
760	800	117.20	105.00	95.00	85.00	75.00	65.00	55.00	45.00	35.00	25.00	15.30
800	840	125.20	112.70	101.40	91.40	81.40	71.40	61.40	51.40	41.40	31.40	21.40
840	880	133.20	120.70	108.20	97.80	87.80	77.80	67.80	57.80	47.80	37.80	27.80
880	920	141.30	128.70	116.20	104.20	94.20	84.20	74.20	64.20	54.20	44.20	34.20
920	960	150.90	136.70	124.20	111.70	100.60	90.60	80.60	70.60	60.60	50.60	40.60
960	1,000	160.50	145.50	132.20	119.70	107.20	97.00	87.00	77.00	67.00	57.00	47.00
1,000	1,040	170.10	155.10	140.20	127.70	115.20	103.40	93.40	83.40	73.40	63.40	53.40
1,040	1,080	179.70	164.70	149.70	135.70	123.20	110.70	99.80	89.80	79.80	69.80	59.80
1,080	1,120	189.30	174.30	159.30	144.30	131.20	118.70	106.20	96.20	86.20	76.20	66.20
1,120	1,160	198.90	183.90	168.90	153.90	139.20	126.70	114.20	102.60	92.60	82.60	72.60
1,160	1,200	208.50	193.50	178.50	163.50	148.50	134.70	122.20	109.70	99.00	89.00	79.00
1,200	1,240	218.10	203.10	188.10	173.10	158.10	143.10	130.20	117.70	105.40	95.40	85.40
1,240	1,280	227.70	212.70	197.70	182.70	167.70	152.70	138.20	125.70	113.20	101.80	91.80

160	164	16.30	7.50	0	0	0	0	0
164	168	16.80	8.10	0	0	0	0	0
168	172	17.40	8.60	0	0	0	0	0
172	176	18.00	9.20	.40	0	0	0	0
176	180	18.60	9.80	1.00	0	0	0	0
180	184	19.30	10.00	1.60	0	0	0	0
184	188	19.90	10.90	2.10	0	0	0	0
188	192	20.60	11.40	2.70	0	0	0	0
192	196	21.20	12.00	3.20	0	0	0	0
196	200	21.80	12.60	3.80	0	0	0	0
200	204	22.50	13.10	4.40	0	0	0	0
204	208	23.10	13.70	4.90	0	0	0	0
208	212	23.80	14.20	5.50	0	0	0	0
212	216	24.40	14.80	6.00	0	0	0	0
216	220	25.00	15.40	6.60	0	0	0	0
220	224	25.70	15.90	7.20	0	0	0	0
224	228	26.30	16.50	7.70	0	0	0	0
228	232	27.00	17.00	8.30	0	0	0	0
232	236	27.60	17.60	8.80	.10	0	0	0
236	240	28.20	18.20	9.40	.70	0	0	0
240	248	29.20	19.20	10.20	1.50	0	0	0
248	256	30.50	20.50	11.40	2.60	0	0	0
256	264	31.80	21.80	12.50	3.70	0	0	0
264	272	33.00	23.00	13.60	4.90	0	0	0
272	280	34.30	24.30	14.70	6.00	0	0	0
280	288	35.60	25.60	15.80	7.10	0	0	0
288	296	36.90	26.90	17.00	8.20	0	0	0
296	304	38.20	28.20	18.20	9.30	.60	0	0
304	312	39.40	29.40	19.40	10.50	1.70	0	0
312	320	40.70	30.70	20.70	11.60	2.80	0	0
320	328	42.00	32.00	22.00	12.70	3.90	0	0
328	336	43.30	33.30	23.30	13.80	5.10	0	0
336	344	44.60	34.60	24.60	14.90	6.20	0	0
344	352	45.80	35.80	25.80	16.10	7.30	0	0
352	360	47.10	37.10	27.10	17.20	8.40	0	0
360	368	48.40	38.40	28.40	18.40	9.50	.80	0

*FOR 6 OR MORE EXEMPTIONS THE INCOME TAX WITHHELD IS 0.

1,280	1,320	237.30	222.30	207.30	192.30	177.30	162.30	147.30	133.70	121.20	108.70	98.20
1,320	1,360	246.90	231.90	216.90	201.90	186.90	171.90	156.90	141.90	129.20	116.70	104.60
1,360	1,400	256.50	241.50	226.50	211.50	196.50	181.50	166.50	151.50	137.20	124.70	112.20
1,400	1,440	266.80	251.10	236.10	221.10	206.10	191.10	176.10	161.10	146.10	132.70	120.20
1,440	1,480	278.00	260.70	245.70	230.70	215.70	200.70	185.70	170.70	155.70	140.70	128.20
1,480	1,520	289.20	271.70	255.30	240.30	225.30	210.30	195.30	180.30	165.30	150.30	136.20
1,520	1,560	300.40	282.90	265.40	249.90	234.90	219.90	204.90	189.90	174.90	159.90	144.90
1,560	1,600	311.60	294.10	276.60	259.50	244.50	229.50	214.50	199.50	184.50	169.50	154.50
1,600	1,640	322.80	305.30	287.80	270.30	254.10	239.10	224.10	209.10	194.10	179.10	164.10
1,640	1,680	334.00	316.50	299.00	281.50	264.00	248.70	233.70	218.70	203.70	188.70	173.70
1,680	1,720	345.20	327.70	310.20	292.70	275.20	258.30	243.30	228.30	213.30	198.30	183.30
1,720	1,760	356.40	338.90	321.40	303.90	286.40	268.90	252.90	237.90	222.90	207.90	192.90
1,760	1,800	367.90	350.10	332.60	315.10	297.60	280.10	262.60	247.50	232.50	217.50	202.50
1,800	1,840	380.70	361.30	343.80	326.30	308.80	291.30	273.80	257.10	242.10	227.10	212.10
1,840	1,880	393.50	373.50	355.00	337.50	320.00	302.50	285.00	267.50	251.70	236.70	221.70
1,880	1,920	406.30	386.30	366.30	348.70	331.20	313.70	296.20	278.70	261.30	246.30	231.30
1,920	1,960	419.10	399.10	379.10	359.90	342.40	324.90	307.40	289.90	272.40	255.90	240.90
1,960	2,000	431.90	411.90	391.90	371.90	353.60	336.10	318.60	301.10	283.50	266.10	250.50
2,000	2,040	444.70	424.70	404.70	384.70	364.80	347.30	329.80	312.30	294.80	277.30	260.10
2,040	2,080	457.50	437.50	417.50	397.50	377.50	358.50	341.00	323.50	306.00	288.50	271.00
2,080	2,120	470.30	450.30	430.30	410.30	390.30	370.30	352.20	334.70	317.20	299.70	282.20
2,120	2,160	484.60	463.10	443.10	423.10	403.10	383.10	363.40	345.90	328.40	310.90	293.40
2,160	2,200	499.00	476.50	455.90	435.90	415.90	395.90	375.90	357.10	339.60	322.10	304.60
2,200	2,240	513.40	490.90	468.70	448.70	428.70	408.70	388.70	368.70	350.80	333.30	315.80
2,240	2,280	527.80	505.30	482.80	461.50	441.50	421.50	401.50	381.50	362.00	344.50	327.00
2,280	2,320	542.20	519.70	497.20	474.70	454.30	434.30	414.30	394.30	374.30	355.70	338.20
2,320	2,360	556.60	534.10	511.60	489.10	467.10	447.10	427.10	407.10	387.10	367.10	349.40
2,360	2,400	571.00	548.50	526.00	503.50	481.00	459.90	439.90	419.90	399.90	379.90	360.60
2,400	2,440	585.40	562.90	540.40	517.90	495.40	472.90	452.70	432.70	412.70	392.70	372.70
2,440	2,480	599.80	577.30	554.80	532.30	509.80	487.30	465.50	445.50	425.50	405.50	385.50
2,480	2,520	614.20	591.70	569.20	546.70	524.20	501.70	479.20	458.30	438.30	418.30	398.30
2,520	2,560	628.60	606.10	583.60	561.10	538.60	516.10	493.60	471.10	451.10	431.10	411.10
2,560	2,600	643.00	620.50	598.00	575.50	553.00	530.50	508.00	485.50	463.90	443.90	423.90
2,600	2,640	657.40	634.90	612.40	589.90	567.40	544.90	522.40	499.90	477.40	456.70	436.70
2,640	2,680	671.80	649.30	626.80	604.30	581.80	559.30	536.80	514.30	491.80	469.50	449.50
2,680	2,720	686.20	663.70	641.20	618.70	596.20	573.70	551.20	528.70	506.20	483.70	462.30
2,720	2,760	700.60	678.10	655.60	633.10	610.60	588.10	565.60	543.10	520.60	498.10	475.60
$2,760 and over		36 percent of the excess over $2,760 plus										
		707.80	685.30	662.80	640.30	617.80	595.30	572.80	550.30	527.80	505.30	482.80

AICO-UTILITY Line Form No. 70-746—1/1/72

Objectives

Instructions offered here are for the preparation of the short Federal Income Tax Form 1040A and the long Form 1040. Use is made of actual forms and tables in preparing reports. Since a federal income tax return may be quite involved and require several schedules, this chapter deals primarily with simple returns and problems in which the income is mostly wages or salaries.

Problems are structured so that they may be used for any year—not just for the year used in the illustrations and examples.

8 FEDERAL INCOME TAXES

Income taxes represent the greater part of the revenue of the federal government. This income, in addition to that of other sources, is needed to meet the costs of the many services demanded by the public and the functions it performs.

Since any business concern and most individuals are obligated to pay an income tax, it is very important that accurate records be kept of all transactions that affect income and expenses. What forms and schedules the taxpayer will use will depend on his source and amount of income. The Internal Revenue Service furnishes special forms and schedules to meet special needs. For example, there are special forms for corporations, partnerships, and farmers to use in preparing their income tax returns. Also, special schedules are provided for deductions, dividend and interest income, profit and loss from a profession or business, sale or exchange of property, supplementary income, and others.

Income tax returns can become very complex involving

several forms and schedules. Since every taxpayer must file either Form 1040A or Form 1040, the purpose of this chapter is to enable anyone to fill out either one. These forms change slightly each year, but the basic information required remains the same. Also, tax laws, rates, and allowance limitations change from time to time, but the general procedure is basically the same. Forms and rates for the year 1972 are used in this text—the latest information available at the time of writing.

FORM 1040A

Form 1040A is referred to as "the short form" and may be used if:

1. your income is from salary, wages, tips, etc.,
2. you did not receive more than $200 of dividends or $200 of interest income,
3. you did not itemize your deductions.

If a taxpayer meets these requirements, he may use this form. The order in which it should be prepared is numbered (in circles) on the illustrations and examples that follow.

example A: Mr. John E. Brown is a clerk and earns a salary of $7200. He has no other income. He is single with no dependents. Income tax withheld as reported on his W-2 Form is $1142.40.

Instructions† for Completing Form 1040A (Pages 154–155)

(1) Use label if IRS (Internal Revenue Service) sent one; otherwise *type* or print name as shown.
(2) Enter social security number.
(3) Indicate marital status.
(4) One exemption† is allowed for each taxpayer and each

NOTE: There are 12 tables (up to 12 exemptions) which prohibits including them all in this text.

†Any office of the Internal Revenue Service will be glad to furnish any information upon request in the preparation of your income tax and assist you in its preparation if necessary. Complete instructions are provided for all forms needed for an income tax return.

dependent. (An additional exemption is allowed if you are 65 or over or are blind.) Line 9 information is obtained from line 25 on the reverse side—page 2 of the form.

(5) Lines 11 to 14 inclusive—enter information as requested.

(6) Instructions between lines 14 and 15 indicate the method of determining your tax. Since line 14 is LESS than $10,000, use the tax table as indicated in the instructions. In this case, refer to Table 1 (1 exemption, page 174). Enter this amount on line 19, page 2 of the form.

(7) Complete this section as indicated.

(8) Any dependents listed here must be listed on line 9.

(9) It is important to complete this section. It is a plan whereby the federal government will return money to your state, county, and city or township.

(10) Sign and date.

(11) Attach your forms W-2 and any statements or other forms required.

example B: Mr. Ralph E. Hayes is a bookkeeper and receives an annual salary of $12,600. He also receives $200 in dividends and $36.60 in interest. He is married. (His wife's name is Mary O.) They have one dependent child, Helen. Income tax withheld amounted to $1478.40.

Instructions for Completing Form 1040A (Pages 156, 157)

(1), (2,), (3), (4)—Follow the same procedure as indicated for example A. Notice that the wife's social security number must be entered as well as her husband's even though she may have had no income for the year since this is a joint return.

(5) line 11—enter information as required.
line 12a—enter total amount of dividends (cannot be more than $200 to use this form).
line 12b—the first $100 of dividends you receive from

EXAMPLE A

Short Form 1040A U.S. Individual Income Tax Return

Department of the Treasury
Internal Revenue Service

1972

Please print or type

(1) First name and initial (if joint return, use first names and middle initials of both): John E. — Last name: Brown

Your social security number (Husband's, if joint return): 666 52 0819 (2)

Wife's number, if joint return:

Present home address (Number and street (including apartment number) or rural route): General Delivery

City, town or post office, State and ZIP code: Chipley, Florida 32428

Occupation — Yours: clerk — Wife's:

Filing Status—check only one:

(3)
1 [X] Single
2 [ ] Married filing joint return
3 [ ] Married filing separately. If wife (husband) is also filing, give her (his) social security number and first name here.
4 [ ] Unmarried Head of Household
5 [ ] Widow(er) with dependent child (Enter year of death of husband (wife) ▶ 19)

Exemptions

	Regular	65 or over	Blind	
6 Yourself . . .	[X]	[ ]	[ ]	Enter number of boxes checked ▶ 1
7 Wife (husband) .	[ ]	[ ]	[ ]	

8 First names of your dependent children *who lived with you* — Enter number ▶

9 Number of other dependents (from line 25) . . . . ▶

10 Total exemptions claimed . . . . ▶ 1 (4)

Attach Copy B of Form W-2 here
Attach Check or Money Order here

11	Wages, salaries, tips, etc. (attach Form W-2 to front. If unavailable, attach explanation)	11	7200	00
12a	Dividends (if over $200, use Form 1040—see instructions) $.......... 12b Less Exclusion $.......... Balance ▶	12c		
13	Interest income (if over $200, use Form 1040)	13		
14	Total lines 11, 12c, and 13 (Adjusted Gross Income)	14	7200	00

(5)

- **If line 14 is $20,000 or less and you want IRS to figure your tax, see instructions on page 3.**
- **If line 14 is under $10,000, find tax in Tables 1–12 and enter on line 19. Skip lines 15 through 18.**
- **If line 14 is $10,000 or more go to line 15.**

(6)

15	If line 14 is $10,000 or more, enter 15% of line 14 but not more than $2,000 ($1,000 if line 3 was checked) .	15		
16	Subtract line 15 from line 14	16		
17	Multiply total number of exemptions claimed on line 10 by $750	17		
18	Taxable income (subtract line 17 from line 16) . . . . (Figure tax on amount on line 18 using Tax Rate Schedule X, Y, or Z, and enter tax on line 19.)	18		

Line	Description	Line	Amount	Cents	
19	Tax, check if from: [X] Tax Tables 1–12, or [ ] Tax Rate Schedule X, Y, or Z	19	937	00	
20	Credit for contributions to candidates for public office (see instructions on page 5)	20			
21	Income tax (subtract line 20 from line 19). If less than zero, enter zero	21	937	00	(7)
22	Total Federal income tax withheld (attach Form W–2 to front)	22	1142	40	
23	If line 21 is larger than line 22, enter BALANCE DUE IRS. Pay in full with return. Write social security number on check or money order and make payable to Internal Revenue Service ▶	23			
24	If line 22 is larger than line 21, enter REFUND. ▶	24	205	40	

Other Dependents

(a) NAME	(b) Relationship	(c) Months lived in your home. If born or died during year, write B or D.	(d) Did dependent have income of $750 or more?	(e) Amount YOU furnished for dependent's support. If 100% write ALL.	(f) Amount furnished by OTHERS including dependent.	
				$	$	(8)

25 **Total number of dependents listed in column (a). Enter here and on line 9** ▶

Revenue Sharing

26 Print or type the location of your principal place of residence at end of year (not necessarily the same as your post office address).

(a) State	(b) County	(c) Locality. If you lived inside the boundaries of an incorporated city, town, etc., enter its name; if not, check here ▶ [X]	(d) Township (see instructions on page 5)	
Florida	Washington			(9)

27 Enter the number of persons included on line 10 who (1) are filing a return of their own; or, (2) did not live at your principal place of residence at the end of the year ▶

For IRS use only—Leave blank

Under penalties of perjury, I declare that I have examined this return, including accompanying schedules and statements, and to the best of my knowledge and belief it is true, correct, and complete. Declaration of preparer (other than taxpayer) is based on all information of which he has any knowledge.

Sign here

▶ Your signature ____ Date ____

▶ Wife's (husband's) signature (if filing jointly, BOTH must sign even if only one had income)

▶ Preparer's signature (other than taxpayer) ____ Date ____ (10)

Address (and ZIP Code) ____ **Preparer's Emp. Ident. or Soc. Sec. No.**

☆ U.S. GOVERNMENT PRINTING OFFICE:1972—O-458-057 94-1149624

EXAMPLE B

Short Form 1040A U.S. Individual Income Tax Return

Department of the Treasury
Internal Revenue Service

1972

Please print or type

(1) First name and initial (if joint return, use first names and middle initials of both): Ralph E. & Mary O.
Last name: Hayes

Present home address (Number and street (including apartment number) or rural route): 2506 Rincon Road

City, town or post office, State and ZIP code: Rockway, California 94602

Your social security number (Husband's, if joint return): 517 | 03 | 1381 (2)

Wife's number, if joint return: 566 | 65 | 0823

Occupation — Yours: bookkeeper; Wife's: housewife

(3) **Filing Status**—check only one:

1 ☐ Single
2 ☒ Married filing joint return
3 ☐ Married filing separately. If wife (husband) is also filing, give her (his) social security number and first name here.
4 ☐ Unmarried Head of Household
5 ☐ Widow(er) with dependent child (Enter year of death of husband (wife) ▶ 19)

Exemptions

	Regular	65 or over	Blind	Enter number of boxes checked ▶
6 Yourself . . .	☒	☐	☐	
7 Wife (husband) .	☒	☐	☐	2

8 First names of your dependent children *who lived with you*: Helen — Enter number ▶ 1

9 Number of other dependents (from line 25) . . . . ▶

10 Total exemptions claimed . . . . ▶ 3 (4)

Attach Copy B of Form W-2 here
Attach Check or Money Order here

Line	Description	Line	Amount	
11	Wages, salaries, tips, etc. (attach Form W-2 to front. If unavailable, attach explanation)	11	12,600	00
12a	Dividends (if over \$200, use Form 1040—see instructions) \$ 200.00 **12b** Less Exclusion \$ 100.00 **Balance ▶**	12c	100	00
13	Interest income (if over \$200, use Form 1040)	13	37	00 (5)
14	Total lines 11, 12c, and 13 (Adjusted Gross Income)	14	12,737	00

- **If line 14 is \$20,000 or less and you want IRS to figure your tax, see instructions on page 3.**
- **If line 14 is under \$10,000, find tax in Tables 1–12 and enter on line 19. Skip lines 15 through 18.**
- **If line 14 is \$10,000 or more go to line 15.**

(6)

Line	Description	Line	Amount	
15	If line 14 is \$10,000 or more, enter 15% of line 14 but not more than \$2,000 (\$1,000 if line 3 was checked) .	15	1,910	55
16	Subtract line 15 from line 14	16	10,826	45
17	Multiply total number of exemptions claimed on line 10 by \$750	17	2,250	00
18	Taxable income (subtract line 17 from line 16) . . . . (Figure tax on amount on line 18 using Tax Rate Schedule X, Y, or Z, and enter tax on line 19.)	18	8,576	45

Line	Description	No.	Amount	¢
19	Tax, check if from: [ ] Tax Tables 1–12, or [X] Tax Rate Schedule X, Y, or Z	19	1,506	82
20	Credit for contributions to candidates for public office (see instructions on page 5)	20		
21	Income tax (subtract line 20 from line 19). If less than zero, enter zero	21	1,506	82
22	Total Federal income tax withheld (attach Form W–2 to front)	22	1,478	40
23	If line 21 is larger than line 22, enter BALANCE DUE IRS. Pay in full with return. Write social security number on check or money order and make payable to Internal Revenue Service ▶	23	28	42
24	If line 22 is larger than line 21, enter REFUND. ▶	24		

(7)

Other Dependents

(a) NAME	(b) Relationship	(c) Months lived in your home. If born or died during year, write B or D.	(d) Did dependent have income of $750 or more?	(e) Amount YOU furnished for dependent's support. If 100% write ALL.	(f) Amount furnished by OTHERS including dependent.
				$	$

(8)

25 Total number of dependents listed in column (a). Enter here and on line 9 ▶

Revenue Sharing

26 Print or type the location of your principal place of residence at end of year (not necessarily the same as your post office address).

(a) State	(b) County	(c) Locality. If you lived inside the boundaries of an incorporated city, town, etc., enter its name; if not, check here ▶ [ ]	(d) Township (see instructions on page 5)
Calif.	Sonoma		

(9)

27 Enter the number of persons included on line 10 who (1) are filing a return of their own; or, (2) did not live at your principal place of residence at the end of the year ▶

For IRS use only—Leave blank

Under penalties of perjury, I declare that I have examined this return, including accompanying schedules and statements, and to the best of my knowledge and belief it is true, correct, and complete. Declaration of preparer (other than taxpayer) is based on all information of which he has any knowledge.

(10)

Sign here

▶ Your signature — Date

▶ Wife's (husband's) signature (if filing jointly, BOTH must sign even if only one had income)

▶ Preparer's signature (other than taxpayer) — Date

Address (and ZIP Code) — **Preparer's Emp. Ident. or Soc. Sec. No.**

☆ U.S. GOVERNMENT PRINTING OFFICE: 1972—O–458–057 94-1149624

qualifying domestic corporations is not taxed. If husband and wife each had dividends from qualifying domestic corporations, each can exclude the first $100. If one receives less than $100 in dividends, the other still cannot exclude more than $100.

line 13—Although, according to our example, interest equals $36.60, you may record amounts rounded to the nearest dollar if you wish.

line 14—Equals total of lines 11, 12c, and 13. This is your *Adjusted Gross Income* for tax purposes.

(6) Read and follow instructions shown between lines 14 and 15. In this case, since line 14 is more than $10,000, complete lines 15 to 18 inclusive as per instructions. Instructions under line 18 indicate that the tax is to be taken from the Tax Rate Schedule X, Y, or Z and entered on line 19.

(7) Turn to page No. 180 and use Tax Rate Schedule Y. Since line 18 ($8,576.45) is more than $8000 but less $12,000, the tax equals $1380 + 22% of any amount over $8000. Then tax = $1380 + (.22 x $576.45) = $1506.82.

line 22—this was given in the example.

line 23 or 24—Complete as indicated.

(8), (9), and (10)—Complete as indicated.

(11) Attach W-2 Forms and any statements or other forms if necessary.

EXERCISE 35

I. Refer to example A, page 154 and answer the following questions.

1. Why did Mr. Brown use the Tax Table instead of the Tax Rate Schedule X?

2. If. Mr. Brown's salary as reported on line 11 and line 14 is $18,820, what is his tax?

3. If. Mr. Brown supported his mother and thereby qualified as "head of household," (a) What is his tax? (b) If he had 2 exemptions, is he required to have federal income tax withheld from his paycheck for both? If not, what is the advantage or disadvantage?

II. Refer to example B, page 156 and answer the following questions.

1. If line 14 is $14,737, what is the amount for line 15?

2. If line 14 is $9350, what is Mr. Hayes' tax?

III. Verify the answers to the following problems.

1. Compute the tax for a married taxpayer filing a joint return with an income (line 14) of (a) $7500, (b) $9825.

answers (a) $758, (b) $1162

2. Compute the tax for a single taxpayer with a taxable income (line 18) of (a) $10,500, (b) $8260

answers (a) $2225, (b) $1655

IV. **1.** Mr. James, a married man whose wife does not work, filed a joint return with a total *taxable* income of $7588. Compute his total federal income tax using the tax rate schedule.

2. Harry Parker is single but supports his mother who lives with him. His total *taxable* income is $11,090. What is his federal income tax? Mr. Parker qualifies as "head of household."

3. Mr. Turner is 65 years of age, his wife is 60. He earns $10,400 annually. He received $150 in dividends and $42 interest. Compute his federal income tax if he filed a joint return.

4. Henry Adams, a single taxpayer, earned $8260 last year. He also received $72 in interest. (a) Compute his tax. (b) If federal income tax withheld was $986, how much did he still owe in taxes?

FORM 1040

This form, sometimes referred to as the "long form," must be used by all taxpayers unless they qualify for and use Form 1040A. Sometimes even though a person may use the "short form," it may be to his advantage to use the "long form" and itemize deductions. The Internal Revenue Service stresses this fact and lists the conditions when it is an advantage.

Form 1040 consists of many schedules. The schedules used will depend on the sources of income and deductions of the taxpayer. It is beyond the scope of this text to discuss in detail and illustrate these schedules. However, information derived from them is shown in the illustrations shown on the following pages and will be used in examples and exercises to determine the taxable income.

General Instructions in Completing Form 1040

Line 15 (total income for tax purposes)† is the total of lines 11–14 inclusive. Line 17 equals the total income as reported on line 15 *less* the total of any adjustments listed on page 2, Part II, line 50. Line 12c (dividends)—the same rules apply as mentioned for Form 1040A.

Exemptions: Every taxpayer is entitled to 1 exemption for himself and 1 for each dependent. For every exemption he may exclude from taxable income a certain amount. For the year 1972, it was $750.00.

Deductions: A taxpayer may itemize his deductions or take the standard deduction (15% of the adjusted gross income not to exceed $2000). This is an important decision to make. Naturally, the choice should be the one that results in the lowest taxable income. If deductions are itemized they are classified as follows:

1. Medical and dental—Only a certain percent is allowed, the amount depending on the adjusted gross income.
2. Taxes—Real estate, state and local income, personal property, other.
3. Contributions—Church, organizations, charities, etc.
4. Interest expenses—(Interest charges on installment accounts, loans, etc.)
5. Casualties or theft (losses)—Insurance coverage is deducted.
6. Child and dependent care
7. Miscellaneous—Example would be professional dues.

†By definition, gross income is the sum of all income but, for tax purposes, some income is not taxable—Social Security Payments for example.

Summary: In preparing any income tax return, the goal is to determine the amount of taxable income. This is the base on which the tax is computed.

Taxable Income = Adjusted Gross Income Less Deductions and Exemptions

1. List all taxable sources of income and total.
2. Subtract adjustments if any to obtain adjusted gross income.
3. Follow instructions directly under line 17 (between line 17 and line 18).
   (a) If you do not itemize deductions and line 17 is less than $10,000, it is your taxable income and the tax is taken from the tax tables.
   (b) If you itemize deductions OR if line 17 is $10,000 or more, then you must compute your tax as outlined in Part III (lines 51–55 inclusive).

Estimated Earnings: Many taxpayers derive all of their income from sources other than wages, salaries, etc. as reported on line 11. These taxpayers are required to complete Form 1040 SE, Estimated Tax Declaration and vouchers which are attached to their tax return. The information on this form indicates what the taxpayer estimates his taxable income will be for the next year. He pays $\frac{1}{4}$ of this amount 4 times a year (April 15, June 15, September 15, and January 15). If the total amount paid is less than 80% of the actual tax due there is a penalty unless justified to the satisfaction of the IRS.

example A: Alice Meier earns $7880.00 a year as a saleslady. She is single with no dependents. She also receives $132.67 in interest and $520.00 from a trust fund yearly. She itemized her deductions which amounted to $1423.00.

example B: Although this example may not apply to you, it is one in which several sources of income are included in addition to other considerations that must be taken into account in filling out Form 1040.

EXAMPLE A

Form **1040** US Department of the Treasury / Internal Revenue Service
Individual Income Tax Return **1972**

For the year January 1–December 31, 1972, or other taxable year beginning ...., 1972, ending ...., 19 ......

Please print or type

First name and initial (If joint return, use first names and middle initials of both)	Last name	Your social security number (~~Husband's, if joint return~~)
Alice E.	Meier	542 55 0819
Present home address (Number and street, including apartment number, or rural route)		Wife's number, if joint return
P.O. Box X		
City, town or post office, State and ZIP code	Occupation	Yours: saleslady
Napa, California 94558		Wife's:

Filing Status—check only one:

1 [x] Single
2 [ ] Married filing joint return (even if only one had income)
3 [ ] Married filing separately. If wife (husband) is also filing give her (his) social security number and first name here.
4 [ ] Unmarried Head of Household
5 [ ] Widow(er) with dependent child (Enter year of death of husband (wife) ▶ 19)

Exemptions Regular / 65 or over / Blind

6 Yourself . . . . . . [x] Regular [ ] 65 or over [ ] Blind
7 Wife (husband) . . . [ ] [ ] [ ]
Enter number of boxes checked ▶ 1
8 First names of your dependent children who lived with you ______
Enter number ▶
9 Number of other dependents (from line 32) . . . ▶
10 Total exemptions claimed . . . . ▶ 1

Please attach Copy B of Form W-2 here

Income

11	Wages, salaries, tips, and other employee compensation. (Attach Form W-2 to front. If unavailable, attach explanation) .	11	7880	00
12a	Dividends (see pages 6 and 13 of instr.) $ ...... 12b Less exclusion $ ...... Balance . ▶	12c		
	(If gross dividends and other distributions are over $200, list in Part I of Schedule B.)			
13	Interest income. [If $200 or less, enter total without listing in Schedule B / If over $200, enter total and list in Part II of Schedule B] . .	13	132	67
14	Income other than wages, dividends, and interest (from line 45)	14	520	00
15	Total (add lines 11, 12c, 13 and 14)	15	8532	67
16	Adjustments to income (such as "sick pay," moving expenses, etc. from line 50) .	16		
17	Subtract line 16 from line 15 (adjusted gross income)	17	8532	67

- **Caution:** *If you have unearned income and you could be claimed as a dependent on your parent's return, see boxed instruction on page 7, under the heading "Tax-Credits-Payments." Check this block* [ ].
- *If you do not itemize deductions and line 17 is under $10,000, find tax in Tables and enter on line 18.*
- *If you itemize deductions or line 17 is $10,000 or more, go to line 51 to figure tax.*

Write soc. sec. no. on Check or Money Order. Attach here

Tax, Payments and Credits

Line	Description		Amount		Line	Amount	
18	Tax, check if from: [ ] Tax Tables 1–12, [ ] Schedule D; [X] Tax Rate Schedule X, Y, or Z; [ ] Schedule G or [ ] Form 4726				18	1196	32
19	Total credits (from line 61)				19		
20	Income tax (subtract line 19 from line 18)				20		
21	Other taxes (from line 67)				21		
22	Total (add lines 20 and 21)				22	1196	32
23	Total Federal income tax withheld (attach Forms W–2 or W–2P to front)	23	1344	00			
24	1972 Estimated tax payments (include amount allowed as credit from 1971 return)	24					
25	Amount paid with Form 4868, Application for Automatic Extension of Time to File U.S. Individual Income Tax Return	25					
26	Other payments (from line 71)	26					
27	Total (add lines 23, 24, 25, and 26)				27	1344	00

Bal. Due or Refund

Line	Description			Line	Amount	
28	If line 22 is larger than line 27, enter BALANCE DUE IRS	Pay in full with return. Make check or money order payable to Internal Revenue Service	▶	28		
29	If line 27 is larger than line 22, enter amount OVERPAID		▶	29	147	68
30	Line 29 to be **REFUNDED TO YOU**		▶	30	147	68
31	Line 29 to be credited on 1973 estimated tax	31				

Foreign Accounts

Did you, at any time during the taxable year, have any interest in or signature or other authority over a bank, securities, or other financial account in a foreign country (except in a U.S. military banking facility operated by a U.S. financial institution)? ▶ [ ] Yes [X] No

If "Yes," attach Form 4683. (For definitions, see Form 4683.)

Note: Be sure to complete Revenue Sharing (lines 33 and 34) on next page.

Sign here

Under penalties of perjury, I declare that I have examined this return, including accompanying schedules and statements, and to the best of my knowledge and belief it is true, correct, and complete. Declaration of preparer (other than taxpayer) is based on all information of which he has any knowledge.

Your signature — Date

Preparer's signature (other than taxpayer) — Date

Wife's (husband's) signature (if filing jointly, BOTH must sign even if only one had income)

Address (and ZIP Code) — **Preparer's Emp. Ident. or Soc. Sec. No.**

EXAMPLE A

Other Dependents

(a) NAME	(b) Relationship	(c) Months lived in your home. If born or died during year, write B or D.	(d) Did dependent have income of $750 or more?	(e) Amount YOU furnished for dependent's support. If 100% write ALL.	(f) Amount furnished by OTHERS including dependent.
				$	$

32 Total number of dependents listed in column (a). Enter here and on line 9 . . . . ▶

Revenue Sharing

33 Print or type the location of your principal place of residence at end of year (not necessarily the same as your post office address).

(a) State	(b) County	(c) Locality. If you lived inside the boundaries of an incorporated city, town, etc., enter its name; if not, check here ▶	(d) Township (see instructions on page 8)
California	Napa	[x]	

34 Enter the number of persons included on line 10 who (1) are filing a return of their own; or, (2) did not live at your principal place of residence at the end of the year . . . . ▶

For IRS use only—Leave blank

PART I.—Income other than Wages, Dividends, and Interest

35 Business income (or loss) (attach Schedule C)	35		
36 Net gain (or loss) from sale or exchange of capital assets (attach Schedule D)	36		
37 Net gain (or loss) from Supplemental Schedule of Gains and Losses (attach Form 4797)	37		
38 Pensions and annuities, rents and royalties, partnerships, estates or trusts, etc. (attach Schedule E)	38	520	00
39 Farm income (or loss) (attach Schedule F)	39		
40 Fully taxable pensions and annuities (not reported on Schedule E—see instructions on page 8)	40		
41 50% of capital gain distributions (not reported on Schedule D)	41		
42 State income tax refunds **(caution—see instructions on page 8)**	42		
43 Alimony	43		
44 Other (state nature and source)	44		
45 **Total** (add lines 35 through 44). **Enter here and on line 14** ▶	45	520	00

PART II.—Adjustments to Income

46 "Sick pay" if included in income (attach Form 2440 or other required statement)	46		
47 Moving expense (attach Form 3903)	47		
48 Employee business expense (attach Form 2106 or other statement)	48		
49 Payments as a self-employed person to a retirement plan, etc. (see Form 4848)	49		
50 **Total adjustments** (add lines 46, 47, 48, and 49). **Enter here and on line 16** ▶	50		

PART III.—Tax Computation **(Do not use this part if you use Tax Tables 1–12 to find your tax.)**

	Line	Dollars	Cents
51 Adjusted gross income (from line 17)	51	8532	67
52 (a) If you itemize deductions, enter total from Schedule A, line 40 and attach Schedule A (b) If you do not itemize deductions, enter 15% of line 51, but do NOT enter more than $2,000. ($1,000 if line 3 is checked)	52	1423	00
53 Subtract line 52 from line 51	53	7109	67
54 Multiply total number of exemptions claimed on line 10, by $750	54	750	00
55 **Taxable income.** Subtract line 54 from line 53	55	6359	67

(Figure your tax on the amount on line 55 by using Tax Rate Schedule X, Y or Z, or if applicable, the alternative tax from Schedule D, income averaging from Schedule G, or maximum tax from Form 4726.) **Enter tax on line 18.**

PART IV.—Credits

	Line	Dollars	Cents
56 Retirement income credit (attach Schedule R)	56		
57 Investment credit (attach Form 3468)	57		
58 Foreign tax credit (attach Form 1116)	58		
59 Credit for contributions to candidates for public office—see instructions on page 9	59		
60 **Work Incentive Program credit (attach Form 4874)**	60		
61 **Total credits** (add lines 56, 57, 58, 59, and 60). **Enter here and on line 19** ▶	61		

PART V.—Other Taxes

	Line	Dollars	Cents
62 Self-employment tax (attach Schedule SE)	62		
63 Tax from recomputing prior-year investment credit (attach Form 4255)	63		
64 Minimum tax (see instructions on page 10). Check here ☐, if Form 4625 is attached	64		
65 Social security tax on tip income not reported to employer (attach Form 4137)	65		
66 Uncollected employee social security tax on tips (from Forms W–2)	66		
67 **Total** (add lines 62, 63, 64, 65, and 66). **Enter here and on line 21** ▶	67		

PART VI.—Other Payments

	Line	Dollars	Cents
68 Excess FICA tax withheld (two or more employers—see instructions on page 10)	68		
69 Credit for Federal tax on special fuels, nonhighway gasoline and lubricating oil (attach Form 4136)	69		
70 Credit from a Regulated Investment Company (attach Form 2439)	70		
71 **Total** (add lines 68, 69, and 70). **Enter here and on line 26** ▶	71		

☆ U.S. GOVERNMENT PRINTING OFFICE: 1972—O-458-048 ☆ GPO 793-501

EXAMPLE B

Form **1040** US Department of the Treasury / Internal Revenue Service
Individual Income Tax Return 1972

For the year January 1–December 31, 1972, or other taxable year beginning ...., 1972, ending ...., 19 ......

Please print or type

First name and initial (If joint return, use first names and middle initials of both)	Last name	Your social security number (Husband's, if joint return)
Harry F. & Ruth B.	Warren	560 01 3256
Present home address (Number and street, including apartment number, or rural route)		Wife's number, if joint return
52 Ryan Way		573 33 3060
City, town or post office, State and ZIP code	Occupation — Yours	salesman
Havens, New York 10067	Occupation — Wife's	housewife/writer

Filing Status—check only one:

1 ☐ Single
2 ☒ Married filing joint return (even if only one had income)
3 ☐ Married filing separately. If wife (husband) is also filing give her (his) social security number and first name here.
4 ☐ Unmarried Head of Household
5 ☐ Widow(er) with dependent child (Enter year of death of husband (wife) ▶ 19)

Exemptions Regular / 65 or over / Blind — Enter number of boxes checked

6 Yourself . . . . . ☒ ☐ ☐
7 Wife (husband) . . . ☒ ☐ ☐ ▶ 2
8 First names of your dependent children who lived with you Donald — Enter number ▶ 1
9 Number of other dependents (from line 32) . . . ▶
10 Total exemptions claimed . . . . ▶ 3

Please attach Copy B of Form W–2 here

Income

11	Wages, salaries, tips, and other employee compensation. (Attach Form W–2 to front. If unavailable, attach explanation)	11	9,471 28
12a	Dividends (see pages 6 and 13 of instr.) $ 730.00 12b Less exclusion $ 200.00 Balance ▶ (If gross dividends and other distributions are over $200, list in Part I of Schedule B.)	12c	530 00
13	Interest income. [If $200 or less, enter total without listing in Schedule B; If over $200, enter total and list in Part II of Schedule B]	13	1,105 42
14	Income other than wages, dividends, and interest (from line 45)	14	3,416 25
15	Total (add lines 11, 12c, 13 and 14)	15	14,522 95
16	Adjustments to income (such as "sick pay," moving expenses, etc. from line 50)	16	
17	Subtract line 16 from line 15 (adjusted gross income)	17	14,522 95

- **Caution:** *If you have unearned income and you could be claimed as a dependent on your parent's return, see boxed instruction on page 7, under the heading "Tax-Credits-Payments." Check this block* ☐.
- *If you do not itemize deductions and line 17 is under $10,000, find tax in Tables and enter on line 18.*
- *If you itemize deductions or line 17 is $10,000 or more, go to line 51 to figure tax.*

Write soc. sec. no. on Check or Money Order. Attach here

Tax, Payments and Credits

Line	Description	Line	Amount		Line	Amount	
18	Tax, check if from: [ ] Tax Tables 1–12, Schedule D [X] Tax Rate Schedule X, Y, or Z [ ] Schedule G or [ ] Form 4726				18	1,880	05
19	Total credits (from line 61)				19	13	08
20	Income tax (subtract line 19 from line 18)				20	1,866	97
21	Other taxes (from line 67)				21	215	01
22	Total (add lines 20 and 21)				22	2,081	98
23	Total Federal income tax withheld (attach Forms W–2 or W–2P to front)	23	1,040	00			
24	1972 Estimated tax payments (include amount allowed as credit from 1971 return)	24	900	00			
25	Amount paid with Form 4868, Application for Automatic Extension of Time to File U.S. Individual Income Tax Return	25					
26	Other payments (from line 71)	26					
27	Total (add lines 23, 24, 25, and 26)				27	1,940	00

Bal. Due or Refund

Line	Description	Line	Amount		Line	Amount	
28	If line 22 is larger than line 27, enter BALANCE DUE IRS — Pay in full with return. Make check or money order payable to Internal Revenue Service ▶				28	141	98
29	If line 27 is larger than line 22, enter amount OVERPAID ▶				29		
30	Line 29 to be **REFUNDED TO YOU** ▶				30		
31	Line 29 to be credited on 1973 estimated tax	31					

Foreign Accounts

Did you, at any time during the taxable year, have any interest in or signature or other authority over a bank, securities, or other financial account in a foreign country (except in a U.S. military banking facility operated by a U.S. financial institution)? ▶ [ ] Yes [X] No
If "Yes," attach Form 4683. (For definitions, see Form 4683.)

Note: Be sure to complete Revenue Sharing (lines 33 and 34) on next page.

Sign here

Under penalties of perjury, I declare that I have examined this return, including accompanying schedules and statements, and to the best of my knowledge and belief it is true, correct, and complete. Declaration of preparer (other than taxpayer) is based on all information of which he has any knowledge.

▶ Your signature — Date

▶ Preparer's signature (other than taxpayer) — Date

▶ Wife's (husband's) signature (if filing jointly, BOTH must sign even if only one had income)

Address (and ZIP Code) — **Preparer's Emp. Ident. or Soc. Sec. No.**

EXAMPLE B

Other Dependents

(a) NAME	(b) Relationship	(c) Months lived in your home. If born or died during year, write B or D.	(d) Did dependent have income of $750 or more?	(e) Amount YOU furnished for dependent's support. If 100% write ALL.	(f) Amount furnished by OTHERS including dependent.
				$	$

32 Total number of dependents listed in column (a). Enter here and on line 9 . . . . ▶

Revenue Sharing

33 Print or type the location of your principal place of residence at end of year (not necessarily the same as your post office address).

(a) State	(b) County	(c) Locality. If you lived inside the boundaries of an incorporated city, town, etc., enter its name; if not, check here ▶ ☐	(d) Township (see instructions on page 8)
New York	Allegany	Havens	

34 Enter the number of persons included on line 10 who (1) are filing a return of their own; or, (2) did not live at your principal place of residence at the end of the year . . . . ▶

For IRS use only—Leave blank

PART I.—Income other than Wages, Dividends, and Interest

Line	Description	No.	Amount	Cents
35	Business income (or loss) (attach Schedule C)	35	2,866	75
36	Net gain (or loss) from sale or exchange of capital assets (attach Schedule D)	36	116	00
37	Net gain (or loss) from Supplemental Schedule of Gains and Losses (attach Form 4797)	37		
38	Pensions and annuities, rents and royalties, partnerships, estates or trusts, etc. (attach Schedule E)	38	660	00
39	Farm income (or loss) (attach Schedule F)	39	(226	50) LOSS
40	Fully taxable pensions and annuities (not reported on Schedule E—see instructions on page 8)	40		
41	50% of capital gain distributions (not reported on Schedule D)	41		
42	State income tax refunds **(caution—see instructions on page 8)**	42		
43	Alimony	43		
44	Other (state nature and source)	44		
45	**Total** (add lines 35 through 44). **Enter here and on line 14** ▶	45	3,416	25

PART II.—Adjustments to Income

Line	Description	No.	Amount	Cents
46	"Sick pay" if included in income (attach Form 2440 or other required statement)	46		
47	Moving expense (attach Form 3903)	47		
48	Employee business expense (attach Form 2106 or other statement)	48		
49	Payments as a self-employed person to a retirement plan, etc. (see Form 4848)	49		
50	**Total adjustments** (add lines 46, 47, 48, and 49). **Enter here and on line 16** ▶	50		

PART III.—Tax Computation (Do not use this part if you use Tax Tables 1–12 to find your tax.)

Line	Description	Line	Dollars	Cents
51	Adjusted gross income (from line 17)	51	14,522	95
52	(a) If you itemize deductions, enter total from Schedule A, line 40 and attach Schedule A (b) If you do not itemize deductions, enter 15% of line 51, but do NOT enter more than $2,000. ($1,000 if line 3 is checked)	52	2,000	00*
53	Subtract line 52 from line 51	53	12,522	95
54	Multiply total number of exemptions claimed on line 10, by $750	54	2,250	00
55	**Taxable income.** Subtract line 54 from line 53	55	10,272	95

(Figure your tax on the amount on line 55 by using Tax Rate Schedule X, Y or Z, or if applicable, the alternative tax from Schedule D, income averaging from Schedule G, or maximum tax from Form 4726.) **Enter tax on line 18.**

PART IV.—Credits

Line	Description	Line	Dollars	Cents
56	Retirement income credit (attach Schedule R)	56		
57	Investment credit (attach Form 3468)	57	13	08
58	Foreign tax credit (attach Form 1116)	58		
59	Credit for contributions to candidates for public office—see instructions on page 9	59		
60	**Work Incentive Program credit (attach Form 4874)**	60		
61	**Total credits** (add lines 56, 57, 58, 59, and 60). **Enter here and on line 19** ▶	61	13	08

PART V.—Other Taxes

Line	Description	Line	Dollars	Cents
62	Self-employment tax (attach Schedule SE)	62	215	01
63	Tax from recomputing prior-year investment credit (attach Form 4255)	63		
64	**Minimum tax (see instructions on page 10). Check here ☐, if Form 4625 is attached**	64		
65	Social security tax on tip income not reported to employer (attach Form 4137)	65		
66	Uncollected employee social security tax on tips (from Forms W–2)	66		
67	**Total** (add lines 62, 63, 64, 65, and 66). **Enter here and on line 21** ▶	67	215	01

PART VI.—Other Payments

Line	Description	Line	Dollars	Cents
68	Excess FICA tax withheld (two or more employers—see instructions on page 10)	68		
69	Credit for Federal tax on special fuels, nonhighway gasoline and lubricating oil (attach Form 4136)	69		
70	**Credit from a Regulated Investment Company (attach Form 2439)**	70		
71	**Total** (add lines 68, 69, and 70). **Enter here and on line 26** ▶	71		

☆ U.S. GOVERNMENT PRINTING OFFICE: 1972—O-458-048 ☆ GPO 793-501

*Deductions were not itemized.

Harry F. Warren and his wife, Ruth B. Warren, both work. They have 1 dependent child, Donald. Mr. Warren earned $9471.28 as a salesman. Mrs. Warren is self-employed and earned $2866.75 as a writer. They also received a net profit of $116.00 on the sale of capital assets, $660.00 from a trust fund, interest income of $1105.42, and dividends of $730.00. They qualify for a $200 exclusion on the dividends. They also had a loss of $226.50 from a small farm they operate. They took the standard deduction since it was more than their itemized deductions.

Before completing the following exercises, review the calculations shown on Form 1040, pages 160 and 161.

EXERCISE 36

1. Refer to Form 1040 for Example A, page 162. (a) If she had not itemized her deductions, what would her tax have been? (b) Why did she itemize her deductions?

2. Refer to Form 1040 for Example B, page 166. (a) Why was the standard deduction taken? (b) If line 51 had been $10,620 and the total of itemized deductions were $1750, which method (52a or 52b) should the taxpayer choose? Why? (c) Assume line 17 (adjusted gross income) amounted to $9625 and itemized deductions amounted to $1525.00. Would you use the tax table or complete Part III of Form 1040 and use the tax rate schedule instead? Why or why not? Prove it. Assume that there are no amounts in lines 12, 13, and 14.

3. Using the appropriate tax rate *schedule* (page 180), determine the income tax for the following:

	Taxable Income	Status
(a)	$ 6,750	married, filing separately
(b)	7,825	married, filing jointly
(c)	10,462	single
(d)	9,800	head of household
(e)	8,750	single

4. How many exemptions are there for the following?

	Status	Number of Dependents
(a)	man and wife, joint return	3
(b)	man, 67 years of age and wife, 60 years of age, joint return	none
(c)	man and wife, both 65 years joint return	none
(d)	single taxpayer	none
(e)	single but head of household	2

5. Mr. and Mrs. Jones have an adjusted gross income of $12,620. If they take the standard deduction, what is their income tax if they file a joint return? They have 3 dependent children.

6. Mr. Owens owned a building that he rented for $225.00 a month. Operating expenses were as follows: depreciation $968.42, repairs $397.72, taxes $743.54, other expenses (painting, insurance, etc.) $1028.58. (a) What was his net income or loss? (b) On what schedule should this be determined? (c) Where on Form 1040 would this amount be recorded?

7. Compute the tax for the following (a joint return).

Mr. and Mrs. Allen received income as follows: wages $10,326.00, dividends of $350.00 for Mr. Allen (separate property) and dividends of $225.00 for Mrs. Allen (separate property), rentals $622.00, no adjustments to income. Deductions totaled $1565.00. They have 1 dependent child.

8. (a) An unmarried man earned $17,800 a year and was the sole support of his mother. How much was his federal income tax if he took the standard deduction? He qualified as "head of household." (b) If earnings amounted to $12,500, what would his tax equal?

9. A single taxpayer earns $8300 in wages and $120 interest. (a) Which form should he use? (b) What is his tax?

10. A single taxpayer earns $12,500.00 a year, $37.50 in interest, and $200.00 in dividends. (a) What form should he use? (b) What is his tax? (c) If this taxpayer had deductions amounting to $2100, which form should he use? Why or why not? Compute his tax, using itemized deductions of $2100.

EXERCISE 37 Chapter Summary and Review of Payroll

1. If overtime is figured at one and a half times the regular rate, compute the overtime rates on the following regular hourly rates: $3.56, $4.85, and $5.25.

2. Calculate the hourly rate on (a) a weekly salary of $360, (b) a monthly salary of $715. (Carry answers to 3 places if necessary.)

3. J. Bradley earns $850 a month. Last month he worked 15 hours overtime. At time and a half for overtime, what are his total earnings? (Carry hourly rate correct to 3 decimals.)

4. Compute the net earnings for J. Bradley in problem 3, including his overtime. Use the 1973–77 social security tax rate and the 1973 federal income withholding tax tables. Mr. Bradley is single and claims 1 exemption.

5. Janet Davis works on a commission of 6% on all sales made. If her sales amounted to $2120 for the week, (a) What were her gross earnings? (b) What is her hourly rate on the basis of a 40-hour week?

6. Record federal income tax withholding for the following:

	Wages	Marital Status	Exemptions	Pay Period
(a)	$ 95.60	single	1	week
(b)	158.50	married	2	week
(c)	125.00	single	none	week
(d)	726.00	single	2	month
(e)	835.00	married	2	month

7. Henry Word earns $920.00 a month as a carpenter. This is his only income. He is buying a car on which the interest charges are $525.00 for the year. Automobile license fee is $18.00. He paid out $192.00 in sales taxes and $26.32 interest on a loan. He gave $25.00 to the Boy Scouts and $150.00 to his church. Deductible medical expenses (including dental work) amounted to $478.50. He is single and has no dependents. (a) Should he itemize his deductions? Why or why not? (b) Which form would be best for him to use? (c) If he used the short form, what is his tax?

8. Mary Allen earns $8900 a year as a typist and has no other income.

She is single but qualifies as "head of household" with 1 dependent. Itemized deductions totaled $1565.20. (a) Which tax form should she use? (b) Compute her tax using the long form and taking itemized deductions. (c) Compute her tax by using the short form in which case she would not use itemized deductions.

1972 Tax Tables

For persons with incomes under $10,000 using Short Form 1040A.

The standard deduction and deduction for exemptions have been taken into account in determining the tax shown in these Tables.

The Tables show the **lower** tax after taking into account both the percentage standard deduction and the low income allowance except in the case of married persons filing separately. For married persons filing separate returns, the tables show the tax figured on the percentage standard deduction and on the low income allowance.

Select the Tax Table that covers the total number of exemptions on Short Form 1040A, line 10. On the appropriate table, read down the income columns until you find the line covering the income you entered on Short Form 1040A, line 14. Then read across to the column heading describing your filing status. If you checked line 5, use the column for "Married filing joint return." Enter the tax you find there on Short Form 1040A, line 19.

Married persons filing separate returns: Choose either the low income allowance or percentage standard deduction to figure your tax; but if one uses the percentage standard deduction, then both must use it. **If you are a married person living apart** from your spouse,

Table 1 —Returns claiming ONE exemption (and not itemizing deductions)

If the amount on Form 1040A, line 14, is—		And you are—				If the amount on Form 1040A, line 14, is—		And you are—				If the amount on Form 1040A, line 14, is—		And you are—			
				Married filing separate return claiming—						Married filing separate return claiming—						Married filing separate return claiming—	
At least	But less than	Single, not head of household	Head of household	Low income allowance	%Standard deduction	At least	But less than	Single, not head of household	Head of household	Low income allowance	%Standard deduction	At least	But less than	Single, not head of household	Head of household	Low income allowance	%Standard deduction
		Your tax is—						Your tax is—						Your tax is—			
$0	$875	$0	$0	$0	$0	$2,750	$2,775	$102	$100	$203	$242	$6,250	$6,300	$737	$703	$883	$818
875	900	0	0	0	1	2,775	2,800	106	103	207	245	6,300	6,350	748	712	894	828
900	925	0	0	0	4	2,800	2,825	109	107	211	249	6,350	6,400	758	722	905	837
925	950	0	0	0	7	2,825	2,850	113	110	215	253	6,400	6,450	769	731	916	846
950	975	0	0	0	10	2,850	2,875	117	114	219	256	6,450	6,500	779	741	927	856
975	1,000	0	0	0	13	2,875	2,900	121	117	223	260	6,500	6,550	790	750	938	865
1,000	1,025	0	0	0	15	2,900	2,925	124	121	227	263	6,550	6,600	800	760	949	875
1,025	1,050	0	0	0	18	2,925	2,950	128	124	231	267	6,600	6,650	811	769	960	884
1,050	1,075	0	0	0	21	2,950	2,975	132	128	236	271	6,650	6,700	821	779	971	894
1,075	1,100	0	0	0	24	2,975	3,000	136	131	240	274	6,700	6,750	832	788	982	905
1,100	1,125	0	0	0	27	3,000	3,050	141	137	246	280	6,750	6,800	842	798	993	916
1,125	1,150	0	0	0	30	3,050	3,100	149	144	255	287	6,800	6,850	853	807	1,004	927
1,150	1,175	0	0	0	33	3,100	3,150	157	152	263	294	6,850	6,900	863	817	1,015	938
1,175	1,200	0	0	0	36	3,150	3,200	165	160	272	301	6,900	6,950	874	826	1,026	949
1,200	1,225	0	0	0	39	3,200	3,250	173	168	280	309	6,950	7,000	884	836	1,037	960
1,225	1,250	0	0	0	42	3,250	3,300	181	176	289	316	7,000	7,050	895	845	1,048	971
1,250	1,275	0	0	0	45	3,300	3,350	189	184	297	324	7,050	7,100	905	855	1,059	982
1,275	1,300	0	0	0	48	3,350	3,400	197	192	306	333	7,100	7,150	916	864	1,070	993
1,300	1,325	0	0	0	51	3,400	3,450	205	200	315	341	7,150	7,200	926	874	1,081	1,004
1,325	1,350	0	0	0	54	3,450	3,500	213	208	324	349	7,200	7,250	937	883	1,092	1,015
1,350	1,375	0	0	0	57	3,500	3,550	221	216	334	357	7,250	7,300	947	893	1,103	1,026
1,375	1,400	0	0	0	60	3,550	3,600	229	224	343	365	7,300	7,350	958	902	1,114	1,037
1,400	1,425	0	0	2	63	3,600	3,650	238	232	353	373	7,350	7,400	968	912	1,125	1,048
1,425	1,450	0	0	5	66	3,650	3,700	246	240	362	381	7,400	7,450	979	921	1,136	1,059
1,450	1,475	0	0	9	69	3,700	3,750	255	248	372	389	7,450	7,500	989	931	1,149	1,070

1,475	**1,500**	0	0	12	72
1,500	**1,525**	0	0	16	75
1,525	**1,550**	0	0	19	79
1,550	**1,575**	0	0	23	82
1,575	**1,600**	0	0	26	85
1,600	**1,625**	0	0	30	88
1,625	**1,650**	0	0	33	91
1,650	**1,675**	0	0	37	94
1,675	**1,700**	0	0	40	98
1,700	**1,725**	0	0	44	101
1,725	**1,750**	0	0	47	104
1,750	**1,775**	0	0	51	107
1,775	**1,800**	0	0	54	110
1,800	**1,825**	0	0	58	114
1,825	**1,850**	0	0	61	117
1,850	**1,875**	0	0	65	120
1,875	**1,900**	0	0	68	123
1,900	**1,925**	0	0	72	126
1,925	**1,950**	0	0	76	130
1,950	**1,975**	0	0	79	133
1,975	**2,000**	0	0	83	136
2,000	**2,025**	0	0	87	139
2,025	**2,050**	0	0	91	142
2,050	**2,075**	2	2	94	145
2,075	**2,100**	5	5	98	149
2,100	**2,125**	9	9	102	152
2,125	**2,150**	12	12	106	156
2,150	**2,175**	16	16	109	159
2,175	**2,200**	19	19	113	162
2,200	**2,225**	23	23	117	166
2,225	**2,250**	26	26	121	169
2,250	**2,275**	30	30	124	173
2,275	**2,300**	33	33	128	176
2,300	**2,325**	37	37	132	179
2,325	**2,350**	40	40	136	183
2,350	**2,375**	44	44	139	186
2,375	**2,400**	47	47	143	190
2,400	**2,425**	51	51	147	193
2,425	**2,450**	54	54	151	196
2,450	**2,475**	58	58	155	200
2,475	**2,500**	61	61	159	203
2,500	**2,525**	65	65	163	207
2,525	**2,550**	68	68	167	210
2,550	**2,575**	72	72	171	213
2,575	**2,600**	76	75	175	217
2,600	**2,625**	79	79	179	220
2,625	**2,650**	83	82	183	224
2,650	**2,675**	87	86	187	227
2,675	**2,700**	91	89	191	231
2,700	**2,725**	94	93	195	234
2,725	**2,750**	98	96	199	238

3,750	**3,800**	263	256	381	397
3,800	**3,850**	272	264	391	405
3,850	**3,900**	280	272	400	413
3,900	**3,950**	289	280	410	421
3,950	**4,000**	297	288	419	429
4,000	**4,050**	306	296	429	438
4,050	**4,100**	315	305	438	446
4,100	**4,150**	324	314	448	454
4,150	**4,200**	334	323	457	462
4,200	**4,250**	343	332	467	470
4,250	**4,300**	353	341	476	478
4,300	**4,350**	362	350	486	486
4,350	**4,400**	372	359	495	494
4,400	**4,450**	381	368	505	502
4,450	**4,500**	391	377	514	510
4,500	**4,550**	400	386	524	518
4,550	**4,600**	410	395	533	526
4,600	**4,650**	419	404	543	534
4,650	**4,700**	429	413	552	543
4,700	**4,750**	438	422	562	551
4,750	**4,800**	448	431	571	559
4,800	**4,850**	457	440	581	567
4,850	**4,900**	467	449	590	575
4,900	**4,950**	476	458	600	583
4,950	**5,000**	486	467	609	591
5,000	**5,050**	495	476	619	599
5,050	**5,100**	505	485	628	607
5,100	**5,150**	514	494	638	615
5,150	**5,200**	524	503	647	623
5,200	**5,250**	533	512	657	631
5,250	**5,300**	543	521	666	639
5,300	**5,350**	552	530	676	647
5,350	**5,400**	562	539	685	656
5,400	**5,450**	571	548	696	664
5,450	**5,500**	581	557	707	672
5,500	**5,550**	590	566	718	680
5,550	**5,600**	600	575	729	688
5,600	**5,650**	609	584	740	697
5,650	**5,700**	619	593	751	706
5,700	**5,750**	628	602	762	716
5,750	**5,800**	638	611	773	725
5,800	**5,850**	647	620	784	734
5,850	**5,900**	657	629	795	744
5,900	**5,950**	666	638	806	753
5,950	**6,000**	676	647	817	762
6,000	**6,050**	685	656	828	772
6,050	**6,100**	695	665	839	781
6,100	**6,150**	706	674	850	790
6,150	**6,200**	716	684	861	800
6,200	**6,250**	727	693	872	809

7,500	**7,550**	1,000	940	1,161	1,081
7,550	**7,600**	1,010	950	1,174	1,092
7,600	**7,650**	1,021	959	1,186	1,103
7,650	**7,700**	1,031	969	1,199	1,114
7,700	**7,750**	1,042	978	1,211	1,125
7,750	**7,800**	1,052	988	1,224	1,136
7,800	**7,850**	1,063	997	1,236	1,149
7,850	**7,900**	1,073	1,007	1,249	1,161
7,900	**7,950**	1,084	1,016	1,261	1,174
7,950	**8,000**	1,094	1,026	1,274	1,186
8,000	**8,050**	1,105	1,035	1,286	1,199
8,050	**8,100**	1,116	1,046	1,299	1,211
8,100	**8,150**	1,128	1,057	1,311	1,224
8,150	**8,200**	1,140	1,068	1,324	1,236
8,200	**8,250**	1,152	1,079	1,336	1,249
8,250	**8,300**	1,164	1,090	1,349	1,261
8,300	**8,350**	1,176	1,101	1,361	1,274
8,350	**8,400**	1,188	1,112	1,374	1,286
8,400	**8,450**	1,200	1,123	1,386	1,299
8,450	**8,500**	1,212	1,134	1,399	1,311
8,500	**8,550**	1,224	1,145	1,411	1,324
8,550	**8,600**	1,236	1,156	1,424	1,336
8,600	**8,650**	1,248	1,167	1,436	1,349
8,650	**8,700**	1,260	1,177	1,449	1,361
8,700	**8,750**	1,270	1,187	1,461	1,374
8,750	**8,800**	1,280	1,196	1,474	1,386
8,800	**8,850**	1,290	1,205	1,486	1,399
8,850	**8,900**	1,301	1,215	1,499	1,411
8,900	**8,950**	1,311	1,224	1,511	1,424
8,950	**9,000**	1,321	1,233	1,524	1,436
9,000	**9,050**	1,331	1,243	1,536	1,449
9,050	**9,100**	1,341	1,252	1,549	1,461
9,100	**9,150**	1,352	1,261	1,561	1,474
9,150	**9,200**	1,362	1,271	1,574	1,486
9,200	**9,250**	1,372	1,280	1,586	1,499
9,250	**9,300**	1,382	1,289	1,599	1,511
9,300	**9,350**	1,392	1,299	1,611	1,524
9,350	**9,400**	1,403	1,308	1,624	1,536
9,400	**9,450**	1,413	1,317	1,637	1,549
9,450	**9,500**	1,423	1,327	1,651	1,561
9,500	**9,550**	1,433	1,336	1,665	1,574
9,550	**9,600**	1,443	1,346	1,679	1,586
9,600	**9,650**	1,454	1,355	1,693	1,599
9,650	**9,700**	1,464	1,364	1,707	1,611
9,700	**9,750**	1,474	1,374	1,721	1,624
9,750	**9,800**	1,484	1,383	1,735	1,637
9,800	**9,850**	1,494	1,392	1,749	1,651
9,850	**9,900**	1,505	1,402	1,763	1,665
9,900	**9,950**	1,515	1,411	1,777	1,679
9,950	**10,000**	1,525	1,420	1,791	1,693

Table 2 —Returns claiming TWO exemptions (and not itemizing deductions)

If the amount on Form 1040A, line 14, is—		And you are—				
					Married filing separate return claiming—	
At least	But less than	Single, not head of household	Head of household	Married* filing joint return	Low income allowance	%Standard deduction
		Your tax is—				
$0	$1,775	$0	$0	$0	$0	$0
1,775	1,800	0	0	0	0	3
1,800	1,825	0	0	0	0	6
1,825	1,850	0	0	0	0	9
1,850	1,875	0	0	0	0	12
1,875	1,900	0	0	0	0	15
1,900	1,925	0	0	0	0	18
1,925	1,950	0	0	0	0	21
1,950	1,975	0	0	0	0	24
1,975	2,000	0	0	0	0	27
2,000	2,025	0	0	0	0	29
2,025	2,050	0	0	0	0	32
2,050	2,075	0	0	0	0	35
2,075	2,100	0	0	0	0	38
2,100	2,125	0	0	0	0	41
2,125	2,150	0	0	0	0	44
2,150	2,175	0	0	0	2	47
2,175	2,200	0	0	0	5	50
2,200	2,225	0	0	0	9	53
2,225	2,250	0	0	0	12	56
2,250	2,275	0	0	0	16	59
2,275	2,300	0	0	0	19	62
2,300	2,325	0	0	0	23	65
2,325	2,350	0	0	0	26	68
2,350	2,375	0	0	0	30	71

If the amount on Form 1040A, line 14, is—		And you are—				
					Married filing separate return claiming—	
At least	But less than	Single, not head of household	Head of household	Married* filing joint return	Low income allowance	%Standard deduction
		Your tax is—				
$3,700	$3,750	$134	$130	$130	$238	$253
3,750	3,800	141	137	137	246	260
3,800	3,850	149	144	144	255	268
3,850	3,900	157	152	151	263	275
3,900	3,950	165	160	159	272	282
3,950	4,000	173	168	166	280	289
4,000	4,050	181	176	174	289	297
4,050	4,100	189	184	181	297	304
4,100	4,150	197	192	189	306	311
4,150	4,200	205	200	196	315	319
4,200	4,250	213	208	204	324	327
4,250	4,300	221	216	211	334	335
4,300	4,350	229	224	219	343	343
4,350	4,400	238	232	226	353	352
4,400	4,450	246	240	234	362	360
4,450	4,500	255	248	241	372	368
4,500	4,550	263	256	249	381	376
4,550	4,600	272	264	256	391	384
4,600	4,650	280	272	264	400	392
4,650	4,700	289	280	271	410	400
4,700	4,750	297	288	279	419	408
4,750	4,800	306	296	286	429	416
4,800	4,850	315	305	294	438	424
4,850	4,900	324	314	302	448	432
4,900	4,950	334	323	310	457	440

If the amount on Form 1040A, line 14, is—		And you are—				
					Married filing separate return claiming—	
At least	But less than	Single, not head of household	Head of household	Married* filing joint return	Low income allowance	%Standard deduction
		Your tax is—				
$6,850	$6,900	$706	$674	$634	$850	$773
6,900	6,950	716	684	644	861	784
6,950	7,000	727	693	653	872	795
7,000	7,050	737	703	663	883	806
7,050	7,100	748	712	672	894	817
7,100	7,150	758	722	682	905	828
7,150	7,200	769	731	691	916	839
7,200	7,250	779	741	701	927	850
7,250	7,300	790	750	710	938	861
7,300	7,350	800	760	720	949	872
7,350	7,400	811	769	729	960	883
7,400	7,450	821	779	739	971	894
7,450	7,500	832	788	748	982	905
7,500	7,550	842	798	758	993	916
7,550	7,600	853	807	767	1,004	927
7,600	7,650	863	817	777	1,015	938
7,650	7,700	874	826	786	1,026	949
7,700	7,750	884	836	796	1,037	960
7,750	7,800	895	845	805	1,048	971
7,800	7,850	905	855	815	1,059	982
7,850	7,900	916	864	824	1,070	993
7,900	7,950	926	874	834	1,081	1,004
7,950	8,000	937	883	843	1,092	1,015
8,000	8,050	947	893	853	1,103	1,026
8,050	8,100	958	902	862	1,114	1,037

2,375	**2,400**	0	0	0	33	74
2,400	**2,425**	0	0	0	37	78
2,425	**2,450**	0	0	0	40	81
2,450	**2,475**	0	0	0	44	84
2,475	**2,500**	0	0	0	47	87
2,500	**2,525**	0	0	0	51	90
2,525	**2,550**	0	0	0	54	94
2,550	**2,575**	0	0	0	58	97
2,575	**2,600**	0	0	0	61	100
2,600	**2,625**	0	0	0	65	103
2,625	**2,650**	0	0	0	68	106
2,650	**2,675**	0	0	0	72	109
2,675	**2,700**	0	0	0	76	113
2,700	**2,725**	0	0	0	79	116
2,725	**2,750**	0	0	0	83	119
2,750	**2,775**	0	0	0	87	122
2,775	**2,800**	0	0	0	91	125
2,800	**2,825**	2	2	2	94	129
2,825	**2,850**	5	5	5	98	132
2,850	**2,875**	9	9	9	102	135
2,875	**2,900**	12	12	12	106	138
2,900	**2,925**	16	16	16	109	141
2,925	**2,950**	19	19	19	113	145
2,950	**2,975**	23	23	23	117	148
2,975	**3,000**	26	26	26	121	151
3,000	**3,050**	32	32	32	126	156
3,050	**3,100**	39	39	39	134	163
3,100	**3,150**	46	46	46	141	170
3,150	**3,200**	53	53	53	149	177
3,200	**3,250**	60	60	60	157	184
3,250	**3,300**	67	67	67	165	190
3,300	**3,350**	74	74	74	173	197
3,350	**3,400**	81	81	81	181	204
3,400	**3,450**	89	88	88	189	211
3,450	**3,500**	96	95	95	197	218
3,500	**3,550**	104	102	102	205	224
3,550	**3,600**	111	109	109	213	232
3,600	**3,650**	119	116	116	221	239
3,650	**3,700**	126	123	123	229	246

4,950	**5,000**	343	332	318	467	448
5,000	**5,050**	353	341	326	476	457
5,050	**5,100**	362	350	334	486	465
5,100	**5,150**	372	359	342	495	473
5,150	**5,200**	381	368	350	505	481
5,200	**5,250**	391	377	358	514	489
5,250	**5,300**	400	386	366	524	497
5,300	**5,350**	410	395	374	533	505
5,350	**5,400**	419	404	382	543	513
5,400	**5,450**	429	413	390	552	521
5,450	**5,500**	438	422	398	562	529
5,500	**5,550**	448	431	406	571	537
5,550	**5,600**	457	440	414	581	545
5,600	**5,650**	467	449	422	590	553
5,650	**5,700**	476	458	430	600	562
5,700	**5,750**	486	467	438	609	570
5,750	**5,800**	495	476	446	619	578
5,800	**5,850**	505	485	454	628	586
5,850	**5,900**	514	494	463	638	594
5,900	**5,950**	524	503	471	647	602
5,950	**6,000**	533	512	480	657	610
6,000	**6,050**	543	521	488	666	618
6,050	**6,100**	552	530	497	676	626
6,100	**6,150**	562	539	505	685	634
6,150	**6,200**	571	548	514	696	642
6,200	**6,250**	581	557	522	707	650
6,250	**6,300**	590	566	531	718	658
6,300	**6,350**	600	575	539	729	666
6,350	**6,400**	609	584	548	740	675
6,400	**6,450**	619	593	556	751	683
6,450	**6,500**	628	602	565	762	691
6,500	**6,550**	638	611	573	773	700
6,550	**6,600**	647	620	582	784	710
6,600	**6,650**	657	629	590	795	719
6,650	**6,700**	666	638	599	806	729
6,700	**6,750**	676	647	607	817	740
6,750	**6,800**	685	656	616	828	751
6,800	**6,850**	695	665	625	839	762

8,100	**8,150**	968	912	872	1,125	1,048
8,150	**8,200**	979	921	881	1,136	1,059
8,200	**8,250**	989	931	891	1,149	1,070
8,250	**8,300**	1,000	940	900	1,161	1,081
8,300	**8,350**	1,010	950	910	1,174	1,092
8,350	**8,400**	1,021	959	919	1,186	1,103
8,400	**8,450**	1,031	969	929	1,199	1,114
8,450	**8,500**	1,042	978	938	1,211	1,125
8,500	**8,550**	1,052	988	948	1,224	1,136
8,550	**8,600**	1,063	997	957	1,236	1,149
8,600	**8,650**	1,073	1,007	967	1,249	1,161
8,650	**8,700**	1,083	1,016	976	1,261	1,174
8,700	**8,750**	1,092	1,024	984	1,274	1,186
8,750	**8,800**	1,101	1,032	992	1,286	1,199
8,800	**8,850**	1,110	1,040	1,000	1,299	1,211
8,850	**8,900**	1,121	1,050	1,008	1,311	1,224
8,900	**8,950**	1,131	1,059	1,016	1,324	1,236
8,950	**9,000**	1,141	1,068	1,024	1,336	1,249
9,000	**9,050**	1,151	1,078	1,033	1,349	1,261
9,050	**9,100**	1,161	1,087	1,041	1,361	1,274
9,100	**9,150**	1,172	1,096	1,049	1,374	1,286
9,150	**9,200**	1,182	1,106	1,057	1,386	1,299
9,200	**9,250**	1,192	1,115	1,065	1,399	1,311
9,250	**9,300**	1,202	1,124	1,073	1,411	1,324
9,300	**9,350**	1,212	1,134	1,081	1,424	1,336
9,350	**9,400**	1,223	1,143	1,089	1,436	1,349
9,400	**9,450**	1,233	1,152	1,097	1,449	1,361
9,450	**9,500**	1,243	1,162	1,105	1,461	1,374
9,500	**9,550**	1,253	1,171	1,113	1,474	1,386
9,550	**9,600**	1,263	1,181	1,121	1,486	1,399
9,600	**9,650**	1,274	1,190	1,129	1,499	1,411
9,650	**9,700**	1,284	1,199	1,138	1,511	1,424
9,700	**9,750**	1,294	1,209	1,146	1,524	1,436
9,750	**9,800**	1,304	1,218	1,154	1,536	1,449
9,800	**9,850**	1,314	1,227	1,162	1,549	1,461
9,850	**9,900**	1,325	1,237	1,170	1,561	1,474
9,900	**9,950**	1,335	1,246	1,178	1,574	1,486
9,950	**10,000**	1,345	1,255	1,186	1,586	1,499

Table 3 —Returns claiming THREE exemptions (and not itemizing deductions) (Continued)

If the amount on Form 1040A, line 14, is—		And you are—				
					Married filing separate return claiming—	
At least	But less than	Single, not head of house-hold	Head of house-hold	Married* filing joint return	Low income allow-ance	%Stand-ard deduc-tion
		Your tax is—				
$0	$2,650	$0	$0	$0	$0	$0
2,650	2,675	0	0	0	0	2
2,675	2,700	0	0	0	0	5
2,700	2,725	0	0	0	0	8
2,725	2,750	0	0	0	0	11
$2,750	$2,775	$0	$0	$0	$0	$14
2,775	2,800	0	0	0	0	17
2,800	2,825	0	0	0	0	20
2,825	2,850	0	0	0	0	23
2,850	2,875	0	0	0	0	26
$2,875	$2,900	$0	$0	$0	$0	$29
2,900	2,925	0	0	0	2	32
$2,925	$2,950	$0	$0	$0	$5	$35
2,950	2,975	0	0	0	9	38
2,975	3,000	0	0	0	12	41
3,000	3,050	0	0	0	18	45
3,050	3,100	0	0	0	25	51
3,100	3,150	0	0	0	32	57
3,150	3,200	0	0	0	39	63
3,200	3,250	0	0	0	46	69
3,250	3,300	0	0	0	53	75
3,300	3,350	0	0	0	60	81
3,350	3,400	0	0	0	67	88

If the amount on Form 1040A, line 14, is—		And you are—				
					Married filing separate return claiming—	
At least	But less than	Single, not head of house-hold	Head of house-hold	Married* filing joint return	Low income allow-ance	%Stand-ard deduc-tion
		Your tax is—				
4,700	4,750	173	168	166	280	270
4,750	4,800	181	176	174	289	277
4,800	4,850	189	184	181	297	285
4,850	4,900	197	192	189	306	292
4,900	4,950	205	200	196	315	299
4,950	5,000	213	208	204	324	306
5,000	5,050	221	216	211	334	314
5,050	5,100	229	224	219	343	322
5,100	5,150	238	232	226	353	330
5,150	5,200	246	240	234	362	338
5.200	5.250	255	248	241	372	346
–5,250	$5,300	$263	$256	$249	$381	$354
5,300	5,350	272	264	256	391	362
5,350	5,400	280	272	264	400	371
5,400	5,450	289	280	271	410	379
5,450	5,500	297	288	279	419	387
5,500	5,550	306	296	286	429	395
5,550	5,600	315	305	294	438	403
5,600	5,650	324	314	302	448	411
5,650	5,700	334	323	310	457	419
5,700	5,750	343	332	318	467	427
5,750	5,800	353	341	326	476	435
5,800	5,850	362	350	334	486	443
5,850	5,900	372	359	342	495	451

If the amount on Form 1040A, line 14, is—		And you are—				
					Married filing separate return claiming—	
At least	But less than	Single, not head of house-hold	Head of house-hold	Married* filing joint return	Low income allow-ance	%Stand-ard deduc-tion
		Your tax is—				
7,500	7,550	685	656	616	828	751
7,550	7,600	695	665	625	839	762
7,600	7,650	706	674	634	850	773
$7,650	$7,700	$716	$684	$644	$861	$784
7,700	7,750	727	693	653	872	795
7,750	7,800	737	703	663	883	806
7,800	7,850	748	712	672	894	817
7,850	7,900	758	722	682	905	828
7,900	7,950	769	731	691	916	839
7,950	8,000	779	741	701	927	850
8,000	8,050	790	750	710	938	861
8,050	8,100	800	760	720	949	872
8,100	8,150	811	769	729	960	883
8,150	8,200	821	779	739	971	894
8,200	8,250	832	788	748	982	905
8,250	8,300	842	798	758	993	916
8,300	8,350	853	807	767	1,004	927
8,350	8,400	863	817	777	1,015	938
8,400	8,450	874	826	786	1,026	949
8,450	8,500	884	836	796	1,037	960
8,500	8,550	895	845	805	1,048	971
8,550	8,600	905	855	815	1,059	982
8,600	8,650	916	864	824	1,070	993
8,650	8,700	926	874	834	1,081	1,004

3,400	**3,450**	0	0	0	74	94
3,450	**3,500**	0	0	0	81	101
3,500	**3,550**	0	0	0	89	107
3,550	**3,600**	4	4	4	96	113
3,600	**3,650**	11	11	11	104	120
3,650	**3,700**	18	18	18	111	126
3,700	**3,750**	25	25	25	119	132
3,750	**3,800**	32	32	32	126	139
3,800	**3,850**	39	39	39	134	145
3,850	**3,900**	46	46	46	141	152
3,900	**3,950**	53	53	53	149	159
3,950	**4,000**	60	60	60	157	166
4,000	**4,050**	67	67	67	165	172
4,050	**4,100**	74	74	74	173	179
4,100	**4,150**	81	81	81	181	186
4,150	**4,200**	89	88	88	189	193
4,200	**4,250**	96	95	95	197	200
4,250	**4,300**	104	102	102	205	206
4,300	**4,350**	111	109	109	213	213
4,350	**4,400**	119	116	116	221	220
4,400	**4,450**	126	123	123	229	227
4,450	**4,500**	134	130	130	238	234
4,500	**4,550**	141	137	137	246	241
4,550	**4,600**	149	144	144	255	249
4,600	**4,650**	157	152	151	263	256
4,650	**4,700**	165	160	159	272	263
5,900	**5,950**	381	368	350	505	459
5,950	**6,000**	391	377	358	514	467

6,000	**6,050**	400	386	366	524	476
6,050	**6,100**	410	395	374	533	484
6,100	**6,150**	419	404	382	543	492
6,150	**6,200**	429	413	390	552	500
6,200	**6,250**	438	422	398	562	508
6,250	**6,300**	448	431	406	571	516
6,300	**6,350**	457	440	414	581	524
6,350	**6,400**	467	449	422	590	532
6,400	**6,450**	476	458	430	600	540
6,450	**6,500**	486	467	438	609	548
6,500	**6,550**	495	476	446	619	556
6,550	**6,600**	505	485	454	628	564
6,600	**6,650**	514	494	463	638	572
6,650	**6,700**	524	503	471	647	581
6,700	**6,750**	533	512	480	657	590
6,750	**6,800**	543	521	488	666	600
6,800	**6,850**	552	530	497	676	609
6,850	**6,900**	562	539	505	685	619
6,900	**6,950**	571	548	514	696	628
6,950	**7,000**	581	557	522	707	638
7,000	**7,050**	590	566	531	718	647
7,050	**7,100**	600	575	539	729	657
7,100	**7,150**	609	584	548	740	666
7,150	**7,200**	619	593	556	751	676
7,200	**7,250**	628	602	565	762	685
7,250	**7,300**	638	611	573	773	696
7,300	**7,350**	647	620	582	784	707
7,350	**7,400**	657	629	590	795	718

7,400	**7,450**	666	638	599	806	729
7,450	**7,500**	676	647	607	817	740
8,700	**8,750**	935	882	842	1,092	1,015
8,750	**8,800**	944	890	850	1,103	1,026
8,800	**8,850**	953	898	858	1,114	1,037
8,850	**8,900**	962	906	866	1,125	1,048
8,900	**8,950**	971	914	874	1,136	1,059
8,950	**9,000**	980	922	882	1,149	1,070
9,000	**9,050**	988	930	890	1,161	1,081
9,050	**9,100**	997	938	898	1,174	1,092
9,100	**9,150**	1,006	946	906	1,186	1,103
9,150	**9,200**	1,015	954	914	1,199	1,114
9,200	**9,250**	1,024	962	922	1,211	1,125
9,250	**9,300**	1,033	970	930	1,224	1,136
9,300	**9,350**	1,042	978	938	1,236	1,149
9,350	**9,400**	1,051	987	947	1,249	1,161
9,400	**9,450**	1,060	995	955	1,261	1,174
9,450	**9,500**	1,069	1,003	963	1,274	1,186
9,500	**9,550**	1,078	1,011	971	1,286	1,199
9,550	**9,600**	1,087	1,019	979	1,299	1,211
9,600	**9,650**	1,096	1,027	987	1,311	1,224
9,650	**9,700**	1,104	1,035	995	1,324	1,236
9,700	**9,750**	1,114	1,044	1,003	1,336	1,249
9,750	**9,800**	1,124	1,053	1,011	1,349	1,261
9,800	**9,850**	1,134	1,062	1,019	1,361	1,274
9,850	**9,900**	1,145	1,072	1,027	1,374	1,286
9,900	**9,950**	1,155	1,081	1,035	1,386	1,299
9,950	**10,000**	1,165	1,090	1,043	1,399	1,311

1972 Tax Rate Schedules

If you do not use one of the Tax Tables, figure your tax on the amount on Form 1040, line 55, by using the appropriate Tax Rate Schedule on this page. Enter tax on Form 1040, line 18.

SCHEDULE X—Single Taxpayers Not Qualifying for Rates in Schedule Y or Z

If the amount on Form 1040, line 55, is: / Enter on Form 1040, line 18:

Not over $500 ..... 14% of the amount on line 55.

Over—	But not over—		of excess over—
$500	$1,000	$70+15%	$500
$1,000	$1,500	$145+16%	$1,000
$1,500	$2,000	$225+17%	$1,500
$2,000	$4,000	$310+19%	$2,000
$4,000	$6,000	$690+21%	$4,000
$6,000	$8,000	$1,110+24%	$6,000
$8,000	$10,000	$1,590+25%	$8,000
$10,000	$12,000	$2,090+27%	$10,000
$12,000	$14,000	$2,630+29%	$12,000
$14,000	$16,000	$3,210+31%	$14,000
$16,000	$18,000	$3,830+34%	$16,000
$18,000	$20,000	$4,510+36%	$18,000
$20,000	$22,000	$5,230+38%	$20,000
$22,000	$26,000	$5,990+40%	$22,000
$26,000	$32,000	$7,590+45%	$26,000
$32,000	$38,000	$10,290+50%	$32,000
$38,000	$44,000	$13,290+55%	$38,000
$44,000	$50,000	$16,590+60%	$44,000
$50,000	$60,000	$20,190+62%	$50,000
$60,000	$70,000	$26,390+64%	$60,000
$70,000	$80,000	$32,790+66%	$70,000
$80,000	$90,000	$39,390+68%	$80,000
$90,000	$100,000	$46,190+69%	$90,000
$100,000	....	$53,090+70%	$100,000

SCHEDULE Y—Married Taxpayers and Certain Widows and Widowers

If you are a married person living apart from your wife (husband), see page 5 of the instructions to see if you can be considered to be "unmarried" for purposes of using Schedule X or Z.

Married Taxpayers Filing Joint Returns and Certain Widows and Widowers (See page 5)

If the amount on Form 1040, line 55, is: / Enter on Form 1040, line 18:

Not over $1,000 14% of the amount on line 55.

Over—	But not over—		of excess over—
$1,000	$2,000	$140+15%	$1,000
$2,000	$3,000	$290+16%	$2,000
$3,000	$4,000	$450+17%	$3,000
$4,000	$8,000	$620+19%	$4,000
$8,000	$12,000	$1,380+22%	$8,000
$12,000	$16,000	$2,260+25%	$12,000
$16,000	$20,000	$3,260+28%	$16,000
$20,000	$24,000	$4,380+32%	$20,000
$24,000	$28,000	$5,660+36%	$24,000
$28,000	$32,000	$7,100+39%	$28,000
$32,000	$36,000	$8,660+42%	$32,000
$36,000	$40,000	$10,340+45%	$36,000
$40,000	$44,000	$12,140+48%	$40,000
$44,000	$52,000	$14,060+50%	$44,000
$52,000	$64,000	$18,060+53%	$52,000
$64,000	$76,000	$24,420+55%	$64,000
$76,000	$88,000	$31,020+58%	$76,000
$88,000	$100,000	$37,980+60%	$88,000
$100,000	$120,000	$45,180+62%	$100,000
$120,000	$140,000	$57,580+64%	$120,000
$140,000	$160,000	$70,380+66%	$140,000
$160,000	$180,000	$83,580+68%	$160,000
$180,000	$200,000	$97,180+69%	$180,000
$200,000	....	$110,980+70%	$200,000

Married Taxpayers Filing Separate Returns

If the amount on Form 1040, line 55, is: / Enter on Form 1040, line 18:

Not over $500 ..... 14% of the amount on line 55.

Over—	But not over—		of excess over—
$500	$1,000	$70+15%	$500
$1,000	$1,500	$145+16%	$1,000
$1,500	$2,000	$225+17%	$1,500
$2,000	$4,000	$310+19%	$2,000
$4,000	$6,000	$690+22%	$4,000
$6,000	$8,000	$1,130+25%	$6,000
$8,000	$10,000	$1,630+28%	$8,000
$10,000	$12,000	$2,190+32%	$10,000
$12,000	$14,000	$2,830+36%	$12,000
$14,000	$16,000	$3,550+39%	$14,000
$16,000	$18,000	$4,330+42%	$16,000
$18,000	$20,000	$5,170+45%	$18,000
$20,000	$22,000	$6,070+48%	$20,000
$22,000	$26,000	$7,030+50%	$22,000
$26,000	$32,000	$9,030+53%	$26,000
$32,000	$38,000	$12,210+55%	$32,000
$38,000	$44,000	$15,510+58%	$38,000
$44,000	$50,000	$18,990+60%	$44,000
$50,000	$60,000	$22,590+62%	$50,000
$60,000	$70,000	$28,790+64%	$60,000
$70,000	$80,000	$35,190+66%	$70,000
$80,000	$90,000	$41,790+68%	$80,000
$90,000	$100,000	$48,590+69%	$90,000
$100,000	....	$55,490+70%	$100,000

SCHEDULE Z—Unmarried (or legally separated) Taxpayers Who Qualify as Heads of Household (See page 5)

If the amount on Form 1040, line 55, is: / Enter on Form 1040, line 18:

Not over $1,000 14% of the amount on line 55.

Over—	But not over—		of excess over—
$1,000	$2,000	$140+16%	$1,000
$2,000	$4,000	$300+18%	$2,000
$4,000	$6,000	$660+19%	$4,000
$6,000	$8,000	$1,040+22%	$6,000
$8,000	$10,000	$1,480+23%	$8,000
$10,000	$12,000	$1,940+25%	$10,000
$12,000	$14,000	$2,440+27%	$12,000
$14,000	$16,000	$2,980+28%	$14,000
$16,000	$18,000	$3,540+31%	$16,000
$18,000	$20,000	$4,160+32%	$18,000
$20,000	$22,000	$4,800+35%	$20,000
$22,000	$24,000	$5,500+36%	$22,000
$24,000	$26,000	$6,220+38%	$24,000
$26,000	$28,000	$6,980+41%	$26,000
$28,000	$32,000	$7,800+42%	$28,000
$32,000	$36,000	$9,480+45%	$32,000
$36,000	$38,000	$11,280+48%	$36,000
$38,000	$40,000	$12,240+51%	$38,000
$40,000	$44,000	$13,260+52%	$40,000
$44,000	$50,000	$15,340+55%	$44,000
$50,000	$52,000	$18,640+56%	$50,000
$52,000	$64,000	$19,760+58%	$52,000
$64,000	$70,000	$26,720+59%	$64,000
$70,000	$76,000	$30,260+61%	$70,000
$76,000	$80,000	$33,920+62%	$76,000
$80,000	$88,000	$36,403+63%	$80,000
$88,000	$100,000	$41,440+64%	$88,000
$100,000	$120,000	$49,120+66%	$100,000
$120,000	$140,000	$62,320+67%	$120,000
$140,000	$160,000	$75,720+68%	$140,000
$160,000	$180,000	$89,320+69%	$160,000
$180,000	....	$103,120+70%	$180,000

Objectives

Fire, automobile, and life insurance are included in this chapter. Each section includes background information, illustrations, and examples followed by problems to which they apply.

Fire insurance covers problems in calculating premiums, multiple carriers, limitations of liability, use of coinsurance clauses, short-term policies, and cancellation of short-term policies. The section on automobile insurance deals primarily with the manner in which premiums and cancellations are calculated. Life insurance covers types of policies and costs.

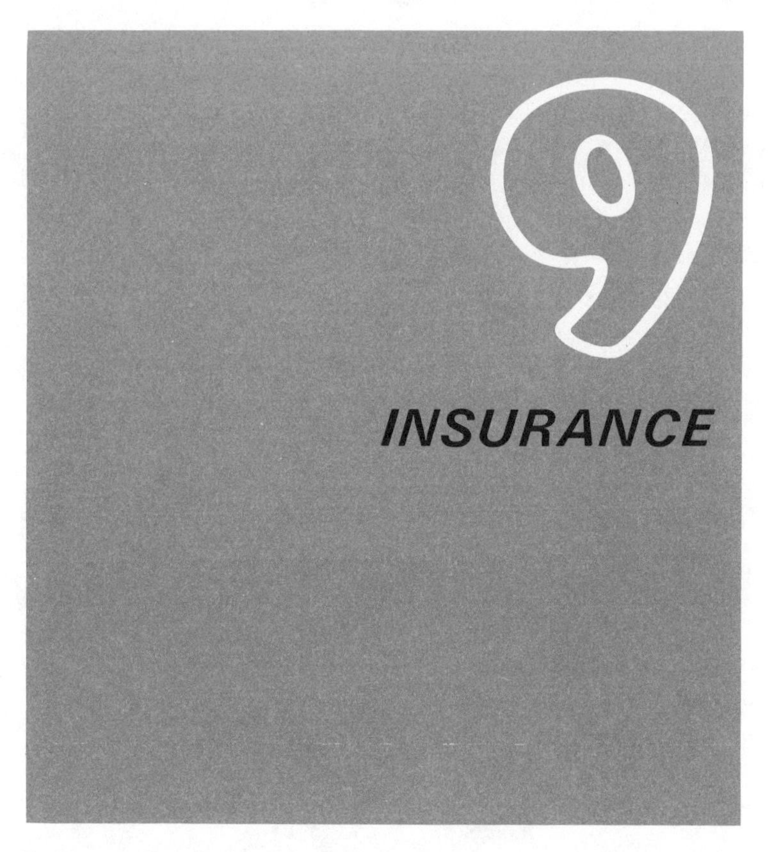

Insurance has become one of the largest and most successful businesses in the world. It has grown out of the needs of society, both of individuals and business enterprises, to be compensated for losses that might otherwise result in financial difficulty. Through the use of insurance, the burden of financial loss is distributed among those who share the same type of risk. Losses may result from injuries, death, poor health, unemployment, fire, water, theft, and many other situations too numerous to mention here.

The *insurance policy* is a written agreement between the *insurer* (insurance company—also called "carrier"), and the *insured* (individual, organization, or business). It is a contract whereby the insurer guarantees that an indemnity will be paid for a specific loss, in accordance with the terms of the insurance policy, in return for a payment called a premium. The face amount of a policy represents the amount of insurance carried.

Many types of insurance are available today. With the increased demand for insurance, many perils are covered in "package policies" or multiperil policies. Multiperil policies for fire and liability have become increasingly popular. However, under many circumstances, policies are written to cover fire losses independent of other types of perils. The manner in which rates are quoted and premiums and cancellations computed is similar for fire and casualty insurance policies.

This chapter deals with some of the mathematical problems and means of solving them that arise in the fire, automobile, and life insurance businesses.

FIRE INSURANCE

Fire insurance provides for protection against fire losses or losses that may be directly related to a fire as a primary cause of the damages sustained, such as water damage from extinguishing the fire. Rates vary, depending on the risk involved, loss experience, and expense factors. They also vary between companies due to their policies and extent of operation. However, many rates must be approved by the insurance commission or insurance department of the state.

There are several factors that must be considered when determining a rate. In the case of fire insurance, rates depend on the type of structure, location, proximity of fire departments and type of equipment used, fire hazards, and water supply. When commercial buildings are concerned, the character of contents and the kind and size of the business are important considerations. A building with an inside sprinkler system may realize a saving of from 70% to 90% credit.

There is also a difference in rates between owner-occupied and tenant-occupied buildings. This applies to rates on contents as well. If the owner occupies the building, the rate may be less than if a tenant occupies it by 2¢ to 11¢ or more per $100.

Most fire insurance policies on private residences carry a deductible clause. Of these, the great majority have $100 deductible clauses with the rest divided between $50 and $250 or $500 clauses. It is general practice for commercial building policies to include a $100 deductible clause. In case of fire loss the insured is reimbursed for the total covered loss less $50, $100, $250, or $500 depending on the provisions of the policy.

The fire rate is based on a unit of insurance and is quoted on $100 increments of insurance, i.e., $0.04 per $100 for a term of 1 year. Typical rates in a protected fire area are $0.10 per $100 for fire, $0.04 for extended coverage,† and $0.05 for an all-risk clause on a private residence or non-commercial property. In unprotected areas the rates may be as much as 200% to 500% more than those in protected areas.

Very few fires result in a total loss. In well-protected areas, it is estimated that many losses are less than 10% of the total value of the property. In unprotected or poorly protected areas the percent of total losses is very high.

PREMIUMS

Policies are generally written in multiples of $100 or $1000. Usually, no policy is written for a sum that includes a fraction of a dollar.

Most fire insurance policies are written for 1 or 3 years. However, under some circumstances, policies are written for 2, 4, or 5 years. Periods less than 1 year are discussed later in this unit.

Finding the Annual Premium

Annual Premium = Rate × Amount of Insurance

example: A building is insured for $25,000 for 1 year at a rate of $0.40 per $100 of coverage. Find the annual premium.

solution: 250 × $0.40 = $100 annual premium

example: If a home is insured for $30,000 at $1.56 per $1000, what is the annual premium?

solution: 30 × $1.56 = $46.80 annual premium

†*Extended coverage* covers insurance against losses due to windstorm, hail, explosion, riot and civil commotion, damage by aircraft or by vehicle, and smoke damage.

Finding the Premium for More Than a Year

Policies written for a year are called "annual policies" and use a basic rate, whereas policies for more than a year are called "term policies" and are written at a discount rate. The following schedule is representative.

Term (Years)	Annual Rate Factor†
2	1.85
3	2.7

example: A building is insured for $25,000. If the rate is $0.40 per $100 of coverage, what are (a) the annual premium and the premium for 3 years? (b) What is the saving to the insured on a 3-year policy?

solution: (a) annual premium 250 × $0.40 = $100
3-year premium $100 × 2.7 = $270

(b) premium for 3 years on an annual basis
$100 × 3 = $300
amount saved $300 − $270 = $ 30‡

EXERCISE 38

I. Find the premium on the following:

No.	Amount of Insurance	Rate	Terms (Years)
1.	$12,500	$0.18 per $ 100	1
2.	36,000	0.113 per $ 100	3
3.	4,200	0.52 per $ 100	2
4.	15,000	1.863 per $1000	3
5.	7,500	2.86 per $1000	2
6.	22,000	1.19 per $1000	3

†From Insurance Services Office, San Francisco, California (1973). Policies at a reduced rate are no longer written for 4 and 5 years.

‡Interest on the second- or third-year payments is not taken into account here. If it was, then the amount saved would be less by the amount of the interest that could be earned on these payments for 1 and 2 years, respectively.

II. How much did the insured "save" in problems 3, 4, 5, and 6 above over a straight annual rate basis?

INDEMNITY PAID ON AN ORDINARY FIRE INSURANCE POLICY

Insurance companies pay only the amount of insurance carried (face of the policy) or the amount of the loss, whichever is lower, but never more than the property is worth, assuming replacement cost coverage exists and compliance with coinsurance requirements is met. (Coinsurance is discussed later in this chapter.) For example:

Amount of Insurance	Amount of Loss	Amount Paid
$20,000	$5,000	$5,000
26,000	30,000	26,000
10,000	10,000	10,000

MULTIPLE CARRIERS

Insurance on a particular piece of property may be divided between two or more companies. This may be done for several reasons. The insured may prefer this arrangement if the amount carried is large. The insured may wish to favor more than one insurance company for business reasons. The total risk may be too great for any one company to assume. In any event, when any insurance coverage is alloted to more than one company, each company must bear any losses in the same proportion that its coverage is to the total insurance carried. In addition, it is important that all policies read exactly alike or there may be a penalty in case of loss.

example: A warehouse and its contents, worth $125,000, was insured in Company A for $50,000 at 69¢ per $100, Company B for $30,000 at 45¢ per $100, and Company C for $20,000 at 75¢ per $100. (a) Find the premium paid to each company and the total premium. (b) If a fire caused a 50% loss ($62,500), how much must each company pay the insured?

solution: (a) Company A

annual premium $= \$500 \times 0.69 = \$345$

Company B

annual premium $= \$300 \times 0.45 = \$135$

Company C

annual premium $= \$200 \times 0.75 = \underline{\$150}$

total annual premium $= \$630$

(b) The total insurance carried is \$50,000 + \$30,000 + \$20,000 = \$100,000. The percent of total insurance carried by each company is:

Company A

$\$50,000 \div \$100,000 = 0.50$ or 50%

Company B

$\$30,000 \div \$100,000 = 0.30$ or 30%

Company C

$\$20,000 \div \$100,000 = 0.20$ or 20%

Therefore, the amount of the loss paid by each company is as follows:

Company A

50% of \$62,500 *or* $.50 \times 62,500 = \$31,250$

Company B

30% of \$62,500 *or* $.30 \times 62,500 = \$18,750$

Company C

20% of \$62,500 *or* $.20 \times 62,500 = \underline{\$12,500}$

total loss $= \$62,500$

EXERCISE 39

1. A home valued at \$30,000 is insured for \$24,000. (a) If fire caused damages of \$16,500, how much did the insurance company pay? (b) If the damages amounted to \$25,000, how much did the insurance company pay?

2. A building worth \$75,000 was insured for \$60,000 at \$0.2065 per \$100. (a) What was the annual premium? (b) How much did the insurance company pay on a loss of \$61,000 due to fire and water damage?

3. A building was insured for \$75,000. The insurance was distributed as follows: Company A, \$15,000; Company B, \$10,000; Company C, \$12,500; and Company D, \$37,500. If the building was damaged by fire to the extent of \$52,000, how much did each company pay?

4. In problem 3, if the damages amounted to \$78,000, how much did each company pay?

COINSURANCE

Most fires result in partial losses. Consequently, property owners are inclined to carry no more insurance than they think is necessary. This practice naturally reduces the amount of premiums received by the insurance companies since the property is insured for only a portion of its value. To encourage policy holders to carry more insurance, the insurance companies offer (at reduced rates) policies with a *coinsurance* clause.

When a policy contains a coinsurance clause, the insured must carry a specified percent (80% is common) of the replacement cost value or actual cash value of the property in order to be reimbursed for the loss or the amount of insurance carried, whichever is the lowest figure. However, if the amount of insurance carried is less than that required by the coinsurance clause, the insured must bear the proportionate difference between the actual loss and the insurance allowed under the terms of the policy. In this sense, the property owner is sharing the risk with the insurance company.

Coinsurance clauses are very common in commercial insurance but not in private or noncommercial insurance except in certain states. The 80% or 90% clauses are common, although many policies contain 100% clauses and some may be less than 80%. Rates vary when a coinsurance clause is used—the higher the coinsurance percentage, the lower the rate per \$100 of coverage.

The formula for computing the amount paid on losses by the insurance company is

$$\frac{\text{Amount of Insurance in Force}}{(\text{Coinsurance }\%)\ (\text{Value of Property})} \times \text{Loss}$$

The value of the property may be the actual cash value or replacement cost, depending on how the policy is written.

example: Property valued at \$75,000 with an 80% coinsurance clause is insured for \$60,000. Fire damages amounted to \$42,500. Find the amount of payment by the insurance company.

solution:
$$\frac{\$60{,}000}{80\% \text{ of } \$75{,}000} \times \$42{,}500 = \frac{\$60{,}000}{\$60{,}000} \times \$42{,}500$$
$$= \$42{,}500$$

example: Property worth \$16,000 is insured for \$12,800. If the loss due to fire amounted to \$14,400 and the policy contained an 80% coinsurance clause, what was the amount of indemnity (amount paid by the insurance company)?

solution:
$$\frac{\$12{,}800}{80\% \text{ of } \$16{,}000} \times \$14{,}400 = \frac{\$12{,}800}{\$12{,}800} \times \$14{,}400$$
$$= \$14{,}400$$

But \$14,400 is greater than the insurance carried. The insurance company will pay no more than the face of the policy, i.e., \$12,800. Therefore, the amount of indemnity is \$12,800.

example: An apartment house valued at \$150,000 is insured for \$100,000. Loss due to fire was \$62,500. If the policy contained an 80% coinsurance clause, what was the amount of indemnity?

solution:
$$\frac{\$100{,}000}{80\% \text{ of } \$150{,}000} \times \$62{,}500$$
$$= \frac{\$100{,}000}{\$120{,}000} \times \$62{,}500$$
$$= \tfrac{5}{6} \times \$62{,}500 = \$52{,}083.33$$

In this case the property owner bears the remainder of the loss, i.e., \$62,500.00 − \$52,083.33 = \$10,416.67.

EXERCISE 40

I. Find the amount paid to the insured in each of the following examples under a policy with a coinsurance clause as stated.

No.	Value of Property	Insurance	Fire Loss	Coinsurance Percent
1.	$57,000	$40,000	$15,000	100
2.	28,000	16,000	13,000	80
3.	37,500	30,000	27,000	90
4.	12,000	10,000	12,000	80
5.	7,000	5,000	4,000	90
6.	8,500	6,200	6,800	100

II. Applications:

1. A motel valued at $125,000 was insured for $90,000. Fire caused damage amounting to $95,000. If the insurance contained a 90% coinsurance clause, how much did the insurance company pay on the loss?

2. A building worth $60,000 was insured for $\frac{4}{5}$ of its value. A fire loss amounted to $20,000. What was the amount of indemnity if the policy contained a 90% coinsurance clause?

3. The A. C. Electric Company insured its warehouse worth $90,000 for $50,000 and the contents worth $200,000 for $150,000. The insurance policy contained an 80% coinsurance clause. If fire destroyed $\frac{2}{3}$ of the building and $\frac{1}{2}$ of the contents, (a) what was the amount of settlement and (b) what was the total loss to the property owner?

4. A hotel worth $90,000 was completely destroyed by fire. Insurance with a 90% coinsurance clause was carried by three companies as follows: Company A, $36,000; Company B, $18,000; and Company C, $9,000. (a) Find the amount of the claim paid by each company. (b) What part of the loss did the property owner share?

→ $27,000 loss − total paid by co.

5. Hipple Canvas Company insured their shop for $55,000 with an 80% coinsurance clause. Fire caused damages amounting to $37,500. If the shop was worth $75,000, how much was the insurance settlement?

$34375.00

6. A piece of property was insured for $4000 with a 70% coinsurance clause. The property was worth $8000. If damages, due to fire, amounted to $6000, what did the insurance company pay?

SHORT-TERM POLICIES AND CANCELLATION

Short-term policies (less than 1 year) for fire and casualty insurance, with the exception of the automobile,† are

SHORT-RATE TABLE

Days Policy in Force	Percent of 1-Year Premium	Days Policy in Force	Percent of 1-Year Premium	Days Policy in Force	Percent of 1-Year Premium
1	5	95–98	37	219–223	69
2	6	99–102	38	224–228	70
3– 4	7	103–105	39	229–232	71
5– 6	8	106–109	40	233–237	72
7– 8	9	110–113	41	238–241	73
9–10	10	114–116	42	242–246 (8 months)	74
11–12	11	117–120	43	247–250	75
13–14	12	121–124 (4 months)	44	251–255	76
15–16	13	125–127	45	256–260	77
17–18	14	128–131	46	261–264	78
19–20	15	132–135	47	265–269	79
21–22	16	136–138	48	270–273 (9 months)	80
23–25	17	139–142	49	274–278	81
26–29	18	143–146	50	279–282	82
30–32 (1 month)	19	147–149	51	283–287	83
33–36	20	150–153 (5 months)	52	288–291	84
37–40	21	154–156	53	292–296	85
41–43	22	157–160	54	297–301	86
44–47	23	161–164	55	302–305 (10 months)	87
48–51	24	165–167	56	306–310	88
52–54	25	168–171	57	311–314	89
55–58	26	172–175	58	315–319	90
59–62 (2 months)	27	176–178	59	320–323	91
63–65	28	179–182 (6 months)	60	324–328	92
66–69	29	183–187	61	329–332	93
70–73	30	188–191	62	333–337 (11 months)	94
74–76	31	192–196	63	338–342	95
77–80	32	197–200	64	343–346	96
81–83	33	201–205	65	347–351	97
84–87	34	206–209	66	352–355	98
88–91 (3 months)	35	210–214 (7 months)	67	356–360	99
92–94	36	215–218	68	361–365 (12 months)	100

†Many automobile insurance policies are written for 3 to 6 months with rates quoted on a 3- and 6-month basis. Short-rate tables are also available for 2- and 3-year policies.

written at a higher rate than those for annual policies. For example, a short-term policy for 4 months costs 44% of the annual premium and a 6-month policy costs 60% of the annual premium. The daily charges increase as the periods become shorter.

The standard Short-Rate Table shown here is used to determine the premium charge for short-term policies. It is also used when the insured party cancels a policy.

Finding the Premium on Short-Term Policies

example: Mr. Easton, while on a new assignment, lived in temporary quarters for 3 months before moving to a permanent residence. During that time he insured his household furnishings for $20,000 at a rate of $0.207 per $100. What was the cost of the insurance?

solution: annual premium = $200 × 0.207 = $41.40
rate for 3 months (Short-Rate Table) = 35%

cost of insurance
= 35% of $41.40 = .35 × $41.40 = $14.49

In cases where a party cancels a short-term policy but reinsures with the same carrier, the general practice is to prorate the insurance. For example, if Mr. Easton in the example purchased a new policy for a year or more from the same carrier after he was established in his permanent residence, his insurance on the furnishings for the 3-month period would be $\frac{1}{4}$ or 25% of the annual premium instead of 35%.

example: Mr. and Mrs. Thomas insured a piano valued at $1500 for the full amount for 15 days. What was the cost of this insurance if the rate was $1.25 per $100?

solution: annual premium = $15.00 × $1.25 = $18.75
rate for 15 days (Short-Rate Table) = 13%

cost of insurance = 13% of $18.75
= .13 × 18.75 = $2.44

Cancellation of Short-Term Policies

The cancellation clause in a standard fire insurance policy specifies that the policy holder may cancel at any time. The insurance company may also cancel the policy but must give the insured 5 days notice (10 days to mortgagee) in writing so that other insurance may be obtained and become effective on the date of cancellation. The manner in which refunds are handled is explained as follows.

If the insured cancels: If the insured cancels a fire insurance policy, he receives a refund in accordance with rates shown on the standard Short-Rate Table. The number of days in force must equal the exact number from the date the policy was written up to and including the day it was cancelled.

example: Mr. Axtel cancelled a 1-year policy, costing \$37.00, 33 days after it had been purchased. What was the amount of the refund?

solution: According to the short-rate table, the charge for 33 days is 20% of the annual premium: .20 × \$37.00 = \$7.40. The refund amounts to the difference between this charge and the annual premium: \$37.00 − \$7.40 = \$29.60. Or, since the charge for 33 days is 20%, then the refund is equal to 80% of the annual premium: .80 × \$37.00 = \$29.60.

example: Mr. Hynes purchased a 1-year policy, dated June 17, for his home on which he paid a premium of \$32.00. On September 22 he sold his home and notified his insurance agent that he wished to cancel his policy. (a) What did the insurance cost Mr. Hynes? (b) What was the refund paid by the insurance company?

solution:
annual premium = \$32.00
exact days (July 17–Sept. 22) = 67
rate for 67 days = 29% of annual premium

(a) The premium charge for 67 days is: $.29 \times \$32.00 = \$9.28$. (b) The refund is $\$32.00 - \$9.28 = \$22.72$. Or, since the charge for 67 days is 29%, then the refund equals 71% of the annual premium: $.71 \times \$32.00 = \$22.72$.

If the insurer (insurance company) cancels: Sometimes a policy is cancelled by the insurance company. When this is the case the amount retained by the insurance company is based on a direct prorata basis, using exact days and a 365-day year. It is the ratio of the time the policy was in force to the total time for which the policy was written and paid for. (This ruling applies to any term of a year or more.)

example: Baines & Co. insured its factory on March 3 for a year. The policy was cancelled by the carrier the following October 15. If the annual premium was $130.00, what was the cost of the insurance and the amount of the refund?

solution: days in force (Mar. 3–Oct. 15) = 226

amount of premium retained by the carrier

$$= \$130.00 \times \frac{226}{365} = \$80.49$$

amount refunded to Baines & Co.

$$= \$130.00 - \$80.49 = \$49.51$$

EXERCISE 41

1. What percent of the annual premium should be charged to the policy holder to cover insurance for 39 days? 16 days? 119 days? 87 days?

2. If Maynard & Co. insured merchandise in a warehouse for 90 days, what would be the premium charge on a policy worth $62,000 at $1.16 per $100 per year?

3. Mrs. Hazelton insured her home for a year, effective March 14. On October 3 the home was sold at which time the owner cancelled the insurance. If the premium was $32.50, what was the net premium charge? What was the amount of refund?

4. Mr. Urton insured his housetrailer and contents for $26,000 at a rate of $1.065 per $100 on June 10 for a year. On December 15 the insurance company cancelled the policy. Find the refund due Mr. Urton.

5. The Hoffman Manufacturing Co. insured a building for $750,000 on September 2 for a 1-year term at $0.1075 per $100. On the following February 15 the policy was cancelled by the carrier. What was the amount of the refund?

AUTOMOBILE INSURANCE

Automobile insurance is one of the most common types of casualty insurance. Insurance is necessary in the public interest to cover the many hazards arising out of the ownership, repairs, and use of an automobile.

The need for insurance has increased with the increased number of automobiles in use. An automobile represents an expensive investment which should be protected by insurance. Repairs are also expensive and have increased along with the initial purchase price of an automobile. Insurance is also needed to protect the owner against judgments due to the negligent operation of a vehicle—judgments that could result in financial disaster.

AUTOMOBILE INSURANCE COVERAGE

Liability insurance: Negligence to others in the use of a car can cause bodily injury or property damage. Automobile liability insurance covers these hazards. (1) **Property damage liability insurance** protects the owner against damages to another person's car or property. The basic maximum limit is $5000. Higher limits may be purchased but seldom are since most property damage claims rarely exceed $5000. (2) **Bodily injury or public liability insurance** provides protection for the owner of an automobile against the cost of injuries to other people due to negligence in the operation of his car. This coverage may include medical payments if anyone in the car is injured in an accident. The minimum limits for bodily injury insurance in many states are $5000/$10,000 (5/10). In California the minimum limits are $15,000/$30,000. A coverage of 5/10 means that the owner of a motor vehicle is protected to a maximum payment of $5000 for any one person and to a maximum payment of $10,000 in any one accident in which the owner of the automobile may have injured more than one person. Additional insurance may be purchased with higher maximum limits such as $25,000/$50,000, $50,000/$100,000 or $100,000/$300,000.

Physical damage insurance: There are two kinds of physical damage insurance. (1) **Comprehensive coverage** includes most of the damages to an automobile, including fire and theft but not collision. Hazards such as fire, theft, and windstorms may be insured against separately. (2) **Collision coverage** protects against damage to the car by collision. Usually, a deductible clause of $50 or $100 is written into the agreement. Therefore, in cases of damages due to collision, the owner of the car pays the first $50 or $100 of cost, depending on the amount specified in the insurance policy.

Other insurance: Insurance may be purchased to cover charges for such services as towing, road service, disability, death benefits, use of other cars, and protection against uninsured motorists.

PREMIUMS

Rates used when determining premiums are dependent on several factors which are classified according to the age of the automobile, type of car, use of car (business or pleasure), locality or community in which the car is used primarily, age and sex of the driver, occupation of the driver, and traffic record of the driver. The three primary classifications used to determine rates are: (1) classification by use of car and age, marital status, and sex of driver; (2) territorial area of state in which the car is located; and (3) value of the automobile. These three classifications are then applied to the rate charts such as those shown on pages 206–210 for comprehensive (fire and theft), collision, bodily injury, medical payments, and property damage.

The type of company and the broadness of the policy account for some differences in rates between carriers. Also, some carriers return to their policy holders a part of their net earnings each year. The charts and rates used in this text apply to this type of company.

Classification by Use and Owner of the Automobile

The classifications and rate charts used in this text are representative of a large automobile association and insurance company.

PRIVATE PASSENGER AUTOMOBILE CLASSIFICATION

Class 1 & 4—Use: Pleasure; To Work or School

No business use, no male operator or owner under age 25, no single male principal operator or owner under 30, no single female operator under 21. Class 4 indicates an operator age 65 or over.

1A–4A Not driven to work or school; or owner is a clergyman.

1B–4B Driven to work or school less than 10 miles one way.

1C–4C Driven to work or school 10 or more miles but less than 20 miles one way.

1D–4D Driven to work or school 20 or more miles one way.

Class 2—Male Operator or Owner

Single, Occasional Operator Under Age 25

2A Ages 16 through 20.

2AE Ages 21 through 24.
Class 2A or 2AE rate also applies to a single male, age 16 through 24†, occasional operator, attending college or in military service in the state of residence.

Single, Principal Operator or Owner Under Age 30

2C Ages 16 through 20.

2CE Ages 21 through 24.

2CA Ages 25 through 29, not driven to work or school.

2CB Ages 25 through 29, driven to work or school less than 10 miles one way.

2CC Ages 25 through 29, driven to work or school 10 or more miles but less than 20 miles one way.

2CD Ages 25 through 29, driven to work or school 20 or more miles one way.

†When the use rate (to work or in business) is higher than the age-of-driver rate, always apply the highest total premium rate.

2C3 Ages 25 through 29, business use, annual mileage 12,000 or less.

2CM Ages 25 through 29, business use, annual mileage over 12,000.

Married, Principal Operator or Owner Under Age 25

2D Ages 16 through 20.

2DE Ages 21 through 24.

Apply the Class 2 premium to the automobile owned or principally operated by the male operator. "Principal" means operates automobiles equivalent to 50% or more of one car's total mileage. If two or more automobiles, Class 2 premiums shall apply to the number of automobiles equal to the number of underage male operators.

Class 3 & 4—Business Use

Business use, no male operator under age 25, no single male principal operator or owner under 30, no single female operator under 21. Class 4 indicates an operator 65 or over.

3–43 Annual mileage 12,000 or less.

3M–43M Annual mileage over 12,000.

A Federal Government employee may be rated at the applicable Class 1 rate, as coverage for business use is provided by the Government.

Automobiles owned by a corporation, copartnership or unincorporated association will be rated at the applicable class based on use and age of drivers.

Automobiles owned by a family, copartnership, or corporation, used in farming or ranching and principally garaged on the farm or ranch, will not be considered used in business, and the applicable rate will be based on age of drivers.

Class 5—Female Operator, Single, Principal Operator, Ages 16 through 20

5A Not driven to work or school.

5B Driven to work or school less than 10 miles one way.

5C Driven to work or school 10 or more miles, but less than 20 miles one way.

5D Driven to work or school 20 or more miles one way.

53 Business use, annual mileage 12,000 or less.

53M Business use, annual mileage over 12,000.

Class 5†—Female, Single, Occasional Operator, Ages 16 through 20

†There is no class 6 listed for this company.

5AA	Not driven to work or school.
5BA	Driven to work or school less than 10 miles one way.
5CA	Driven to work or school 10 or more miles, but less than 20 miles one way.
5DA	Driven to work or school 20 or more miles one way.
53A	Business use, annual mileage 12,000 or less.
5MA	Business use, annual mileage over 12,000.

Class 7—Female, Only Operator,† Ages 30 through 64

7A	Not driven to work or school.
7B	Driven to work or school less than 10 miles one way.
7C	Driven to work or school 10 or more miles, but less than 20 miles one way.
7D	Driven to work or school 20 or more miles one way.
73	Business use, annual mileage 12,000 or less.
73M	Business use, annual mileage over 12,000.

TO WORK OR SCHOOL LESS THAN FIVE DAYS A MONTH

Exception to Classes 1, 2, 4, 5, and 7: When an automobile is used five days or less during a month for driving to work or school, it will be rated at the next lower classification.

DISCOUNTS

TWO OR MORE AUTOMOBILES (NET RATE)

When two or more automobiles owned by an individual or husband and wife (or owned and registered jointly to relatives in the same household), are insured on the same members policy and are garaged at the same location and are rated at Class 1, 2, 3, 4, 5, 6, or 7, each automobile shall be subject to a reduction of 15% from the applicable bodily injury, medical payments, property damage, and collision rates.

BUMPER DISCOUNTS

Any 1973 private passenger car which is warranted by the manufacturer to meet or exceed the federal "bumper" standard will qualify for a collision coverage discount. There are three discounts as follows:

†Exception: If there is an additional female in the same household, whose car is also insured in the Bureau, Class 7 is still applicable.

Bumper Standards, Discounts, and Codes

Crash Front Bumper	Speed Rear Bumper	Standard	Discounts	Codes
5 mph	$2\frac{1}{2}$ mph	No damage to safety related items.	10%	1
5 mph	$2\frac{1}{2}$ mph	No damage.	15%	2
5 mph	5 mph	No damage	20%	3

Classification by Territorial Definitions

Each state is divided into areas, often based on county lines. Such areas are usually identified with a number. Rates are prepared for bodily injury, medical payments, property damage, and collision for each area, depending on the past experience of the insurer. For the sake of simplicity, problems are restricted to rates that apply to district or territory 91 on our charts. The following schedule of territories for California is representative.

Territory	***Territory Definitions, California***
50	NORTHERN SAN MATEO COUNTY territory comprises that area enclosed by the outside boundaries of the following townships: Brisbane, Colma, Daly City, Pacifica, and South San Francisco.
51	SAN FRANCISCO COUNTY—Entire county.
52	SAN MATEO COUNTY—(Balance) That portion of San Mateo County not included in Northern San Mateo County.
53	NORTHERN SANTA CLARA COUNTY territory comprises that area enclosed by the outside boundaries of the following townships: Campbell, Palo Alto, San Jose, Saratoga, Fremont, Redwood City, Santa Clara, and Sunnyvale.
60	OAKLAND METROPOLITAN territory comprises that area enclosed by the outside boundaries of the following townships in Alameda County: Alameda, Brooklyn, Eden, Oakland, and Peralta.
61	OAKLAND SUBURBAN territory comprises that area in Contra Costa County situated in Township 8 and that area north of the south boundary of Township 6 and west of the Mount Diablo meridian, but excluding that area bounded on the south by a line which begins at the intersecting point of the Mount Diablo meridian and runs west along Highway 4 until it reaches the Willow Pass Summit meridian where it runs along the meridian to the Bay.
63	STOCKTON METROPOLITAN territory comprises the area enclosed by the outside boundaries of the townships 1N and 2N, and Ranges 6E and 7E.

(*continued*)

Territory	*Territory Definitions, California*
64	SACRAMENTO METROPOLITAN territory comprises that portion of Sacramento County and Yolo County bounded as follows: starting from the intersection of Grant Line Road and Calvine Road, following Calvine Road and its extension west to the line of Range 3E north to the Sacramento River, continuing north along the Sacramento River to the Placer County line, east along the Placer County line to the El Dorado County line, south along the El Dorado County line to its intersection with U.S. Highway 50, west along Route 50 to the American River, west along the American River to County Road E2, south along E2 to its intersection with Calvine Road.
66	FRESNO METROPOLITAN territory comprises that portion of Fresno County bounded as follows: commencing at the point where Blackstone Avenue intersects the San Joaquin River, south on Blackstone to Herndon Avenue, east on Herndon to Fowler Avenue, south on Fowler to Kings Canyon Road, east on Kings Canyon to Armstrong Avenue, south on Armstrong to Jensen Avenue, west on Jensen to Clovis Avenue, south on Clovis to Lincoln Avenue, west on Lincoln to Fig Avenue, north on Fig to North Avenue, west on North to Valentine Avenue, north on Valentine to Whites Bridge Avenue, west on Whites Bridge to Brawley Avenue, north on Brawley to the U.S. Highway 99 Freeway, northwest along U.S. Highway 99 Freeway to its intersection with the San Joaquin River, following the San Joaquin River east to its intersection with Blackstone Avenue.
73	STANISLAUS COUNTY—Entire county.
74	SANTA CLARA COUNTY—(Balance) That portion of Santa Clara County not included in Northern Santa Clara County territory.
75	ALAMEDA COUNTY—(Balance) That portion of Alameda County not included in Oakland Metropolitan territory.
76	CONTRA COSTA COUNTY—(Balance) That portion of Contra Costa County not included in Oakland Suburban territory.
77	MARIN COUNTY—Entire county.
79	REMAINDER OF STATE II comprises that area enclosed by the outside boundaries of the following countries: Fresno (excluding that area in territory 66), Solano (excluding that area in territory 80), Kings, Madera, Mariposa, Merced, Monterey, San Benito, Santa Cruz, Shasta, Tehama.
80	REMAINDER OF STATE III comprises that area enclosed by the outside boundaries of the following counties: Solano—That portion of the county bounded as follows: commencing at the intersection of Highway 12 with the Napa County line, eastward along Highway 12 to its intersection with Highway 21. Then due south following Highway 21 and its extension to the Solano County line (including the city of Benicia and the Benicia Arsenal Military Reservation), then west following the Solano County line to its junction with the Napa County line, then east following the Napa County line to its interesection with Highway 12; and Yolo (excluding that area in territories 64 and 98), Lake, Mendocino, Napa, Sonoma, Sutter, Yuba.

Territory	Territory Definitions, California
91	REMAINDER OF STATE I territory comprises that area enclosed by the outside boundaries of the following counties: Alpine, Amador, Butte, Calaveras, Colusa, Del Norte, El Dorado, Glenn, Humboldt, Lassen, Modoc, Nevada, Placer, Plumas, Sierra, Siskiyou, Trinity, and Tuolumne.
98	SACRAMENTO AND SAN JOAQUIN RURAL territory comprises all of the area in San Joaquin County except Stockton metropolitan territory and all of the area in Sacramento County (and Yolo County east of the east line of Range 3E) except Sacramento Metropolitan territory.

Classification by Value of Automobile

It is necessary to know the value of the automobile in order to determine the cost of insurance to cover comprehensive (including fire and theft) and collision. The rate applied will be higher for a policy with a $50-deductible clause than for one with a $100-deductible clause since the insurance company is bearing an additional $50 of the cost of any damages resulting from collision.

Classification charts for automobiles are too extensive to include in this text. Therefore, all problems will indicate the model classification to be applied to the rate charts for comprehensive and collision coverage.

Determining the Premium

On the charts that follow for collision and liability, notice the lower section is marked N E T. Net rates are to be used only when two or more cars are insured. If only one vehicle is insured, use the upper section of the charts.

example: Mrs. Lane, who is 35, owns a four-door Dodge sedan, classified as J1. It is used for pleasure only. She is the only driver in household. Mrs. Lane wants insurance coverage for collision with a $100-deductible clause: comprehensive, including fire and theft; bodily injury for

50/100 thousand, medical payment for $1000, and property damage for $5000. What is the premium cost of each coverage? The total premium?

solution:

Driver classification (page 200)	7A
Territorial classification	91
Automobile classification (value and age)	J1

comprehensive (column J – K, group 1) in line with class 7A	$ 35.00
collision ($100 deductible), page 208 line 7A, col. J1	78.00
bodily injury 50/100	46.00
medical payment $1000 (page 209 line 7A)	13.00
property damage $5000	17.00
total premium	$189.00

example: Henry Jones, not married, received a car for his twenty-first birthday. He purchased insurance for the following coverage: comprehensive, collision with a $100-deductible clause, bodily injury for 20/40 thousand dollars, and property damage for $5000. If the car was classified as H2, what was the cost for each coverage? The total premium?

solution:

Driver classification	2CE
Territorial classification	91
Automobile classification	H2

comprehensive (H–I group, 2nd column)	$ 37.00
collision ($100 deductible)	105.00
bodily injury, 20/40	178.00
property damage $5000	76.00
total premium	$396.00

example: John Barns, age 52, owns two cars, one listed as H3 and the other as L4. They are used for pleasure and to drive to work. What is the total insurance premium for both cars if he carries comprehensive, $100 deductible collision (for the L4 car only), BI 50/100, MI 1, PD 5.

solution:

Driver classification			1A
Territorial classification			91
Automobile classification			H3, L4
comprehensive	H3	$ 25.00	(NET rate)
collision	L4	26.00	(NET rate)
Collision ($100 deductible) for	L4	63.00†	
bodily injury, 50/100		86.00	(NET—43 × 2)
medical payments $1000		22.00	(NET—11 × 2)
property damage $5000		32.00	(NET—16 × 2)
total premium		$254.00	

†Net rate cannot be used here since only one car was covered for collision.

TERRITORY 91

PRIVATE PASSENGER AUTOMOBILES

COMPREHENSIVE-FIRE AND THEFT

	A - G				H - I				J - K				L - M				N - O				P - S				T - Z			
	1	2	3	4	1	2	3	4	1	2	3	4	1	2	3	4	1	2	3	4	1	2	3	4	1	2	3	4
1A 4A 7A	24	20	20	14	29	25	25	17	35	30	30	21	44	37	37	26	55	47	47	33	70	60	60	42	89	76	76	53
2CA 5A 5AA	29	25	25	17	35	30	30	21	42	36	36	25	53	45	45	32	66	56	56	40	84	71	71	50	107	91	91	64
1B 4B 1C 4C 7B 7C	25	21	21	15	30	26	26	18	36	31	31	22	46	39	39	28	57	48	48	34	73	62	62	44	93	79	79	56
2CB 2CC 5B 5BA 5C 5CA	30	26	26	18	36	31	31	22	44	37	37	26	55	47	47	33	69	59	59	41	88	75	75	53	111	94	94	67
1D 4D 3 43 73 73M 7D 3M 43M	26	22	22	16	31	26	26	19	38	32	32	23	48	41	41	29	59	50	50	35	76	65	65	46	96	82	82	58
2C3 2CM 2CD 5D 5DA 53 53A 53M 5MA	31	26	26	19	37	31	31	22	45	38	38	27	57	48	48	34	71	60	60	43	90	77	77	54	115	98	98	69
2A 2AE 2D 2DE	32	27	27	19	38	32	32	23	46	39	39	28	58	49	49	35	73	62	62	44	92	78	78	55	117	99	99	70
2C 2CE	36	31	31	22	43	37	37	26	52	44	44	31	65	55	55	39	81	69	69	49	104	88	88	62	132	112	112	79
ALL CLASSES, F AND T	6	5	5	4	7	6	6	4	8	7	7	5	11	9	9	7	13	11	11	8	17	14	14	10	22	19	19	13

TERRITORY 91

PRIVATE PASSENGER AUTOMOBILES

COLLISION $50 DEDUCTIBLE

	A - G				H - I				J - K				L - M				N - O				P - S				T - Z			
	1	2	3	4	1	2	3	4	1	2	3	4	1	2	3	4	1	2	3	4	1	2	3	4	1	2	3	4
1A 4A	72	62	58	47	89	77	72	58	104	89	84	68	122	105	99	79	143	123	116	93	161	138	130	105	176	151	143	114
1B 1C 4B 4C	73	63	59	47	91	78	74	59	106	91	86	69	124	107	100	81	146	126	118	95	164	141	133	107	180	155	146	117
1D 4D	75	65	61	49	93	80	75	60	108	93	87	70	127	109	103	83	149	128	121	97	167	144	135	109	183	157	148	119
2A	99	85	80	64	123	106	100	80	144	124	117	94	168	144	136	109	197	169	160	128	222	191	180	144	243	209	197	158
2AE	84	72	68	55	103	89	83	67	121	104	98	79	142	122	115	92	166	143	134	108	187	161	151	122	204	175	165	133
2C	150	129	122	98	185	159	150	120	216	186	175	140	254	218	206	165	297	255	241	193	335	288	271	218	366	315	296	238
2CE	142	122	115	92	175	151	142	114	205	176	166	133	240	206	194	156	282	243	228	183	317	273	257	206	347	298	281	226
2D	104	89	84	68	128	110	104	83	150	129	122	98	176	151	143	114	206	177	167	134	232	200	188	151	253	218	205	164
2DE	81	70	66	53	100	86	81	65	116	100	94	75	137	118	111	89	160	138	130	104	180	155	146	117	197	169	160	128
3 43 73M	90	77	73	59	111	95	90	72	130	112	105	85	153	132	124	99	179	154	145	116	201	173	163	131	220	189	178	143
3M 43M	94	81	76	61	116	100	94	75	135	116	109	88	159	137	129	103	186	160	151	121	209	180	169	136	229	197	185	149
5A 5AA	104	89	84	68	128	110	104	83	150	129	122	98	176	151	143	114	206	177	167	134	232	200	188	151	253	218	205	164
2CA	134	115	109	87	166	143	134	108	193	166	156	125	227	195	184	148	266	229	215	173	299	257	242	194	327	281	265	213
5B 5BA 5C 5CA	107	92	87	70	132	114	107	86	154	132	125	100	181	156	147	118	212	182	172	138	238	205	193	155	260	224	211	169
2CB 2CC	138	119	112	90	170	146	138	111	199	171	161	129	233	200	189	151	273	235	221	177	308	265	249	200	336	289	272	218
5D 5DA	109	94	88	71	134	115	109	87	157	135	127	102	184	158	149	120	216	186	175	140	243	209	197	158	266	229	215	173
2CD	140	120	113	91	173	149	140	112	202	174	164	131	237	204	192	154	277	238	224	180	312	268	253	203	341	293	276	222
53 53A	137	118	111	89	169	145	137	110	198	170	160	129	232	200	188	151	272	234	220	177	306	263	248	199	334	287	271	217
2C3	176	151	143	114	217	187	176	141	254	218	206	165	298	256	241	194	349	300	283	227	393	338	318	255	429	369	347	279
53M 5MA	142	122	115	92	175	151	142	114	205	176	166	133	240	206	194	156	282	243	228	183	317	273	257	206	347	298	281	226
2CM	182	157	147	118	225	194	182	146	263	226	213	171	309	266	250	201	362	311	293	235	407	350	330	265	445	383	360	289
7A	70	60	57	46	86	74	70	56	101	87	82	66	118	101	96	77	139	120	113	90	156	134	126	101	171	147	139	111
7B 7C	71	61	58	46	88	76	71	57	103	89	83	67	121	104	98	79	142	122	115	92	159	137	129	103	174	150	141	113
7D	73	63	59	47	90	77	73	59	105	90	85	68	123	106	100	80	144	124	117	94	163	140	132	106	178	153	144	116
73	86	74	70	56	107	92	87	70	125	108	101	81	146	126	118	95	172	148	139	112	193	166	156	125	211	181	171	137
-------NET-------																												
1A 4A	61	53	49	40	76	65	61	49	88	76	71	58	104	89	84	67	122	105	99	79	137	117	111	89	150	128	122	97
1B 1C 4B 4C	62	54	50	40	77	66	63	50	90	77	73	59	105	91	85	69	124	107	100	81	139	120	113	91	153	132	124	99
1D 4D	64	55	52	42	79	68	64	51	92	79	74	60	108	93	88	71	127	109	103	82	142	122	115	93	156	133	126	101
2A	84	72	68	54	105	90	85	68	122	105	99	80	143	122	116	93	167	144	136	109	189	162	153	122	207	178	167	134
2AE	71	61	58	47	88	76	71	57	103	88	83	67	121	104	98	78	141	122	114	92	159	137	128	104	173	149	140	113
2C	128	110	104	83	157	135	128	102	184	158	149	119	216	185	175	140	252	217	205	164	285	245	230	185	311	268	252	202
2CE	121	104	98	78	149	128	121	97	174	150	141	113	204	175	165	133	240	207	194	156	269	232	218	175	295	253	239	192
2D	88	76	71	58	109	94	88	71	128	110	104	83	150	128	122	97	175	150	142	114	197	170	160	128	215	185	174	139
2DE	69	60	56	45	85	73	69	55	99	85	80	64	116	100	94	76	136	117	111	88	153	132	124	99	167	144	136	109
3 43 73M	77	65	62	50	94	81	77	61	111	95	89	72	130	112	105	84	152	131	123	99	171	147	139	111	187	161	151	122
3M 43M	80	69	65	52	99	85	80	64	115	99	93	75	135	116	110	88	158	136	128	103	178	153	144	116	195	167	157	127
5A 5AA	88	76	71	58	109	94	88	71	128	110	104	83	150	128	122	97	175	150	142	114	197	170	160	128	215	185	174	139
2CA	114	98	93	74	141	122	114	92	164	141	133	106	193	166	156	126	226	195	183	147	254	218	206	165	278	239	225	181
5B 5BA 5C 5CA	91	78	74	60	112	97	91	73	131	112	106	85	154	133	125	100	180	155	146	117	202	174	164	132	221	190	179	144
2CB 2CC	117	101	95	77	145	124	117	94	169	145	137	110	198	170	161	128	232	200	188	150	262	225	212	170	286	246	231	185
5D 5DA	93	80	75	60	114	98	93	74	133	115	108	87	156	134	127	102	184	158	149	119	207	178	167	134	226	195	183	147
2CD	119	102	96	77	147	127	119	95	172	148	139	111	201	173	163	131	235	202	190	153	265	228	215	173	290	249	235	189
53 53A	116	100	94	76	144	123	116	94	168	145	136	110	197	170	160	128	231	199	187	150	260	224	211	169	284	244	230	184
2C3	150	128	122	97	184	159	150	120	216	185	175	140	253	218	205	165	297	255	241	193	334	287	270	217	365	314	295	237
53M 5MA	121	104	98	78	149	128	121	97	174	150	141	113	204	175	165	133	240	207	194	156	269	232	218	175	295	253	239	192
2CM	155	133	125	100	191	165	155	124	224	192	181	145	263	226	213	171	308	264	249	200	346	298	281	225	378	326	306	246
7A	60	51	48	39	73	63	60	48	86	74	70	56	100	86	82	65	118	102	96	77	133	114	107	86	145	125	118	94
7B 7C	60	52	49	39	75	65	60	48	88	76	71	57	103	88	83	67	121	104	98	78	135	116	110	88	148	128	120	96
7D	62	54	50	40	77	65	62	50	89	77	72	58	105	90	85	68	122	105	99	80	139	119	112	90	151	130	122	99
73	73	63	60	48	91	78	74	60	106	92	86	69	124	107	100	81	146	126	118	95	164	141	133	106	179	154	145	116

TERRITORY 91

PRIVATE PASSENGER AUTOMOBILES

COLLISION $100 DEDUCTIBLE

	A - G				H - I				J - K				L - M				N - O				P - S				T - Z			
	1	2	3	4	1	2	3	4	1	2	3	4	1	2	3	4	1	2	3	4	1	2	3	4	1	2	3	4
1A 4A	43	37	35	28	62	53	50	40	80	69	65	52	97	83	79	63	115	99	93	75	131	113	106	85	144	124	117	94
1B 1C 4B 4C	44	38	36	29	63	54	51	41	82	71	66	53	99	85	80	64	117	101	95	76	134	115	109	87	147	126	119	96
1D 4D	45	39	36	29	64	55	52	42	83	71	67	54	101	87	82	66	120	103	97	78	136	117	110	88	150	129	122	98
2A	59	51	48	38	86	74	70	56	110	95	89	72	134	115	109	87	159	137	129	103	181	156	147	118	199	171	161	129
2AE	50	43	41	33	72	62	58	47	93	80	75	60	113	97	92	73	133	114	108	86	152	131	123	99	167	144	135	109
2C	89	77	72	58	129	111	104	84	166	143	134	108	202	174	164	131	239	206	194	155	272	234	220	177	300	258	243	195
2CE	85	73	69	55	122	105	99	79	158	136	128	103	191	164	155	124	227	195	184	148	258	222	209	168	284	244	230	185
2D	62	53	50	40	89	77	72	58	115	99	93	75	140	120	113	91	166	143	134	108	189	163	153	123	207	178	168	135
2DE	48	41	39	31	69	59	56	45	90	77	73	59	109	94	88	71	129	111	104	84	147	126	119	96	161	138	130	105
3 43 73M	54	46	44	35	78	67	63	51	100	86	81	65	121	104	98	79	144	124	117	94	164	141	133	107	180	155	146	117
3M 43M	56	48	45	36	81	70	66	53	104	89	84	68	126	108	102	82	150	129	122	98	170	146	138	111	187	161	151	122
5A 5AA	62	53	50	40	89	77	72	58	115	99	93	75	140	120	113	91	166	143	134	108	189	163	153	123	207	178	168	135
2CA	80	69	65	52	115	99	93	75	149	128	121	97	180	155	146	117	214	184	173	139	244	210	198	159	268	230	217	174
5B 5BA 5C 5CA	64	55	52	42	92	79	75	60	118	101	96	77	144	124	117	94	170	146	138	111	194	167	157	126	213	183	173	138
2CB 2CC	82	71	66	53	118	101	96	77	153	132	124	99	185	159	150	120	220	189	178	143	250	215	203	163	275	237	223	179
5D 5DA	65	56	53	42	94	81	76	61	121	104	98	79	146	126	118	95	174	150	141	113	198	170	160	129	217	187	176	141
2CD	83	71	67	54	120	103	97	78	155	133	126	101	188	162	152	122	223	192	181	145	254	218	206	165	279	240	226	181
53 53A	82	71	66	53	118	101	96	77	152	131	123	99	184	158	149	120	219	188	177	142	249	214	202	162	274	236	222	178
2C3	105	90	85	68	151	130	122	98	195	168	158	127	237	204	192	154	281	242	228	183	320	275	259	208	351	302	284	228
53M 5MA	85	73	69	55	122	105	99	79	158	136	128	103	191	164	155	124	227	195	184	148	258	222	209	168	284	244	230	185
2CM	109	94	88	71	157	135	127	102	202	174	164	131	245	211	198	159	291	250	236	189	331	285	268	215	364	313	295	237
7A	42	36	34	27	60	52	49	39	78	67	63	51	94	81	76	61	112	96	91	73	127	109	103	83	140	120	113	91
7B 7C	43	37	35	28	61	52	49	40	79	68	64	51	96	83	78	62	114	98	92	74	130	112	105	85	143	123	116	93
7D	43	37	35	28	63	54	51	41	81	70	66	53	98	84	79	64	116	100	94	75	132	114	107	86	145	125	117	94
73	52	45	42	34	74	64	60	48	96	83	78	62	116	100	94	75	138	119	112	90	157	135	127	102	173	149	140	112
-------NET-------																												
1A 4A	37	31	30	24	53	45	43	34	68	59	55	44	82	71	67	54	98	84	79	64	111	96	90	72	122	105	99	80
1B 1C 4B 4C	37	32	31	25	54	46	43	35	70	60	56	45	84	72	68	54	99	86	81	65	114	98	93	74	125	107	101	82
1D 4D	38	33	31	25	54	47	44	36	71	60	57	46	86	74	70	56	102	88	82	66	116	99	94	75	128	110	104	83
2A	50	43	41	32	73	63	60	48	94	81	76	61	114	98	93	74	135	116	110	88	154	133	125	100	169	145	137	110
2AE	43	37	35	28	61	53	49	40	79	68	64	51	96	82	78	62	113	97	92	73	129	111	105	84	142	122	115	93
2C	76	65	61	49	110	94	88	71	141	122	114	92	172	148	139	111	203	175	165	132	231	199	187	150	255	219	207	166
2CE	72	62	59	47	104	89	84	67	134	116	109	88	162	139	132	105	193	166	156	126	219	189	178	143	241	207	196	157
2D	53	45	43	34	76	65	61	49	98	84	79	64	119	102	96	77	141	122	114	92	161	139	130	105	176	151	143	115
2DE	41	35	33	26	59	50	48	38	77	65	62	50	93	80	75	60	110	94	88	71	125	107	101	82	137	117	111	89
3 43 73M	46	39	37	30	66	57	54	43	85	73	69	55	103	88	83	67	122	105	99	80	139	120	113	91	153	132	124	99
3M 43M	48	41	38	31	69	60	56	45	88	76	71	58	107	92	87	70	128	110	104	83	145	124	117	94	159	137	128	104
5A 5AA	53	45	43	34	76	65	61	49	98	84	79	64	119	102	96	77	141	122	114	92	161	139	130	105	176	151	143	115
2CA	68	59	55	44	98	84	79	64	127	109	103	82	153	132	124	99	182	156	147	118	207	179	168	135	228	196	184	148
5B 5BA 5C 5CA	54	47	44	36	78	67	64	51	100	86	82	65	122	105	99	80	145	124	117	94	165	142	133	107	181	156	147	117
2CB 2CC	70	60	56	45	100	86	82	65	130	112	105	84	157	135	128	102	187	161	151	122	213	183	173	139	234	201	190	152
5D 5DA	55	48	45	36	80	69	65	52	103	88	83	67	124	107	100	81	148	128	120	96	168	145	136	110	184	159	150	120
2CD	71	60	57	46	102	88	82	66	132	113	107	86	160	138	129	104	190	163	154	123	216	185	175	140	237	204	192	154
53 53A	70	60	56	45	100	86	82	65	129	111	105	84	156	134	127	102	186	160	150	121	212	182	172	138	233	201	189	151
2C3	89	77	72	58	128	111	104	83	166	143	134	108	201	173	163	131	239	206	194	156	272	234	220	177	298	257	241	194
53M 5MA	72	62	59	47	104	89	84	67	134	116	109	88	162	139	132	105	193	166	156	126	219	189	178	143	241	207	196	157
2CM	93	80	75	60	133	115	108	87	172	148	139	111	208	179	168	135	247	213	201	161	281	242	228	183	309	266	251	201
7A	36	31	29	23	51	44	42	33	66	57	54	43	80	69	65	52	95	82	77	62	108	93	88	71	119	102	96	77
7B 7C	37	31	30	24	52	44	42	34	67	58	54	43	82	71	66	53	97	83	78	63	111	95	89	72	122	105	99	79
7D	37	31	30	24	54	46	43	35	69	60	56	45	83	71	67	54	99	85	80	64	112	97	91	73	123	106	99	80
73	44	38	36	29	63	54	51	41	82	71	66	53	99	85	80	64	117	101	95	77	133	115	108	87	147	127	119	95

Revised
2-1-73

TERRITORY 91

PRIVATE PASSENGER AND COMMERCIAL VEHICLES

		BODILY INJURY					MEDICAL PAYMENTS			PROPERTY DAMAGE			
		15/30	20/40	25/50	50/100	100/300	1	2	5	5	10	25	50
1A 4A		43	45	46	51	55	13	15	18	19	21	23	24
1B 4B		49	52	54	59	63	15	17	20	22	24	26	28
1C 4C		60	63	65	72	78	15	17	20	27	30	32	34
1D 4D 3 43 73M		75	79	82	90	97	16	18	21	34	37	41	43
2A		110	116	120	131	142	25	27	30	49	54	59	61
2AE		85	90	93	101	110	25	27	30	38	42	46	48
2C		217	229	237	259	281	25	27	30	97	107	116	121
2CE		169	178	184	202	219	25	27	30	76	84	91	95
2D		117	123	127	139	151	25	27	30	52	57	62	65
2DE		73	77	80	87	94	25	27	30	33	36	40	41
3M 43M		80	84	87	95	103	16	18	21	36	40	43	45
5A 5AA		51	54	56	61	66	17	19	22	33	36	40	41
2CA		57	60	62	68	73	19	21	24	29	32	35	36
5B 5BA		58	61	63	69	75	20	22	25	37	41	44	46
2CB		64	68	70	77	83	22	24	27	32	35	38	40
5C 5CA		69	72	75	82	89	20	22	25	44	48	53	55
2CC		76	81	83	91	99	22	24	27	39	43	47	49
5D 5DA 53 53A		84	89	92	100	109	21	23	26	55	61	66	69
2CD 2C3		94	99	102	112	121	23	25	28	47	52	56	59
53M 5MA		88	93	96	105	114	21	23	26	57	63	68	71
2CM		98	104	107	117	127	23	25	28	50	55	60	63
7A		38	40	42	46	49	13	15	18	17	19	20	21
7B		45	47	49	53	58	15	17	20	20	22	24	25
7C		56	59	61	66	72	15	17	20	25	28	30	31
7D 73		72	76	79	86	93	16	18	21	32	35	38	40
1A 4A	NET	37	38	39	43	47	11	13	15	16	18	20	20
1B 4B	NET	42	44	46	50	54	13	14	17	19	20	22	24
1C 4C	NET	51	54	55	61	66	13	14	17	23	26	27	29
1D 4D 3 43 73M	NET	64	67	70	77	82	14	15	18	29	31	35	37
2A	NET	94	99	102	111	121	21	23	26	42	46	50	52
2AE	NET	72	77	79	86	94	21	23	26	32	36	39	41
2C	NET	184	195	201	220	239	21	23	26	82	91	99	103
2CE	NET	144	151	156	172	186	21	23	26	65	71	77	81
2D	NET	99	105	108	118	128	21	23	26	44	48	53	55
2DE	NET	62	65	68	74	80	21	23	26	28	31	34	35
3M 43M	NET	68	71	74	81	88	14	15	18	31	34	37	38
5A 5AA	NET	43	46	48	52	56	14	16	19	28	31	34	35
2CA	NET	48	51	53	58	62	16	18	20	25	27	30	31
5B 5BA	NET	49	52	54	59	64	17	19	21	31	35	37	39
2CB	NET	54	58	60	65	71	19	20	23	27	30	32	34
5C 5CA	NET	59	61	64	70	76	17	19	21	37	41	45	47
2CC	NET	65	69	71	77	84	19	20	23	33	37	40	42
5D 5DA 53 53A	NET	71	76	78	85	93	18	20	22	47	52	56	59
2CD 2C3	NET	80	84	87	95	103	20	21	24	40	44	48	50
53M 5MA	NET	75	79	82	89	97	18	20	22	48	54	58	60
2CM	NET	83	88	91	99	108	20	21	24	43	47	51	54
7A	NET	32	34	36	39	42	11	13	15	14	16	17	18
7B	NET	38	40	42	45	49	13	14	17	17	19	20	21
7C	NET	48	50	52	56	61	13	14	17	21	24	26	26
7D 73	NET	61	65	67	73	79	14	15	18	27	30	32	34
6		47	49	51	56	61	14	16	19	21	23	25	26
8		64	68	70	77	83	15	17	20	29	32	35	36

Uninsured Motorists Premium--$11 each car

CANCELLING POLICIES

Automobile insurance may be cancelled by the insured or insurer. The same rules apply as those used for fire insurance.

If the insurer cancels: The law permits the insurance company to retain no more than the exact prorata amount of the premium for the number of days the policy has been in force, based on a 365-day year.

example: Mr. Wilson insured his automobile for 1 year. After it had been in effect 135 days the insurance company cancelled the policy. If the premium paid was $160.50, (a) how much was retained by the insurance company and (b) how much was refunded to Mr. Wilson?

solution: (a) $\frac{135}{365} \times \$160.50 = \$59.36$
retained by the insurance company

(b) $\$160.50 - \$59.36 = \$101.14$
refund to Mr. Wilson

If the insured cancels: A short-rate table is used to compute the refund and the amount retained by the insurer by most, but not all, carriers.

example: Assume that Mr. Wilson cancelled his policy instead of the company. (a) How much did the insurance company retain and (b) how much was Mr. Wilson's refund?

solution: (a) Refer to the Short-Rate Table, page 192
The charge for 135 days is 47% of the annual premium.
$\$160.50 \times .47 = \$75.44$
retained by insurance company

(b) $\$160.50 - \$75.44 = \$85.06$
refund to Mr. Wilson

EXERCISE 42

1. How much, more or less, does a 23-year-old male driver have to pay compared to one 18 years of age for $50-deductible collision insurance on a car classified as A2? Both drivers are single and are the principal operators.

2. Mr. Ames lost control of his automobile and, as a consequence, caused damage to another person's property to the extent of $500.00. His own car damages amounted to $136.50. How much did his insurance company pay if he carried collision with $50 deductible and $5000 property damage coverage?

3. How much would a person save on coverage for collision, with $100 deductible as compared to $50 deductible, on a car classified as K2 if the owner was a young woman 25 years of age and used the car primarily to drive to work—a distance of 8 miles each way?

4. Compare the cost of automobile insurance for a single man 25 years of age to that for a man of the same age but married if each was the principal operator. Each car was classified as N3 and was used for pleasure only. Insurance coverage included comprehensive, $100-deductible collision, 50/100 bodily injury, and $5000 property damage.

5. Mr. Pepper's annual insurance cost on his automobile was $173.50. If the policy was cancelled 115 days after it was in force, how much was refunded to him (a) if he cancelled the policy or (b) if the insurance company cancelled the policy?

6. Mr. Johnson purchased automobile insurance costing $155.00 on March 10 and cancelled it on September 6 when he sold the car. How much was his refund?

7. Mr. Hayes, 45 years of age, owns a car classified as P4 that he uses for business only. He drives approximately 18,000 miles a year. What is the difference between $50 and $100 deductible collision insurance?

8. Mr. and Mrs. Smith (both more than 65 years of age) own two cars. They use the cars for pleasure only. One car is classified as L2 and the other as N3. What is the total insurance for both cars if they carry comprehensive, $100 deductible collision, 50/100 BI, $2000 MP, and $5000 PD?

LIFE INSURANCE

In general, the purpose of life insurance is to provide financial assistance to the dependents of the insured in the event of his death. The kind of insurance program purchased depends on the income, indebtedness, amount of security desired for dependents, number of dependents, and the age of the insured.

The first life insurance policies on record were written for very short periods of time—6 months or a year. Today, almost any type of life insurance desired may be purchased.

TYPES OF LIFE INSURANCE POLICIES

Most life insurance policies can be classified under four main groups.

Term insurance policies provide protection for the survivors (beneficiaries) of the insured for a limited time only such as 1, 2, 3, 4, 5, or 10 years. It is therefore temporary. If death of the insured occurs during the term, the full amount of insurance carried (face of policy) is paid to the beneficiaries. If the insured is still living when the insurance expires, he is no longer insured. However, before the insurance is terminated, it may, under certain conditions, be converted to another form of life insurance. Term insurance has the advantage of being the least expensive insurance. However, since it is purchased for a limited time only, if the insured wishes to extend the time, he must pay a higher premium to carry the same amount of insurance.

Straight life insurance policies offer permanent insurance protection during the life of the insured. This type of insurance requires payment of a specified premium each year until the death of the insured when the face of the policy is paid to the beneficiary. This insurance has an advantage over term insurance because it can be borrowed against and usually has a cash surrender value after the second year.

Limited-payment life insurance policies require that premiums be paid for a fixed number of years. If the policy is a 20-year payment policy, the insured pays premiums for 20 years or until his death, whichever occurs first. If the insured is living at the end of 20 years, no more payments are made but he is insured for the rest of his life. At his death, the face of the policy is paid to his beneficiary. Policies of this type may also be written to mature at a specified age such as 65. An advantage of this type of coverage is that premiums are paid during a person's most productive years.

Endowment life insurance policies provide some savings in addition to the protection covered by life insurance. It is the most expensive type of life insurance. Such policies are usually purchased for an anticipated need at some future time such as a supplement to retirement income. Like the limited-payment life insurance policy, the endowment policy is written for a specified number of years or to mature at some designated age such as 60 or 65.

In the event of the insured's death, the face of the policy is paid to the beneficiary. However, if the insured lives beyond his endowment, he may choose one of several options when the policy matures. He may elect to receive the face amount of the policy in a lump sum, or as periodic equal payments with any balance remaining at his death paid to his beneficiary or estate. Or the money may become income for minor children to be paid monthly or annually.

Cash Value: Any life insurance policy except a *term* insurance policy has a cash value after 2 or 3 years. At that time or later the policy may be converted to a new one without making additional payments or the policy may be surrendered for its cash value.

COST OF INSURANCE—PREMIUMS

The cost of insurance depends on the type of policy, the amount of coverage desired, and the age and sex of the insured. The type of policy purchased depends, in turn, on the kind of coverage desired that also fits the buyer's budget.

Rates are based on tables of *life expectancy*. Obviously, life expectancy decreases with age which, in turn, increases the cost of insurance. Rates are slightly higher for males since females have a greater life expectancy. Rates also depend on the type of insurance company—participating or nonparticipating. A participating company is one in which the policy holder shares in the earnings of the company by way of yearly dividends. A nonparticipating insurance company does not share its earnings with policy holders. For this reason, net costs, not gross costs, should be the basis for comparison of rates between companies.

The following table is representative of rates of a non-participating life insurance company for the four types of insurance policies discussed. Since females have a longer life expectancy, their rates are slightly lower than those shown. The rates shown below the table for periods other than annual may differ slightly between companies.

ANNUAL PREMIUMS PER $1000 INSURANCE*

Age	5-Year Term	10-Year Term	Straight Life	20-Payment Life	20-Year Endowment
20	4.94	4.95	11.58	19.35	41.47
25	4.99	5.00	13.40	21.95	41.64
30	5.09	5.18	16.02	24.81	42.00
35	5.62	6.08	19.20	28.14	42.70
40	7.07	7.95	23.11	32.14	43.97
45	9.51	11.08	28.05	37.28	46.02
50	13.50	15.91	34.16	42.80	48.62
55	19.41	—	42.46	49.69	—
60	28.98	—	53.76	58.77	—

*Semiannual rate is 51% of annual rate, quarterly rate is 26% of annual rate, and monthly rate is 9% of annual rate.

example: Find the annual premium on a $3000 straight life insurance policy that was purchased at age 30.

solution:

1. Locate 30 under "age" column. Follow this line across to column headed "Straight Life".
2. The number shown, 16.02, is the annual premium on a $1000 policy.
3. Premium for $3000 = 3 × $16.02 = $48.06.

example: How much less would a policy for 20-payment life insurance cost than a 20-year endowment at age 45?

solution: annual premium for 20-year endowment = \$46.02
annual premium for 20-payment life = 37.28
difference = \$ 8.74

example: How much more are the premiums for a year on a \$1000 20-year endowment at age 25 if paid quarterly instead of annually?

solution: Rates for semiannual, quarterly, and monthly premium payments are listed under the premium tables.

annual premium = \$41.64
quarterly premium = \$41.64 × .26 = \$10.83

premium for 1 year on quarterly basis = \$10.83 × 4 = \$43.32

amount saved if paid annually = \$43.32 \$41.64 = \$1.68

EXERCISE 43

1. Which is the least expensive policy for a person 40 years of age for 10 years? (a) How much would the insured pay in premiums during that time, based on \$1000? (b) How much more would a straight life insurance policy cost for the same period of time?

2. Find the annual premium on a \$5000 20-payment life insurance policy that is issued at age 20? At 30?

3. How much less does a 20-year-old person pay than one 40 years of age for a 20-payment life policy? For a 20-year endowment policy?

4. J. Hampton, age 25, wished to buy a 20-pay life insurance policy for \$4000. What would his premium payments be for a year if they are made quarterly? Semiannually? Annually?

5. Compare the total premiums paid on a \$10,000 life insurance policy for a 10-year term with straight life for 10 years at (a) age 20 and (b) age 40.

6. Mr. Ecklund purchased a 20-payment life policy for $5000 at age 35. He died 10 years later. How much more did the insurance company pay the beneficiaries than he had paid in premiums?

7. Harold Mays, at age 20, bought a 20-payment life insurance policy for $5000. If he was living at the end of that time, how much did he pay in premiums? How much less was the amount paid in premiums than the face of the policy?

8. In problem 7, if Mr. Mays was 60 years of age, how much more would he have paid in premiums over the face of the policy? Why did he pay more when in problem 7 he paid less than the face of the policy?

9. Mr. Mason, 40 years old, wishes to have additional insurance protection for his daughter, who is 15 years of age, for 10 years in order to cover her education expenses in the event of his death. What would term insurance cost on $5000 for that time if he paid premiums annually? Semiannually?

10. If Mr. Mason, in problem 9, died after 5 years, how much more would the insurance company pay his daughter than he had paid in premiums on an annual basis?

EXERCISE 44 Chapter Summary

1. Find the premium for each of the following fire insurance policies.

	Amount of Insurance	Rate	Terms (Years)
(a)	$16,500	$0.72 per $100	1
(b)	32,000	0.55 per $1000	3
(c)	154,500	0.96 per $100	1
(d)	94,000	0.24 per $100	2
(e)	540,000	0.19 per $100	3

2. Moore & Sons insured an apartment building against fire for $350,000. It was distributed as follows: Company A carried $\frac{1}{4}$ of this amount at $0.21 per $100, Company B carried $\frac{1}{5}$ of the remainder at $0.19 per $100, and Company C carried the balance at

$0.175 per $100. What was the premium paid to each company and what was the total premium cost for 1 year?

3. If, in problem 2, a fire caused damages amounting to 10% of the total insurance carried, what was the payment made by each carrier? Assume that there were no coinsurance requirements.

4. Find the payment made by the insurance company for each of the following.

	Value of Property	Amount of Insurance	Fire Loss	Coinsurance Clause
(a)	$215,000	$150,000	$86,000	80%
(b)	80,000	70,000	72,000	90%
(c)	67,500	50,000	7,100	100%
(d)	34,000	25,000	13,600	80%

5. Johnson Bros. insured the contents of its warehouse against fire for 60 days. If the annual premium was $32.50, what was the net cost of the insurance?

6. A store was insured against fire loss for $350,000 at $0.95 per $100; the contents were insured for $500,000 at $0.86 per $100. The effective date of the policy was January 3. On July 6 of the same year the store owners cancelled the policy. Find the refund paid by the insurance company. (Assume that this was not a leap year.)

7. Mr. Hopkins purchased fire insurance on his home on October 14 which he cancelled on December 19 of the same year. The policy was written for $22,000 at $0.12 per $100. What was the premium charge for the period that the policy was in effect?

8. Mr. Hayley insured his shop, valued at $40,000, for $\frac{4}{5}$ of its value. How much would the insurance company pay if damages resulting from a fire amounted to $33,000? To $22,500? (No coinsurance clause was included.)

9. The Fox Realty Co. carries $20,000 fire insurance on a piece of property worth $30,000. The policy contains an 80% coinsurance clause. How much would the insurance company pay for loss from fire amounting to $12,000?

10. If the annual premium on a fire insurance policy was $40.00, how much of the premium was returned to the insured if: (a) the insurance was cancelled by the policy holder after 90 days or (b) the insurance was cancelled by the carrier after 90 days?

11. Mr. Rowan is 35 years of age and married. He drives his automobile to work, a distance of 5 miles each way. He carries comprehensive, $100-deductible collision, 100/300 bodily injury, $5000 medical, $25,000 property damage, and liability insurance on the uninsured motorist. If his car is in class F4, how much does his insurance cost?

12. An automobile policy with an annual premium of $155.00 is cancelled by the insured 105 days after it had been in force. How much was refunded to the insured?

13. After 96 days the carrier cancelled an automobile insurance policy costing $178.00. How much did the carrier retain?

14. John Hope, married and age 30, uses his automobile to drive to work, a distance of 10 miles each way. He carries comprehensive, $100-deductible collision, 50/100 bodily injury, $2000 medical, and $5000 property damage. If his car is classed as G3, how much does his insurance cost?

15. Mr. Nelson, age 65, uses his automobile, class J4, for pleasure only. He carries comprehensive but no collision, 50/100 bodily injury, $2000 medical, and $5000 property damage. He also carries liability insurance for the uninsured motorist. What are the charges for each coverage and the total cost?

16. Find the premium on an automobile, class N2, used for pleasure only by a 25-year-old married woman who carries comprehensive, $50-deductible collision, 50/100 bodily injury, $2000 medical, and $5000 property damage if the car is the second car in the family and qualifies for the net premium cost?

17. Find the annual premium on a $5000 life insurance policy for ages 25, 40, and 60 for: (a) 5-year term, (b) straight life, and (c) 20-year endowment.

18. How much more would a man 35 years of age pay on a 20-pay life policy if premiums were paid (a) quarterly, (b) monthly, or (c) semiannually than if paid annually?

19. B. Burns, who is 20, carries a 10-year term life insurance policy for \$4000. If he is living at the end of that time, how much would he have paid in annual premiums during that time?

20. Compare the cost of a 20-payment life insurance policy for \$5000 for a young man 25 years of age with one 45 years of age. How much in total premiums would each pay as compared to the face of the policy if they lived for the term of the policy?

Objectives

The study of interest—both simple and compound—what it is, how it is calculated, with some simple applications, is the primary purpose of this chapter. The various methods of computing simple interest are included as well as the manner in which maturity value, principal, rate, or time are derived when required. Compound interest is also covered in detail with ample opportunity to practice and use the tables.

10 INTEREST

Interest is the price paid for the use of money. The amount paid depends on the amount borrowed or loaned, the rate charged (expressed as a percent), and the time for which the money is used. Time may be expressed in days, months, or years. The rate (percent) is the charge per annum unless otherwise stated.

Money is borrowed or loaned for many reasons. For example, an individual may wish to establish credit, to take advantage of a desirable market price on a certain commodity, to meet unexpected expenses, or to purchase a car, furniture, or home. Although the borrower may have adequate funds, it may be to his advantage to borrow rather than withdraw funds from his savings account. Similarly, a business concern may borrow against future income to meet current expenses, to expand facilities, to meet emergencies, to take advantage of cash discounts, or for some other purpose.

Banks and many other financial agencies are in the business of lending money. The rates they charge are dependent on many factors. The supply of money, legal restrictions, credit rating of the borrower, type of security, and purpose of the loan all affect the interest rate.

Simple interest is the amount paid on a sum (borrowed or loaned) which remains unchanged for a specified period of time. *Bank discount* is interest paid in advance and is based on the maturity value of a loan rather than the principal. It will be discussed in subsequent chapters. *Compound interest* is computed on a sum which is increased periodically as interest is earned and added to it.

SIMPLE INTEREST

ORDINARY AND ACCURATE INTEREST

Although there are four methods of calculating simple interest when time is expressed in days, only three are in common usage. They are identified as follows and will be referred to in this manner throughout the text.

1. *Ordinary simple interest* is based on a 360-day year and a 30-day month. This method is used most often on installment loans for purchases such as real estate or on other loans which are repaid periodically.

2. *Banker's* or *commercial interest* is based on a 360-day year and exact number of days. The use of this method results in a greater return to the creditor (lending company, agency, or individual) and is, therefore, favored. When no statement is made in a loan agreement, this method generally applies.

3. *Accurate interest*† is based on a 365-day year and exact number of days. This method is always used by the federal government. Also, some banks and other lending agencies use this method, depending on their policy as applied to certain types of loans.

†Accurate interest is sometimes referred to as "exact interest." Only the term "accurate interest" will be used in this text.

4. The fourth method is based on a 365-day year and a 30-day month. This method is never used except by agreement by all parties concerned.

As can be observed, it is necessary to determine the number of days from the time a loan is initiated until the maturity date (date loan is due and payable) in order to determine the interest charge. Whether a 30-day month or the exact number of days is used depends on the method used.

CALCULATING TIME

The 30-Day-per-Month Time Basis

One of two methods may be used to determine the number of days on this basis (sometimes referred to as approximate time). These are illustrated below.

METHODS OF DETERMINING NUMBER OF DAYS

Example	Method I	Method II
Find the time from June 6 to Dec. 15	June 6–Dec. 6 (6 months) = 180 days Dec. 6–Dec. 15 = +9 days Time = 189 days	Month / Day 12 / 15 −6 / −6 6 / 9 180 days + 9 days = 189 days
Find the time from Apr. 20 to Sept. 4	Apr. 20–Sept. 20 (5 months) = 150 days Sept. 4–Sept. 20 = −16 days Time = 134 days	Month / Day 8 / 34* ~~9~~ / ~~4~~ −4 / −20 4 / 14 120 days + 14 days = 134 days
Find the time from Mar. 3, 1974 to Feb. 16, 1975	Mar. 3, 1974–Feb. 3, 1975 (11 months) = 330 days Feb. 3, 1975–Feb. 16, 1975 = +13 days Time = 343 days	Year / Month / Day 9 / 14 / 197~~5~~ / ~~2~~** / 16 −1974 / −3 / −3 / 11 / 13 330 days + 13 days = 343 days

*Since 20 cannot be subtracted from 4, borrow a month (30 days) and proceed as shown.
**Since 3 cannot be subtracted from 2, a year or 12 months was borrowed from 1975.

Exact Time

In determining the exact time (number of days) during the life of a loan, it is necessary to count each day, excluding the first but including the last day. If the month of February is included in the time period and it is a leap year, add an extra day.

example: Find the exact number of days from April 13 to October 10.

solution:

Month	Number of Days
April (13 to 30)	17
May	31
June	30
July	31
August	31
September	30
October	10
Time (days)	180

The method of counting days as illustrated above is awkward and time consuming. Consequently, as in all other types of computational work, tables have been developed when possible to simplify the problem. The following table is one of several that are used to find the number of days between two dates.

EXERCISE 45

Find the approximate time and exact time for the following. Use any method.

1. June 12, 1973 to September 11, 1973

2. February 3, 1976 to June 10, 1976

3. January 31, 1974 to January 1, 1975

4. September 10, 1975 to March 3, 1976

5. July 3, 1975 to December 21, 1975

6. September 17, 1976 to June 9, 1977

7. April 1, 1975 to April 12, 1976

8. May 5, 1972 to October 19, 1972

9. March 29, 1975 to July 27, 1975

10. February 16, 1974 to August 15, 1974

THE NUMBER OF EACH DAY OF THE YEAR

Day of Month	Jan.	Feb.*	Mar.	Apr.	May	June	July	Aug.	Sept.	Oct.	Nov.	Dec.	Day of Month
1	1	32	60	91	121	152	182	213	244	274	305	335	1
2	2	33	61	92	122	153	183	214	245	275	306	336	2
3	3	34	62	93	123	154	184	215	246	276	307	337	3
4	4	35	63	94	124	155	185	216	247	277	308	338	4
5	5	36	64	95	125	156	186	217	248	278	309	339	5
6	6	37	65	96	126	157	187	218	249	279	310	340	6
7	7	38	66	97	127	158	188	219	250	280	311	341	7
8	8	39	67	98	128	159	189	220	251	281	312	342	8
9	9	40	68	99	129	160	190	221	252	282	313	343	9
10	10	41	69	100	130	161	191	222	253	283	314	344	10
11	11	42	70	101	131	162	192	223	254	284	315	345	11
12	12	43	71	102	132	163	193	224	255	285	316	346	12
13	13	44	72	103	133	164	194	225	256	286	317	347	13
14	14	45	73	104	134	165	195	226	257	287	318	348	14
15	15	46	74	105	135	166	196	227	258	288	319	349	15
16	16	47	75	106	136	167	197	228	259	289	320	350	16
17	17	48	76	107	137	168	198	229	260	290	321	351	17
18	18	49	77	108	138	169	199	230	261	291	322	352	18
19	19	50	78	109	139	170	200	231	262	292	323	353	19
20	20	51	79	110	140	171	201	232	263	293	324	354	20
21	21	52	80	111	141	172	202	233	264	294	325	355	21
22	22	53	81	112	142	173	203	234	265	295	326	356	22
23	23	54	82	113	143	174	204	235	266	296	327	357	23
24	24	55	83	114	144	175	205	236	267	297	328	358	24
25	25	56	84	115	145	176	206	237	268	298	329	359	25
26	26	57	85	116	146	177	207	238	269	299	330	360	26
27	27	58	86	117	147	178	208	239	270	300	331	361	27
28	28	59	87	118	148	179	209	240	271	301	332	362	28
29	29		88	119	149	180	210	241	272	302	333	363	29
30	30		89	120	150	181	211	242	273	303	334	364	30
31	31		90		151		212	243		304		365	31

*Add 1 for leap years after February 28.

CALCULATING INTEREST

Simple Interest Formula

$$\text{Interest} = \text{Principal} \times \text{Rate} \times \text{Time}$$

Expressed symbolically,

$$I = Prt$$

where:

I = interest in dollars
P = principal (dollar amount borrowed or loaned)
r = interest rate expressed as a percent
t = time (period during which the borrower used the principal)

The unit of time must correspond to the rate. Therefore, if the rate is understood to be on an annual basis, then the time must be expressed in terms of a year. Rates are quoted as annual rates unless otherwise stated.

example: When time is expressed in *years*: Find the interest on \$300.00 for $2\frac{1}{2}$ years at 5%.

solution: $I = Prt = \$300.00 \times .05 \times 2.5 = \$37.50$

example: When time is expressed in *months*, it must be converted to years, i.e., 7 months = $\frac{7}{12}$ years, 5 months = $\frac{5}{12}$ years, etc. Find the interest on \$500.00 at 4% for 3 months.

solution: $I = \$500.00 \times .04 \times \frac{3}{12} = \$5.00$

example: When time is expressed in *days*, it must be converted to years. Find the interest on \$3000.00 at 3% for 90 days.

solution: If a 360-day year is used (ordinary or commercial interest), then

$$I = \$3000.00 \times .03 \times \frac{90}{360} = \$22.50$$

If a 365-day year is used (accurate interest), then

$$I = \$3000.00 \times .03 \times \frac{90}{365} = \$22.19$$

Simple Interest Formula and Cancellation Method

The formula $I = Prt$ may also be written

$$I = \frac{P \times r \times \text{days}}{360 \text{ or } 365}$$

and the so-called cancellation method is used.

example: Find the interest on \$350.00 at 4.5% for 72 days on a 360-day basis.

solution: In this example the numerator and denominator are divided by common factors to simplify the calculations.

$$I = \frac{35\cancel{0} \times .045 \times \overset{2}{\cancel{72}}}{\cancel{360}} = \$3.15$$

example: Find the *ordinary*, *banker's*, and *accurate* interest on \$1200 at 6% for a loan dated May 16 and due August 16 of the same year.

solution: *Ordinary* interest: $P = \$1200$, $r = .06$, and days (approximate time) $= 90$.

$$I = \frac{\overset{300}{\cancel{1200}} \times .06 \times \cancel{90}}{\underset{4}{\cancel{360}}} = \$18.00$$

Banker's or commercial interest: $P = \$1200$, $r = .06$, and days (exact time) $= 92$.

$$I = \frac{\overset{20}{\cancel{1200}} \times \overset{.01}{\cancel{.06}} \times 92}{\underset{\cancel{60}}{\cancel{360}}} = \$18.40$$

Accurate interest: $P = \$1200$, $r = .06$, and days (exact time) $= 92$.

$$I = \frac{\overset{240}{\cancel{1200}} \times .06 \times 92}{\underset{73}{\cancel{365}}} = \$18.15$$

EXERCISE 46

I. Find the simple interest on the following:

1. $500 for 5 months at 6%
2. $455 for 3 months at $4\frac{1}{2}\%$
3. $600 at $3\frac{1}{2}\%$ for 3 years
4. $4800 at 4% for $4\frac{1}{2}$ years
5. $400 at 4% for a year
6. $1500 at 5% for 19 months
7. $1225 at $4\frac{1}{2}\%$ for 3 months
8. $4000 at 6% for 8 months
9. $1400 at 7% for 15 months
10. $250 at $6\frac{1}{2}\%$ for 2 years and 3 months

II. Find the *ordinary simple* interest (360-day year, 30-day month) on the following:

1. $642 for 60 days at 7%
2. $2000 for 62 days at $3\frac{1}{2}\%$
3. $2400 at $2\frac{3}{4}\%$ for 64 days
4. $720 at $5\frac{1}{2}\%$ for 84 days
5. $500 at 4% for 60 days
6. $750 from June 10 to August 17 at 3%
7. $1200 at 3% from May 12 to July 2
8. $800 at $3\frac{1}{2}\%$ from November 14 to February 12
9. $1640 at 7% from July 19 to September 12
10. $900 at 5% from March 16 to May 21

III. Find the *banker's* or *commercial* interest (360-day year, exact days) on problems 6 to 10 in Section II.

IV. Find the *accurate* interest (365-day year, exact number of days) on the following:

1. $3000 at 3% for 90 days
2. $300 at 5% for 1 year and 16 days
3. $3500 at 4% from January 2 to June 5
4. $7000 at $3\frac{1}{2}\%$ for 72 days
5. $600 at 6% from March 7 to September 25
6. $1200 at $5\frac{1}{2}\%$ from May 12 to October 3

The 60-Day, 6% Method of Calculating Ordinary or Banker's Interest

When a loan is made for 60 days at 6%, the interest is always 1% of the principal. This is due to the fact that $.06 \times 60/360 = .01$.

example: Find the ordinary interest on $575.00 at 6% for 60 days.

solution: $I = \$575.00 \times (.06 \times \frac{60}{360})$
$= \$575.00 \times .01 = \$5.75$

Therefore, the interest at 6% for 60 days on:

$ 326.50	=	$ 3.27
4050.00	=	40.50
39.50	=	.40
127.65	=	1.28

This basic relationship may be used for periods of time that are factors or multiples of 60 days when the rate is 6% or can be broken down into multiples or factors of 60.

When a loan at 6% is made for time that is a multiple or factor of 60: For example, the interest on $500.00 at 6% for:

60 days	=	$ 5.00	(1% of $500)
120 days	=	10.00	(2 × interest for 60 days)
180 days	=	15.00	(3 × interest for 60 days)
90 days	=	7.50	($1\frac{1}{2}$ × interest for 60 days)
40 days	=	3.33	($\frac{2}{3}$ × interest for 60 days)
20 days	=	1.67	($\frac{1}{3}$ × interest for 60 days)
10 days	=	.83	($\frac{1}{6}$ × interest for 60 days)

When a loan at 6% is made for time that may be broken down into multiples or factors of 60: For example, the interest on $1250 at 6% for 92 days can be found as follows:

92 days = 60 + 30 + 2

interest for 60 days	=	$12.50	
interest for 30 days	=	6.25	($\frac{1}{2}$ of $12.50)
interest for 2 days	=	0.42†	($\frac{1}{30}$ of $12.50)
total	=	$19.17	

The interest for 47 days on $600 at 6% can be found as follows:

47 days = 30 + 20 − 3
interest for 60 days = $6.00

interest for 30 days	=	$3.00	($\frac{1}{2}$ of $6.00)
interest for 20 days	=	+2.00	($\frac{1}{3}$ of $6.00)
interest for 3 days	=	−0.30	($\frac{1}{10}$ of 30 days' interest)
total	=	$4.70	

†The interest for 20 days = $\frac{1}{3}$ of $12.50 or $4.19; the interest for 2 days = $\frac{1}{10}$ of this amount or $0.42.

When the interest rate is a multiple or factor of 6% and the time is 60 days: For example, the interest on $800.00 for 60 days at:

$$\begin{aligned} 6\% &= \$8.00 \\ 3\% &= 4.00 \quad (\tfrac{1}{2} \text{ of } 6\%) \\ 2\% &= 2.67 \quad (\tfrac{1}{3} \text{ of } 6\%) \\ 9\% &= 12.00 \quad (1\tfrac{1}{2} \times 6\%) \\ 12\% &= 16.00 \quad (2 \times 6\%) \end{aligned}$$

This principle may be expanded to include rates such as 5%, in which case find the interest for 60 days at 6% and then subtract 1% for 60 days ($\frac{1}{6}$ of 6%). If the rate is 8%, find the interest for 60 days at 6% and then add 2% for 60 days ($\frac{1}{3}$ of 6%), and so forth.

These shortcuts are useful for simple problems. However, they need to be practiced if results are to be dependable. Otherwise, the time and effort to determine the best number combinations for time are not justified and it would be easier to use the interest formula, $I = Prt$.

EXERCISE 47

Find the *ordinary simple* interest for each of the following loans by the 60-day, 6% method.

1. $450 at 6% for 90 days
2. $500 at 3% for 90 days
3. $5400 at 2% for 60 days
4. $1800 at 5% for 120 days
5. $475 at 6% for 26 days
6. $860 at 9% for 30 days
7. $720 at 12% for 90 days
8. $96 at 6% for 126 days
9. $350 at 6% for 27 days
10. $2000 at 3% for 60 days
11. $5500 at 2% for 30 days
12. $198 at 4% for 30 days
13. $1650 at 6% for 180 days
14. $2600 at 3% for 20 days
15. $950 at 5% for 60 days
16. $375 at 4% for 12 days
17. $300 at 6% for 58 days
18. $150 at 6% for 89 days
19. $360 at 6% for 31 days
20. $500 at 3% for 120 days

Simple Interest Tables

The use of tables greatly simplifies the computational work involved in determining interest charges. There are several commercial tables available for this purpose. Those shown here are representative. They list the interest charges per $100 (to 4 decimals) for any number of days, at rates from 3% to 8%, based on a 360-day year and a 365-day year. Time not listed may be obtained by combining time periods shown that add to the time desired or when subtracted one from the other will result in the time desired. This applies to rates as well.

Using the 360-day year table (page 234):

example: Find the interest on $560.00 for 23 days at 4%.

solution: In column headed "Time," find 23 days. In line with 23 days find the number in column headed 4%, i.e., 0.2556.

interest on $100.00
at 4% for 23 days = $0.2556 (from table)
interest on $560.00
at 4% for 23 days = $5.60 × $0.2556
= $1.43136 *or* $1.43

example: Find the interest on $6000.00 at $7\frac{1}{2}$% for 75 days.

solution: Since 75 days is not listed in the tables, any combination of time that results in 75 days may be used, i.e., 2 months + 15 days.

interest on $100.00
at $7\frac{1}{2}$% for 2 months = $1.2500
interest on $100.00
at $7\frac{1}{2}$% for 15 days = +0.3125
interest on $100.00
at $7\frac{1}{2}$% for 75 days = $1.5625

interest on $6000.00
at $7\frac{1}{2}$% for 75 days = 60 × $1.5625
= $93.75

example: Find the interest on \$400.00 at 2% for 12 days.

solution: Since 2% is not included in the table, any combination of rates that result in 2% may be used such as 6 − 4, or 8 − 6.

interest on \$100.00
at 6% for 12 days = \$0.2000
interest on \$100.00
at 4% for 12 days = −0.1333
interest on \$100.00
at 2% for 12 days = \$0.0667

interest on \$400.00
at 2% for 12 days = 4 × \$0.0667
= \$0.2668 *or* \$0.27

Using the 365-day year table (page 235):

example: Find the interest on \$1300.00 at 6% for 19 days.

solution: In column headed "Time," find 19 days. In line with 19 days find the number in column headed 6%, i.e., 0.3123.

interest on \$100.00
at 6% for 19 days = \$0.3123
interest on \$1300.00
at 6% for 19 days = 13 × \$0.3123
= \$4.0599 *or* \$4.06

example: Find interest on \$2575.00 at $5\frac{1}{2}$% for 116 days.

solution: The 116 days is not listed in the table. However, any combination of time periods that equal 116 may be used such as 90 + 26, or 120 − 4.

interest on \$100.00
at $5\frac{1}{2}$% for 90 days = \$1.3562
interest on \$100.00
at $5\frac{1}{2}$% for 26 days = +0.3918
interest on \$100.00
at $5\frac{1}{2}$% for 116 days = \$1.7480

interest on \$2575.00
at $5\frac{1}{2}$% for 116 days = 25.75 × \$1.7480
= \$45.011000
= \$45.01

example: Find the interest on \$300.00 at 10% for 18 days.

solution: Use any combination of rates that will result in a rate of 10% such as 5 + 5, 7 + 3, or 6 + 4.

interest on \$100.00
at 5% for 18 days = \$0.2466
interest on \$100.00
at 10% for 18 days = 2 × \$0.2466
= \$0.4932
interest on \$300.00
at 10% for 18 days = 3 × \$0.4932
= \$1.4796 *or* \$1.48

EXERCISE 48

I. Find the amount of simple interest on each of the following loans by using (a) the 360-day table (page 234) for problems 1 to 5 inclusive and (b) the 365-day table (page 235) for problems 6 to 10 inclusive. (c) Verify problems 2, 5, and 8 by the interest formula.

No.	Principal	Rate (%)	Time (days)
1.	\$460	4	50
2.	100	5	22
3.	300	$4\frac{1}{2}$	36
4.	718	8	28
5.	1000	6	19
6.	850	7	65
7.	2050	5	50
8.	1000	9	26
9.	1460	$6\frac{1}{2}$	110
10.	730	6	70

SIMPLE INTEREST TABLE — Interest on $100, Based on a 360-Day Year

Time (Days)	3%	3½%	4%	4½%	5%	5½%	6%	6½%	7%	7½%	8%
1	.0083	.0097	.0111	.0125	.0139	.0153	.0167	.0181	.0194	.0208	.0222
2	.0167	.0194	.0222	.0250	.0278	.0306	.0333	.0361	.0389	.0417	.0444
3	.0250	.0292	.0333	.0375	.0417	.0458	.0500	.0542	.0583	.0625	.0667
4	.0333	.0389	.0444	.0500	.0556	.0611	.0667	.0722	.0778	.0833	.0889
5	.0417	.0486	.0556	.0625	.0694	.0764	.0833	.0903	.0972	.1042	.1111
6	.0500	.0583	.0667	.0750	.0833	.0917	.1000	.1083	.1167	.1250	.1333
7	.0583	.0681	.0778	.0875	.0972	.1069	.1167	.1264	.1361	.1458	.1556
8	.0667	.0778	.0889	.1000	.1111	.1222	.1333	.1444	.1556	.1667	.1778
9	.0750	.0875	.1000	.1125	.1250	.1375	.1500	.1625	.1750	.1875	.2000
10	.0833	.0972	.1111	.1250	.1389	.1528	.1667	.1806	.1944	.2083	.2222
11	.0917	.1069	.1222	.1375	.1528	.1681	.1833	.1986	.2139	.2197	.2444
12	.1000	.1167	.1333	.1500	.1667	.1833	.2000	.2167	.2333	.2500	.2667
13	.1083	.1264	.1444	.1625	.1806	.1986	.2167	.2347	.2528	.2708	.2889
14	.1167	.1361	.1556	.1750	.1944	.2139	.2333	.2528	.2722	.2917	.3111
15	.1250	.1458	.1667	.1875	.2083	.2292	.2500	.2708	.2917	.3125	.3333
16	.1333	.1556	.1778	.2000	.2222	.2444	.2667	.2889	.3111	.3333	.3556
17	.1417	.1653	.1889	.2125	.2361	.2597	.2833	.3069	.3306	.3542	.3778
18	.1500	.1750	.2000	.2250	.2500	.2750	.3000	.3250	.3500	.3750	.4000
19	.1583	.1847	.2111	.2375	.2639	.2903	.3167	.3431	.3694	.3958	.4222
20	.1667	.1944	.2222	.2500	.2778	.3056	.3333	.3611	.3889	.4167	.4444
21	.1750	.2042	.2333	.2625	.2917	.3208	.3500	.3792	.4083	.4375	.4667
22	.1833	.2139	.2444	.2750	.3056	.3361	.3667	.3972	.4278	.4533	.4889
23	.1917	.2236	.2556	.2875	.3194	.3514	.3833	.4153	.4472	.4792	.5111
24	.2000	.2333	.2667	.3000	.3333	.3667	.4000	.4333	.4667	.5000	.5333
25	.2083	.2431	.2778	.3125	.3472	.3819	.4167	.4514	.4861	.5208	.5556
26	.2167	.2528	.2889	.3250	.3611	.3972	.4333	.4694	.5056	.5417	.5778
27	.2250	.2625	.3000	.3375	.3750	.4125	.4500	.4875	.5250	.5625	.6000
28	.2333	.2722	.3111	.3500	.3889	.4278	.4667	.5056	.5444	.5833	.6222
29	.2417	.2819	.3222	.3625	.4028	.4431	.4833	.5236	.5639	.6042	.6444
(Months)											
1	.2500	.2917	.3333	.3750	.4167	.4583	.5000	.5417	.5833	.6250	.6667
2	.5000	.5833	.6667	.7500	.8333	.9167	1.0000	1.0833	1.1667	1.2500	1.3333
3	.7500	.8750	1.0000	1.1250	1.2500	1.3750	1.5000	1.6250	1.7500	1.8750	2.0000
4	1.0000	1.1667	1.3333	1.5000	1.6667	1.8333	2.0000	2.1667	2.3333	2.5000	2.6667
5	1.2500	1.4583	1.6667	1.8750	2.0833	2.2917	2.5000	2.7083	2.9167	3.1250	3.3333
6	1.5000	1.7500	2.0000	2.2500	2.5000	2.7500	3.0000	3.2500	3.5000	3.7500	4.0000

SIMPLE INTEREST TABLE *continued* Interest on $100, Based on a 365-Day Year

Time (Days)	3%	3½%	4%	4½%	5%	5½%	6%	6½%	7%	7½%	8%
1	.0082	.0096	.0110	.0123	.0137	.0151	.0164	.0178	.0192	.0205	.0219
2	.0164	.0192	.0219	.0247	.0274	.0301	.0329	.0356	.0384	.0411	.0438
3	.0247	.0288	.0329	.0370	.0411	.0452	.0493	.0534	.0575	.0616	.0658
4	.0329	.0384	.0438	.0493	.0548	.0603	.0658	.0712	.0767	.0822	.0877
5	.0411	.0480	.0548	.0616	.0685	.0753	.0822	.0890	.0959	.1027	.1096
6	.0493	.0575	.0658	.0740	.0822	.0904	.0986	.1068	.1151	.1233	.1315
7	.0575	.0671	.0767	.0863	.0959	.1055	.1151	.1247	.1342	.1438	.1534
8	.0658	.0767	.0877	.0986	.1096	.1205	.1315	.1425	.1534	.1654	.1753
9	.0740	.0863	.0986	.1110	.1233	.1356	.1479	.1603	.1726	.1849	.1973
10	.0822	.0959	.1096	.1233	.1370	.1507	.1644	.1781	.1918	.2055	.2192
11	.0904	.1055	.1205	.1356	.1507	.1658	.1808	.1959	.2110	.2260	.2411
12	.0986	.1151	.1315	.1479	.1644	.1808	.1973	.2137	.2301	.2466	.2630
13	.1068	.1247	.1425	.1603	.1781	.1959	.2137	.2315	.2493	.2671	.2849
14	.1151	.1342	.1534	.1726	.1918	.2110	.2301	.2493	.2685	.2877	.3068
15	.1233	.1438	.1644	.1849	.2055	.2260	.2466	.2671	.2877	.3082	.3288
16	.1315	.1534	.1753	.1973	.2192	.2411	.2630	.2849	.3068	.3288	.3507
17	.1397	.1630	.1863	.2096	.2329	.2562	.2795	.3027	.3260	.3493	.3726
18	.1479	.1726	.1973	.2219	.2466	.2712	.2959	.3205	.3452	.3699	.3945
19	.1562	.1822	.2082	.2342	.2603	.2863	.3123	.3384	.3644	.3904	.4164
20	.1644	.1918	.2192	.2466	.2740	.3014	.3288	.3562	.3836	.4110	.4384
21	.1726	.2014	.2301	.2589	.2817	.3164	.3452	.3740	.4027	.4315	.4603
22	.1808	.2110	.2411	.2712	.3014	.3315	.3616	.3918	.4219	.4521	.4822
23	.1890	.2205	.2521	.2836	.3151	.3466	.3781	.4096	.4411	.4726	.5041
24	.1973	.2301	.2630	.2959	.3288	.3616	.3945	.4274	.4603	.4932	.5260
25	.2055	.2397	.2740	.3082	.3425	.3767	.4110	.4452	.4795	.5137	.5479
26	.2137	.2493	.2849	.3205	.3562	.3918	.4274	.4630	.4986	.5342	.5699
27	.2219	.2589	.2959	.3329	.3699	.4068	.4438	.4808	.5178	.5548	.5918
28	.2301	.2685	.3068	.3452	.3836	.4219	.4603	.4986	.5370	.5753	.6137
29	.2384	.2781	.3178	.3575	.3973	.4370	.4767	.5164	.5562	.5959	.6356
30	.2466	.2877	.3288	.3699	.4110	.4521	.4932	.5342	.5753	.6164	.6575
31	.2548	.2973	.3397	.3822	.4247	.4671	.5096	.5521	.5945	.6370	.6795
40	.3288	.3836	.4384	.4932	.5479	.6027	.6575	.7123	.7671	.8219	.8767
50	.4110	.4795	.5479	.6164	.6849	.7534	.8219	.8904	.9589	1.0274	1.0959
60	.4932	.5753	.6575	.7397	.8219	.9041	.9863	1.0685	1.1507	1.2329	1.3151
70	.5753	.6712	.7671	.8630	.9589	1.0548	1.1507	1.2466	1.3425	1.4384	1.5343
80	.6575	.7671	.8767	.9863	1.0959	1.2055	1.3151	1.4247	1.5343	1.6438	1.7534
90	.7397	.8630	.9863	1.1096	1.2329	1.3562	1.4795	1.6027	1.7260	1.8493	1.9726
100	.8219	.9589	1.0959	1.2329	1.3699	1.5069	1.6438	1.7808	1.9178	2.0548	2.1918

II. Find the interest on the following loans by using the appropriate table.

No.	Principal	Rate %	Time Period	Interest
1.	$1300	3	Aug. 10, 1969–Dec. 15, 1969	Commercial
2.	2400	4	Feb. 4, 1973–Mar. 15, 1973	Accurate
3.	725	$6\frac{1}{2}$	Apr. 20, 1972–Oct. 5, 1972	Ordinary
4.	550	8	May 19, 1974–Jan. 20, 1975	Commercial
5.	3000	7	Dec. 20, 1975–Mar. 5, 1976	Accurate
6.	3725	$5\frac{1}{2}$	Oct. 10, 1974–May 21, 1975	Accurate

III. Word problems:

1. Bank A offers all personal unsecured loans on a 365-day year while Bank B uses a 360-day year. If Mr. Smith wishes to borrow $5000 for 90 days at 8%, which bank offers the best proposition? Why? By how much?

2. Using the banker's 60-day method, find the interest on: (a) $1500 at 6% for 30 days, (b) $250 at 3% for 120 days, and (c) $3600 at 6% for 10 days.

3. On May 5, Mr. Howe borrowed $650 at 7% for 90 days. However, when the note was due, Mr Howe could only pay $350 on the principal plus the interest. The bank accepted this payment, cancelled the original note, and made out a new 90-day loan at 7% for $300. (a) When was the last loan due? (b) What was the total interest charge from May 5 until the second loan was paid?

4. A finance company charges 6% on home loans. If a family borrows $12,000, what is the interest for the first month?

5. Find the *ordinary* and *accurate* interest on $2400 for 45 days at $4\frac{1}{2}$%.

6. Find the *ordinary* interest, the *banker's* interest, and the *accurate* interest on $950 at 5% from October 2 to February 11 of the next year?

FINDING MATURITY VALUE, PRINCIPAL, RATE, AND TIME

Maturity Value

When a loan matures it is due and payable in full. The amount to be repaid is called the maturity value and equals the principal plus the interest earned. As a formula, it is expressed as

$$S = P + I$$

where

S = maturity value or future sum
P = principal (amount borrowed or loaned)
I = interest

example: Mr. Adams borrowed \$500.00 at 6% for 6 months. What sum was due at the maturity date?

solution: $I = Prt = \$500.00 \times .06 \times \frac{6}{12} = \$15.00$
$S = P + I = \$500.00 + \$15.00 = \$515.00$ (maturity value)

example: The Peoples Bank loaned The Hipple Co. \$6000 at $6\frac{1}{2}$% for 90 days. If this loan was repaid when due, how much did the bank receive (commercial interest)?

solution: $I = Prt = \$6000.00 \times .065 \times \frac{90}{360} = \$97.50$
$S = P + I = \$6000.00 + \$97.50 = \$6097.50$ (maturity value)

Principal

To find the principal of a loan, it is necessary to know the interest rate, amount of interest, and time period. When these factors are known the principal may be found by substituting these values in the interest formula. If $I = Prt$, then $P = I \div rt$.

example: When time is expressed in *years*: Find the principal necessary to earn \$15 at 5% in 2 years.

solution: $P = I \div rt = \$15 \div (.05 \times 2)$
$= \$15 \div .10 = \$150$ (principal)

example: When time is expressed in *months*: Find the principal necessary to earn \$60 at $4\frac{1}{2}\%$ in 4 months.

solution: $P = I \div rt = \$60 \div (.045 \times \frac{1}{3})$
$= \$60 \div .015 = \$4000$ (principal)

example: When time is expressed in *days*: Find the principal necessary to yield \$25 at 6% in 20 days (ordinary interest).

solution: $P = I \div rt = \$25 \div (.06 \times \frac{20}{360})$
$= \$25 \div \frac{1}{300} = \$7500$ (principal)

Rate

As in the case when finding the principal, use the basic interest formula and substitute known factors to find the rate. $I = Prt$; therefore, $r = I \div Pt$.

example: What rate of interest will earn \$76.80 on on \$640.00 in 3 years?

solution: $r = I \div Pt = \$76.80 \div (640.00 \times 3)$
$= .04$ *or* 4% (interest rate)

example: At what rate must \$720.00 be loaned to yield \$29.70 in 9 months?

solution: $r = I \div Pt = \$29.70 \div (720 \times \frac{9}{12})$
$= .055$ *or* $5\frac{1}{2}\%$ (interest rate)

example: What rate of interest will earn \$6.40 on \$1200.00 in 32 days (use 360-day year)?

solution: $r = I \div Pt = \$6.40 \div (1200.00 \times \frac{32}{360})$
$= .06$ *or* 6% (interest rate)

Time

In the same manner in which the principal and interest rate were determined, time is found by stubstituting the known factors in the basic formula. Time may be expressed in months, in days, or in years. $I = Prt$; therefore, $t = I \div Pr$.

Since interest rates are quoted per year unless otherwise stated, the time is in terms of years. If the result is less than 1, it may be converted to days or months as illustrated in the following examples.

example: How long will it take \$3500.00 to earn \$437.50 at 5%?

solution: $t = I \div Pr = \$437.50 \div (3500.00 \times .05)$
$= 2\frac{1}{2}$ years (time)

example: Find the time required for \$425.00 to earn \$25.50 at 8%.

solution: $t = I \div Pr = \$25.50 \div (425.00 \times .08)$
$= .75$ years (time)

Converted to months, .75 years = .75 × 12 or 9 months. Converted to days (360-day year), .75 years = .75 × 360 or 270 days.

EXERCISE 49

I. Find the principal and maturity value for the following loans at simple ordinary interest.

No.	Interest	Rate (%)	Time
1.	\$ 3.75	$7\frac{1}{2}$	30 days
2.	35.00	$3\frac{1}{2}$	8 months
3.	26.25	6	15 months
4.	8.05	7	18 days
5.	14.50	8	120 days
6.	0.90	5	10 days

II. Find the ordinary interest rate on the following loans.

No.	Principal	Interest	Time
1.	$1200.00	$ 35.10	117 days
2.	700.00	14.00	72 days
3.	3000.00	30.00	6 months
4.	1500.00	367.50	$3\frac{1}{2}$ years
5.	875.00	39.38	1 year
6.	7500.00	87.50	4 months

III. Find the time required to earn the interest charges shown. Convert all fractions of a year to days, using a 360-day year.

No.	Principal	Interest	Rate (%)
1.	$1240.00	$ 18.60	6
2.	1250.00	25.00	4.8
3.	6000.00	37.50	5
4.	1800.00	70.20	6
5.	800.00	35.00	$3\frac{1}{2}$
6.	1545.00	133.90	4

IV. Solve the following. When the time is a fractional part of a year, convert to days.

1. How long will it take the ordinary interest on $1500.00 at 8% to amount to $15.00?
2. In 6 months an investment of $3600.00 earns $96.00. What is the rate of interest?
3. $10,200.00 earned $191.25 from March 3 to June 1, what was the commercial interest rate?
4. How much must be invested to earn $50 in 60 days at 8% (360-day year)?
5. Find the maturity value on a loan of $675 at 4% which was dated July 15 and due September 25? (Use banker's interest.)

6. What principal will earn $16.80 simple interest at 4% in 84 days?
7. Find the commercial interest on the following amounts by the 60-day, 6% method: (a) $160 at 3% for 60 days, (b) $750 at 12% for 30 days, and (c) $90 at 6% for 90 days.
8. Find the accurate interest on $375 at 5% from May 10 to July 15.
9. The interest on a loan of $1500.00 for 90 days amounted to $15.00. What was the rate of interest? (Use a 360-day year.)
10. What was the principal or face of a note if $20.00 were earned in 3 months at $2\frac{1}{2}$%?
11. How long will it take $500 to earn $20.00 at 3% (ordinary interest)?
12. Mr. Haines bought a TV set from the Hi Fi Company for $650 cash or $670 if paid at the end of 3 months. If money is worth 5%, how much would he save by borrowing the money and paying cash (commercial interest)?

COMPOUND INTEREST

Simple interest is computed on a fixed amount or principal for a specified period of time and rate. Compound interest is computed on a principal that changes at stated intervals when interest is added to it. This interval may be any time period during the year such as annually, semiannually, quarterly, monthly, or daily.† When simple interest is added to the principal, it is said to be converted—it becomes part of the principal. The number of conversion periods is the number of times interest is converted during a year.

†The use of computers has brought many changes in banking procedures. The services of many clerks used to be required to calculate and post interest compounded quarterly; the computer can now convert money daily in a fraction of the time. This is the usual practice of banks as well as of savings and loan associations in some areas of the United States. When this system is used, individual accounts are credited with interest earned at the end of each quarter.

The difference between the original principal and the amount it has become at the end of any period of time is called compound interest.

example: Let us assume that $100 is to be converted annually at 6%. What will the compound interest equal at the end of 3 years?

solution:

$100.00	original principal
6.00	interest, first year ($100.00 × .06)
$106.00	principal at end of first year
6.36	interest, second year ($106.00 × .06)
$112.36	principal at end of second year
6.74	interest, third year ($112.36 × .06)
$119.10	principal at end of third year

compound interest = $119.10 − $100.00
= $19.10

example: If a person deposited $100 in a savings account, how much would he have at the end of a year if money is worth† 6% and is compounded (a) semiannually or (b) quarterly? What will the compound interest be in each case?

solution: (a) If the annual interest is 6%, then the interest for half of a year is 3%. In this case there are 2 conversions per year.

$100.00	original principal
3.00	interest, 6 months ($100.00 × .03)
$103.00	principal, end of 6 months
3.09	interest, second 6-month period (103.00 × .03)
$106.09	principal, end of year

compound interest = $106.09 − $100.00
= $6.09

†"Money is worth" is a common expression. It is another way of saying that the rate of interest is (in this case) 6%.

(b) If money is converted quarterly, there are 4 conversion periods. Then the interest rate per conversion period will equal $1\frac{1}{2}\%$ ($6\% \div 4$).

\$100.00	original principal
1.50	interest, first quarter (\$100.00 × .015)
\$101.50	principal, end of first quarter
1.52	interest, second quarter (\$101.50 × .015)
\$103.02	principal, end of second quarter
1.55	interest, third quarter (\$103.02 × .015)
\$104.57	principal, end of third quarter
1.57	interest, fourth quarter (\$104.57 × .015)
\$106.14	principal, end of fourth quarter or the year

compound interest = \$106.14 − \$100.00
= \$6.14

CONVERSION PERIODS

An examination of the examples above illustrates the fact that, at the same rate, money increases as the number of conversion periods increases. In other words, the better investment program is one in which money is most often converted, assuming the annual interest rate is the same.

When computing compound interest, it is important to be able to change the annual rate to the conversion period rate quickly—both when solving a problem as illustrated or when using the Compound Interest Table. For example:

Annual Interest Rate	Conversion Periods per Year	Interest Rate per Conversion Period
8% converted annually	1	8%
8% converted semiannually	2	4%
8% converted quarterly	4	2%
5% converted semiannually	2	$2\frac{1}{2}\%$
5% converted monthly	12	$\frac{5}{12}\%$ *or* $.41\frac{2}{3}\%$

COMPOUND INTEREST TABLES

As can be observed, it becomes a tedious chore to compute the compound interest for sums in excess of 4 or 5 conversion periods. Use of the Compound Interest Tables reduces the arithmetical processes to a minimum. The column headed n represents the number of conversion periods; the interest rates are the rates per conversion period. The table shown on pages 247–250 includes rates from $\frac{1}{2}\%$ to $8\frac{1}{2}\%$ for 50 conversion periods.

example: How much will \$500 amount to in 5 years at 6% compounded quarterly? Find the interest earned.

solution: conversion periods (n) = 20 (4×5 years)
conversion interest rate = $1\frac{1}{2}\%$ ($6\% \div 4$)

Read down column headed n to 20. Follow across the line in which 20 appears to column headed $1\frac{1}{2}\%$. The amount shown = 1.346855. This means that \$1.00 will increase to \$1.346855 in 5 years when compounded quarterly at 6%; the interest earned is \$0.35 approximately. Therefore, to find the sum to which \$500 will increase, multiply \$1.346855 by 500: $500 \times \$1.346855 = \$673.427500$ *or* \$673.43. The interest earned = \$673.43 − \$500.00 = \$173.43.

example: What is the difference between simple interest at 4% and compound interest on \$1000 at 4% compounded quarterly for a year?

solution: Simple interest = $\$1000.00 \times .04 = \$40.00$. With compound interest, $n = 4$, and the conversion interest rate = 1%. According to the table, \$1.00 = \$1.040604 at the end of the year. Therefore, \$1000.00 = \$1040.604 *or* \$1040.60. The difference between compound interest and simple interest is \$0.60.

example: How long, approximately, will it take money to double itself at 6% converted semiannually?

solution: The conversion interest rate = 3%. Under the

column headed 3% in the tables, look for a sum that is equal to $2.00. We find that $1.00 equals $2.03 plus at 24 periods. Since money is converted semiannually, this is 12 years approximately.

example: What simple interest rate will earn the same interest in a year as 5% converted semi-annually?

solution: Refer to the table and read the value given for 2 conversion periods at $2\frac{1}{2}\%$. This value is 1.050625 which means that an annual rate of 5.0625% will earn the same interest as 5% converted semiannually.

NOMINAL AND EFFECTIVE INTEREST RATES

The *effective* rate is the annual rate which will produce the same interest in a year as the *nominal* rate converted a certain number of times. For example, 6% converted semi-annually produces $6.09 per $100. Therefore, 6% is the nominal rate and 6.09% is the effective rate. This is the same as saying that a rate of 6% converted semiannually yields the same interest as a rate of 6.09% on an annual basis. In the last example above, 5% is the nominal rate and 5.0625% is the effective rate.

To make comparisons between different nominal rates, they must be reduced to a common base. Such comparisons can be made when all rates are expressed as effective rates.

EXERCISE 50

1. How many years, approximately, will it take money to double itself at 4% compounded semiannually?

2. If 200 dollars were deposited in a bank at 5% compounded semi-annually, how much will be on deposit at the end of 5 years? How much interest was earned?

3. If the rate of interest is 6%, how many conversion periods are there in 5 years if money is converted semiannually? Quarterly? Monthly? Yearly? What is the interest rate per period of time for each of these conversion periods?

4. How much more will $100 earn at 6% converted quarterly as compared to semiannually for a year?

5. How much does $100 earn at 5% converted annually for 10 years?

6. How many years must $2000 be invested at 4% compounded annually to earn $1746 (nearest whole dollar)?

7. A building and loan association pays interest on deposits at the rate of 3% compounded semiannually. How much, at this rate, would a deposit of $1200 equal at the end of 6 years?

8. A man borrows $500 on September 10, 1971. If he pays interest at 6% converted monthly, how much must he pay on April 10, 1973 (use a 30-day month)?

9. What sum will be needed to repay a loan of $1600 at an interest rate of 5% converted semiannually at the end of 2 years?

10. What is the effective interest rate equivalent to 7% converted semiannually, to 8% converted quarterly, and to 9% converted semiannually (nearest hundredth of 1%)?

11. How much interest will $5000 earn in 15 months at 6% compounded quarterly?

12. An investment of $2000 earned 4% converted quarterly for 3 years, then 5% converted semiannually for the next 2 years. Find the total of the investment fund at the end of the 5 years and the interest earned.

COMPOUND INTEREST TABLE Amount of $1 at Compound Interest: $S = (1 + i)^n$

n	½%	1%	1½%	2%	n
1	1.0050 0000	1.0100 0000	1.0150 0000	1.0200 0000	1
2	1.0100 2500	1.0201 0000	1.0302 2500	1.0404 0000	2
3	1.0150 7513	1.0303 0100	1.0456 7838	1.0612 0800	3
4	1.0201 5050	1.0406 0401	1.0613 6355	1.0824 3216	4
5	1.0252 5125	1.0510 1005	1.0772 8400	1.1040 8080	5
6	1.0303 7751	1.0615 2015	1.0934 4326	1.1261 6242	6
7	1.0355 2940	1.0721 3535	1.1098 4491	1.1486 8567	7
8	1.0407 0704	1.0828 5671	1.1264 9259	1.1716 5938	8
9	1.0459 1058	1.0936 8527	1.1433 8998	1.1950 9257	9
10	1.0511 4013	1.1046 2213	1.1605 4083	1.2189 9442	10
11	1.0563 9583	1.1156 6835	1.1779 4894	1.2433 7431	11
12	1.0616 7781	1.1268 2503	1.1956 1817	1.2682 4179	12
13	1.0669 8620	1.1380 9328	1.2135 5244	1.2936 0663	13
14	1.0723 2113	1.1494 7421	1.2317 5573	1.3194 7876	14
15	1.0776 8274	1.1609 6896	1.2502 3207	1.3458 6834	15
16	1.0830 7115	1.1725 7864	1.2689 8555	1.3727 8571	16
17	1.0884 8651	1.1843 0443	1.2880 2033	1.4002 4142	17
18	1.0939 2894	1.1961 4748	1.3073 4064	1.4282 4625	18
19	1.0993 9858	1.2081 0895	1.3269 5075	1.4568 1117	19
20	1.1048 9558	1.2201 9004	1.3468 5501	1.4859 4740	20
21	1.1104 2006	1.2323 9194	1.3670 5783	1.5156 6634	21
22	1.1159 7216	1.2447 1586	1.3875 6370	1.5459 7967	22
23	1.1215 5202	1.2571 6302	1.4083 7715	1.5768 9926	23
24	1.1271 5978	1.2697 3465	1.4295 0281	1.6084 3725	24
25	1.1327 9558	1.2824 3200	1.4509 4535	1.6406 0599	25
26	1.1384 5955	1.2952 5631	1.4727 0953	1.6734 1811	26
27	1.1441 5185	1.3082 0888	1.4948 0018	1.7068 8648	27
28	1.1498 7261	1.3212 9097	1.5172 2218	1.7410 2421	28
29	1.1556 2197	1.3345 0388	1.5399 8051	1.7758 4469	29
30	1.1614 0008	1.3478 4892	1.5630 8022	1.8113 6158	30
31	1.1672 0708	1.3613 2740	1.5865 2642	1.8475 8882	31
32	1.1730 4312	1.3749 4068	1.6103 2432	1.8845 4059	32
33	1.1789 0833	1.3886 9009	1.6344 7918	1.9222 3140	33
34	1.1848 0288	1.4025 7699	1.6589 9637	1.9606 7603	34
35	1.1907 2689	1.4166 0276	1.6838 8132	1.9998 8955	35
36	1.1966 8052	1.4307 6878	1.7091 3954	2.0398 8734	36
37	1.2026 6393	1.4450 7647	1.7347 7663	2.0806 8509	37
38	1.2086 7725	1.4595 2724	1.7607 9828	2.1222 9879	38
39	1.2147 2063	1.4741 2251	1.7872 1025	2.1647 4477	39
40	1.2207 9424	1.4888 6373	1,8140 1841	2.2080 3966	40
41	1.2268 9821	1.5037 5237	1.8412 2868	2.2522 0046	41
42	1.2330 3270	1.5187 8989	1.8688 4712	2.2972 4447	42
43	1.2391 9786	1.5339 7779	1.8968 7982	2.3431 8936	43
44	1.2453 9385	1.5493 1757	1.9253 3302	2.3900 5314	44
45	1.2516 2082	1.5648 1075	1.9542 1301	2.4378 5421	45
46	1.2578 7892	1.5804 5885	1.9835 2621	2.4866 1129	46
47	1.2641 6832	1.5962 6344	2.0132 7910	2.5363 4351	47
48	1.2704 8916	1.6122 2608	2.0434 7829	2.5870 7039	48
49	1.2768 4161	1.6283 4834	2.0741 3046	2.6388 1179	49
50	1.2832 2581	1.6446 3182	2.1052 4242	2.6915 8803	50

continued

COMPOUND INTEREST TABLE   Amount of $1 at Compound Interest: $S = (1 + i)^n$—(*continued*)

n	2½%	3%	3½%	4%	n
1	1.0250 0000	1.0300 0000	1.0350 0000	1.0400 0000	1
2	1.0506 2500	1.0609 0000	1.0712 2500	1.0816 0000	2
3	1.0768 9063	1.0927 2700	1.1087 1788	1.1248 6400	3
4	1.1038 1289	1.1255 0881	1.1475 2300	1.1698 5856	4
5	1.1314 0821	1.1592 7407	1.1876 8631	1.2166 5290	5
6	1.1596 9342	1.1940 5230	1.2292 5533	1.2653 1902	6
7	1.1886 8575	1.2298 7387	1.2722 7926	1.3159 3178	7
8	1.2184 0290	1.2667 7008	1.3168 0904	1.3685 6905	8
9	1.2488 6297	1.3047 7318	1.3628 9735	1.4233 1181	9
10	1.2800 8454	1.3439 1638	1.4105 9876	1.4802 4428	10
11	1.3120 8666	1.3842 3387	1.4599 6972	1.5394 5406	11
12	1.3448 8882	1.4257 6089	1.5110 6866	1.6010 3222	12
13	1.3785 1104	1.4685 3371	1.5639 5606	1.6650 7351	13
14	1.4129 7382	1.5125 8972	1.6186 9452	1.7316 7645	14
15	1.4482 9817	1.5579 6742	1.6753 4883	1.8009 4351	15
16	1.4845 0562	1.6047 0644	1.7339 8604	1.8729 8125	16
17	1.5216 1826	1.6528 4763	1.7946 7555	1.9479 0050	17
18	1.5596 5872	1.7024 3306	1.8574 8920	2.0258 1652	18
19	1.5986 5019	1.7535 0605	1.9225 0132	2.1068 4918	19
20	1.6386 1644	1.8061 1123	1.9897 8886	2.1911 2314	20
21	1.6795 8185	1.8602 9457	2.0594 3147	2.2787 6807	21
22	1.7215 7140	1.9161 0341	2.1315 1158	2.3699 1879	22
23	1.7646 1068	1.9735 8651	2.2061 1448	2.4647 1554	23
24	1.8087 2595	2.0327 9411	2.2833 2849	2.5633 0416	24
25	1.8539 4410	2.0937 7793	2.3632 4498	2.6658 3633	25
26	1.9002 9270	2.1565 9127	2.4459 5856	2.7724 6978	26
27	1.9478 0002	2.2212 8901	2.5315 6711	2.8833 6858	27
28	1.9964 9502	2.2879 2768	2.6201 7196	2.9987 0332	28
29	2.0464 0739	2.3565 6551	2.7118 7798	3.1186 5145	29
30	2.0975 6758	2.4272 6247	2.8067 9370	3.2433 9751	30
31	2.1500 0677	2.5000 8035	2.9050 3148	3.3731 3341	31
32	2.2037 5694	2.5750 8276	3.0067 0759	3.5080 5875	32
33	2.2588 5086	2.6523 3524	3.1119 4235	3.6483 8110	33
34	2.3153 2213	2.7319 0530	3.2208 6033	3.7943 1634	34
35	2.3732 0519	2.8138 6245	3.3335 9045	3.9460 8899	35
36	2.4325 3532	2.8982 7833	3.4502 6611	4.1039 3255	36
37	2.4933 4870	2.9852 2668	3.5710 2543	4.2680 8986	37
38	2.5556 8242	3.0747 8348	3.6960 1132	4.4388 1345	38
39	2.6195 7448	3.1670 2698	3.8253 7171	4.6163 6599	39
40	2.6850 6384	3.2620 3779	3.9592 5972	4.8010 2063	40
41	2.7521 9043	3.3598 9893	4.0978 3381	4.9930 6145	41
42	2.8209 9520	3.4606 9589	4.2412 5799	5.1927 8391	42
43	2.8915 2008	3.5645 1677	4.3897 0202	5.4004 9527	43
44	2.9638 0808	3.6714 5227	4.5433 4160	5.6165 1508	44
45	3.0379 0328	3.7815 9584	4.7023 5855	5.8411 7568	45
46	3.1138 5086	3.8950 4372	4.8669 4110	6.0748 2271	46
47	3.1916 9713	4.0118 9503	5.0372 8404	6.3178 1562	47
48	3.2714 8956	4.1322 5188	5.2135 8898	6.5705 2824	48
49	3.3532 7680	4.2562 1944	5.3960 6459	6.8333 4937	49
50	3.4371 0872	4.3839 0602	5.5849 2686	7.1066 8335	50

COMPOUND INTEREST TABLE Amount of $1 at Compound Interest: $S = (1 + i)^n$—(*continued*)

n	4½%	5%	5½%	6%	n
1	1.0450 0000	1.0500 0000	1.0550 0000	1.0600 0000	1
2	1.0920 2500	1.1025 0000	1.1130 2500	1.1236 0000	2
3	1.1411 6613	1.1576 2500	1.1742 4138	1.1910 1600	3
4	1.1925 1860	1.2155 0625	1.2388 2465	1.2624 7696	4
5	1.2461 8194	1.2762 8156	1.3069 6001	1.3382 2558	5
6	1.3022 6012	1.3400 9564	1.3788 4281	1.4185 1911	6
7	1.3608 6183	1.4071 0042	1.4546 7916	1.5036 3026	7
8	1.4221 0061	1.4774 5544	1.5346 8651	1.5938 4807	8
9	1.4860 9514	1.5513 2822	1.6190 9427	1.6894 7896	9
10	1.5529 6942	1.6288 9463	1.7081 4446	1.7908 4770	10
11	1.6228 5305	1.7103 3936	1.8020 9240	1.8982 9856	11
12	1.6958 8143	1.7958 5633	1.9012 0749	2.0121 9647	12
13	1.7721 9610	1.8856 4914	2.0057 7390	2.1329 2826	13
14	1.8519 4492	1.9799 3160	2.1160 9146	2.2609 0396	14
15	1.9352 8244	2.0789 2818	2.2324 7649	2.3965 5819	15
16	2.0223 7015	2.1828 7459	2.3552 6270	2.5403 5168	16
17	2.1133 7681	2.2920 1832	2.4848 0215	2.6927 7279	17
18	2.2084 7877	2.4066 1923	2.6214 6627	2.8543 3915	18
19	2.3078 6031	2.5269 5020	2.7656 4691	3.0255 9950	19
20	2.4117 1402	2.6532 9771	2.9177 5749	3.2071 3547	20
21	2.5202 4116	2.7859 6259	3.0782 3415	3.3995 6360	21
22	2.6336 5201	2.9252 6072	3.2475 3703	3.6035 3742	22
23	2.7521 6635	3.0715 2376	3.4261 5157	3.8197 4966	23
24	2.8760 1383	3.2250 9994	3.6145 8990	4.0489 3464	24
25	3.0054 3446	3.3863 5494	3.8133 9235	4.2918 7072	25
26	3.1406 7901	3.5556 7269	4.0231 2893	4.5493 8296	26
27	3.2820 0956	3.7334 5632	4.2444 0102	4.8223 4594	27
28	3.4296 9999	3.9201 2914	4.4778 4307	5.1116 8670	28
29	3.5840 3649	4.1161 3560	4.7241 2444	5.4183 8790	29
30	3.7453 1813	4.3219 4238	4.9339 5129	5.7434 9117	30
31	3.9138 5745	4.5380 3949	5.2580 6861	6.0881 0064	31
32	4.0899 8104	4.7649 4147	5.5472 6238	6.4533 8668	32
33	4.2740 3018	5.0031 8854	5.8523 6181	6.8405 8988	33
34	4.4663 6154	5.2533 4797	6.1742 4171	7.2510 2528	34
35	4.6673 4781	5.5160 1537	6.5138 2501	7.6860 8679	35
36	4.8773 7846	5.7918 1614	6.8720 8538	8.1472 5200	36
37	5.0968 6049	6.0814 0694	7.2500 5008	8.6360 8712	37
38	5.3262 1921	6.3854 7729	7.6488 0283	9.1542 5235	38
39	5.5658 9908	6.7047 5115	8.0694 8699	9.7035 0749	39
40	5.8163 6454	7.0399 8871	8.5133 0877	10.2857 1794	40
41	6.0781 0094	7.3919 8815	8.9815 4076	10.9028 6101	41
42	6.3516 1548	7.7615 8756	9.4755 2550	11.5570 3267	42
43	6.6374 3818	8.1496 6693	9.9966 7940	12.2504 5463	43
44	6.9361 2290	8.5571 5028	10.5464 9677	12.9854 8191	44
45	7.2482 4843	8.9850 0779	11.1265 5409	13.7646 1083	45
46	7.5744 1961	9.4342 5818	11.7385 1456	14.5904 8748	46
47	7.9152 6849	9.9059 7109	12.3841 3287	15.4659 1673	47
48	8.2714 5557	10.4012 6965	13.0652 6017	16.3938 7173	48
49	8.6436 7107	10.9213 3313	13.7838 4948	17.3775 0403	49
50	9.0326 3627	11.4673 9979	14.5419 6120	18.4201 5427	50

COMPOUND INTEREST TABLE Amount of $1 at Compound Interest: $S = (1 + i)^n$—(*continued*)

n	6½%	7%	7½%	8%	n
1	1.0650 0000	1.0700 0000	1.0750 0000	1.0800 0000	1
2	1.1342 2500	1.1449 0000	1.1556 2500	1.1664 0000	2
3	1.2079 4963	1.2250 4300	1.2422 9688	1.2597 1200	3
4	1.2864 6635	1.3107 9601	1.3354 6914	1.3604 8896	4
5	1.3700 8666	1.4025 5173	1.4356 2933	1.4693 2808	5
6	1.4591 4230	1.5007 3035	1.5433 0153	1.5868 7432	6
7	1.5539 8655	1.6057 8148	1.6590 4914	1.7138 2427	7
8	1.6549 9567	1.7181 8618	1.7834 7783	1.8509 3021	8
9	1.7625 7039	1.8384 5921	1.9172 3866	1.9990 0463	9
10	1.8771 3747	1.9671 5136	2.0610 3156	2.1589 2500	10
11	1.9991 5140	2.1048 5195	2.2156 0893	2.3316 3900	11
12	2.1290 9624	2.2521 9159	2.3817 7960	2.5181 7012	12
13	2.2674 8750	2.4098 4500	2.5604 1307	2.7196 2373	13
14	2.4148 7418	2.5785 3415	2.7524 4405	2.9371 9362	14
15	2.5718 4101	2.7590 3154	2.9588 7735	3.1721 6911	15
16	2.7390 1067	2.9521 6375	3.1807 9315	3.4259 4264	16
17	2.9170 4637	3.1588 1521	3.4193 5264	3.7000 1805	17
18	3.1066 5438	3.3799 3228	3.6758 0409	3.9960 1950	18
19	3.3085 8691	3.6165 2754	3.9514 8940	4.3157 0106	19
20	3.5236 4506	3.8696 8446	4.2478 5110	4.6609 5714	20
21	3.7526 8199	4.1405 6237	4.5664 3993	5.0338 3372	21
22	3.9966 0632	4.4304 0174	4.9089 2293	5.4365 4041	22
23	4.2563 8573	4.7405 2986	5.2770 9215	5.8714 6365	23
24	4.5330 5081	5.0723 6695	5.6728 7406	6.3411 8074	24
25	4.8276 9911	5.4274 3264	6.0983 3961	6.8484 7520	25
26	5.1414 9955	5.8073 5292	6.5557 1508	7.3963 5321	26
27	5.4756 9702	6.2138 6763	7.0473 9371	7.9880 6147	27
28	5.8316 1733	6.6488 3836	7.5759 4824	8.6271 0639	28
29	6.2106 7245	7.1142 5705	8.1441 4436	9.3172 7490	29
30	6.6143 6616	7.6122 5504	8.7549 5519	10.0626 5689	30
31	7.0442 9996	8.1451 1290	9.4115 7683	10.8676 6944	31
32	7.5021 7946	8.7152 7080	10.1174 4509	11.7370 8300	32
33	7.9898 2113	9.3253 3975	10.8762 5347	12.6760 4964	33
34	8.5091 5950	9.9781 1354	11.6919 7248	13.6901 3361	34
35	9.0622 5487	10.6765 8148	12.5688 7042	14.7853 4429	35
36	9.6513 0143	11.4239 4219	13.5115 3570	15.9681 7184	36
37	10.2786 3603	12.2236 1814	14.5249 0088	17.2456 2558	37
38	10.9467 4737	13.0792 7141	15.6142 6844	18.6252 7563	38
39	11.6582 8595	13.9948 2041	16.7853 3858	20.1152 9768	39
40	12.4160 7453	14.9744 5784	18.0442 3897	21.7245 2150	40
41	13.2231 1938	16.0226 6989	19.3975 5689	23.4624 8322	41
42	14.0826 2214	17.1442 5678	20.8523 7366	25.3394 8187	42
43	14.9979 9258	18.3443 5475	22.4163 0168	27.3666 4042	43
44	15.9728 6209	19.6284 5959	24.0975 2431	29.5559 7166	44
45	17.0110 9813	21.0024 5176	25.9048 3863	31.9204 4939	45
46	18.1168 1951	22.4726 2338	27.8477 0153	34.4740 8534	46
47	19.2944 1278	24.0457 0702	29.9362 7915	37.2320 1217	47
48	20.5485 4961	25.7289 0651	32.1815 0008	40.2105 7314	48
49	21.8842 0533	27.5299 2997	34.5951 1259	43.4274 1899	49
50	23.3066 7868	29.4570 2506	37.1897 4603	46.9016 1251	50

Objectives

Notes and drafts are direct applications of the use of interest. They represent the "paper" on which transactions involving interest are recorded. This chapter is devoted to the manner in which they are prepared, used, bought and sold, and how proceeds are computed.

11 NOTES AND DRAFTS

The *promissory note* and the *draft* are two of the most common forms of *negotiable instruments*. A negotiable instrument must be written, must be signed by the maker, must have an exact due date—some date that is determinable, must be an unconditional promise to pay a definite sum of money, and must be payable to order or to bearer.

Notes and drafts are usually written on a standard form but may be written on a plain piece of paper if all legal requirements for a negotiable instrument are met.

This chapter deals with the purpose and use of these documents and how bank discounting procedures are applied.

NOTES

A *promissory note* is a written promise to pay a debt in accordance with the terms stated thereon. Such notes are generally used for short-term loans. Although the time may be stated in years or months, the majority of promissory notes are for short periods such as 30, 60, or 90 days.

Notes may be interest bearing or noninterest bearing. Interest is expressed as a percent. When the note bears no interest, it is so indicated. Notes sold to a bank are of two types—those that originate with the borrower and those that the borrower holds from a third party in payment of a debt.

NOTES DISCOUNTED AT THE BANK THAT ORIGINATE WITH THE BORROWER

Many businessmen need cash to meet immediate commitments—commitments that are more convenient to pay at some future time. For example, Mr. Jones needs $500 to pay for expanding his shop to include a new department. He expects new business to bring in enough additional income to meet this expense 6 months hence. If his credit standing is good, his bank will loan him this amount on his personal note after deducting a charge for the service. This charge is called *bank discount* and is calculated in the same way that interest is determined. However, unlike interest, the discount is calculated on the maturity value rather than the principal; it is collected when the loan is made rather than at the maturity date.

Although banks have their own forms, the promissory note shown here is typical. The face of the note is $3000 (the amount borrowed); the term of the note is 90 days; the due date or maturity date is August 4 (90 days after May 6); the maker is James Smith, and the payee is the High Street National Bank.

$ 3000.00 Oakland, California May 6 19

90 days after date, (without grace) I promise to pay to the order of High Street National Bank

Three thousand and no/100 - - - - - - - - - - - - - Dollars

for value received with interest of no per cent per annum from May 6, 19 until paid both principal and interest payable only in LAWFUL MONEY OF THE UNITED STATES.

Payable at High Street National Bank

No. 6607 Due August 4, 19

James Smith

James Smith, 2147 8th Ave., Harperville, Calif.

THE UTILITY LINE FORM NO. 60-012

In this case, James Smith has given his personal note to the High Street National Bank for \$3000 which it has accepted for 90 days. In exchange for this loan, the bank has made a charge. It is general practice for the loan departments of banks to use a discount rate, not an interest rate, for commercial papers of this type. Bank discount is sometimes called interest in advance. Instead of waiting until the maturity date to collect the charge for the loan, the bank collects it at the time the loan is made.

If the discount rate is 5%, then the bank discount = \$3000† × .05 × 90/360‡ = \$37.50. Mr. Smith receives \$3000 − \$37.50 or \$2962.50. This amount (\$2962.50) is called the *proceeds* of the note.

NOTES DISCOUNTED AT THE BANK THAT ARE HELD BY THE PAYEE

Under some circumstances, especially for certain types of business, a company accepts a note in payment of an obligation. In this case, the note may either be held until maturity, at which time payment is made by the maker, or it may be discounted at the bank (converted to cash) at some earlier date. The bank holds the note until maturity and collects from the maker. If the maker fails to meet this obligation, then the payee is held responsible.

\$ 600.00 Chicago, Illinois June 12 19

2 months after date (without grace) I promise to pay to the order of John Brown

Six hundred and no/100 - - - - - - - - - - - - - - - Dollars

for value received with interest of 6% per cent per annum from June 12, 19 until paid both principal and interest payable only in LAWFUL MONEY OF THE UNITED STATES.

Payable at Third National Bank

No. 72 Due August 12, 19

Robert Smith

Robert Smith, Suburb, Illinois

THE UTILITY LINE FORM No. 60-012

†Since this note does not bear interest, the maturity value and face of the note (principal) are the same.

‡If the terms of the note are for 3 months instead of 90 days, then the time or term of discount must be counted in days from May 6 to August 6 inclusive which amounts to 92 days.

The promissory note illustrated is for this type of transaction. This note was given to John Brown (payee) by Robert Smith (maker) in lieu of a money payment for services or merchandise. Mr. Brown accepted this note with the understanding that he would collect the $600 plus interest at 6% from Mr. Smith at the end of 2 months. However, before the note was due, he needed cash for unexpected expenses. Therefore, he could not wait until the maturity date to collect this debt.

If his credit rating and standing with his bank were good, he could have the note discounted. Suppose he took it to his bank on July 15 where it was discounted at $5\frac{1}{2}\%$. The bank held the note until maturity and collected from Mr. Smith. The question is: "What were the proceeds of the note?"

As mentioned previously, discount rates are always applied to the maturity value of a note. When the note bears no interest, the face of the note equals the maturity value. When interest is charged, then the maturity value equals the face of the note plus the interest earned.

When a note is discounted, the time is figured in exact days from the date it is "sold" up to and including the date of maturity. This period of time is called the *term of discount.*

face of note	$600.00	
interest	6.00	(6% of $600.00 for 2 months)
maturity value	$606.00	
date of maturity	August 12	
term of discount	28 days	(July 15–August 12)
rate of discount	$5\frac{1}{2}\%$	
discount	$2.59	($\$606.00 \times .055 \times \frac{28}{360}$)
proceeds	$603.41	($606.00 − $2.59)

INTEREST RATE EQUIVALENT TO DISCOUNT RATE

With the first promissory note, page 254, Mr. Smith paid 5% on $3000 but had the use of only $2962.50. Therefore, he actually paid an interest rate greater than 5% on the money he received. The sum of \$2962.50 represents the principal in calculating the interest rate. Interest rate = \$37.50 ÷ (\$2962.50 × $\frac{90}{360}$) = .0506 = 5.06%. This means that the 5.06% interest rate is equivalent to a 5% discount rate.

Formulas have been derived for use in calculating the interest rate for any given discount rate and for calculating the discount rate for any given interest rate.

(1) To find the interest rate that corresponds to a discount rate

$$r = \frac{d}{1 - dt}$$

where

r = interest rate
d = discount rate
t = time

Substituting in the above example, to find r (formula 1)

$$r = \frac{.05}{1 - (.05)(90/360)} = \frac{.05}{1 - .0125} = \frac{.05}{.9875}$$

$$= .0506 \textit{ or } 5.06\%$$

(2) To find the discount rate that corresponds to an interest rate

$$d = \frac{r}{1 + rt}$$

where

d = discount rate
r = interest rate
t = time

example: Find the discount rate that corresponds to an interest rate of 8%. Let t = 1 year.

solution: $d = r \div (1 + rt) = .08 \div (1 + .08)$

$= .08 \div 1.08 = 7.4\%$

EXERCISE 51

1. Mann Bros. accepted a 3-months note, dated November 11, for \$500 at 6% from Mr. Hope in payment of a debt. On January 2, it was discounted at the First National Bank at $5\frac{1}{2}\%$. Furnish the following information: maker of note, payee, assignee, maturity date, maturity value, term of discount, bank discount, and proceeds.

2. Find the proceeds on each of the following noninterest-bearing notes.

	Face of Note	Time to Run	Date of Note	Discount Date	Rate of Discount
(a)	\$350.00	60 days	July 17	Aug. 6	$4\frac{1}{2}\%$
(b)	1450.00	3 months	Oct. 11	Nov. 22	8%
(c)	653.00	90 days	Mar. 26	May 15	$5\frac{1}{2}\%$
(d)	800.50	5 months	Jan. 5	Mar. 4	6%

3. Mr. Cox held a 60-day note for \$4200 with interest at 5%. This note was discounted at his bank 25 days before maturity at 6%. What were the proceeds?

4. A noninterest-bearing note for \$2400 was discounted 30 days before it was due. If the proceeds were \$2389 what was the discount rate?

5. A 6-months note for \$1500, bearing interest at 6% and dated May 17, was discounted at 8% on September 22. What were the proceeds?

6. Mr. Sams, a building contractor, needed \$45,000 for 120 days for a construction project. The bank accepted his note for this amount at a discount rate of $4\frac{1}{2}\%$. How much did Mr. Sams receive from the bank? What is the equivalent interest rate (to 2 decimals in terms of percent)?

7. Mr. Hays gave Rincon Manufacturing Co. a 90-day note for \$1250 on October 10. If it was discounted at the bank on November 15 at $6\frac{1}{2}\%$, what were the proceeds?

8. If a bank charges $6\frac{1}{2}\%$ for discounting notes, how much can a borrower receive now if he must repay a loan of \$1300 in 3 months?

9. What interest rates are equivalent to the discount rates of 6% and 8% on a sum of money due in 1 year? (Record answers to the nearest tenth of 1%.)

10. Mr. Ames needs $1860 which he plans to repay a year later. How much must he borrow if the bank discounts his note at 7%?

DRAFTS

A *draft* is an order for the payment of money by an individual or bank, called the maker, on another, called the *drawee*. The person or company to whom payment is made is called the *payee*.

Drafts are drawn for a specific purpose and, with the exception of the bank draft, involve the buying and selling of goods. The bank draft is an order by one bank on another bank to pay a third party a specified sum of money. Drafts involving the transfer of goods are sometimes called "trade acceptances" or "bills of exchange". Such drafts are drawn for the amount of a specific purchase by the seller on the buyer and generally specify the place of payment (such as a specific bank).

SIGHT DRAFT

A *sight draft* is a written order for payment of a purchase by the buyer to the seller. It is payable upon presentation. Merchandise in this case generally is not released until the draft is accepted by the buyer and payment is made to the seller—usually to his bank. The bank, acting as a collection agent, charges a small fee (fraction of 1% of the face of the draft) for its services.

TIME DRAFT

As the name implies, payment is not due on a time draft when it is presented. The due date or maturity date is computed in accordance with the terms as stated on the face of the draft. "Two months after date" means that the draft is due and payable 2 months after the date of the draft. However, if the terms specify "2 months after sight," then the maturity date is 2 months after it is accepted. For example:

1. A draft dated October 12 is due and payable 30 days after date. The maturity date is November 11.

2. A draft dated October 12 is due and payable 2 months after date. The maturity date is December 12.

3. A draft dated October 12 is due and payable 30 days after sight. If it is accepted on October 15, the maturity date is 30 days after October 15 or November 14.

4. A draft dated February 3 is due and payable 3 months after sight. It is accepted on February 10. The maturity date is 3 months after February 10 or May 10.

INTEREST-BEARING AND NONINTEREST-BEARING DRAFTS

Drafts may or may not bear interest. Also, they may be discounted ("sold to a bank") in the same manner as promissory notes. However, the bank may also charge a fee which is a small percent of the face of the draft. The face of the draft less the discount and collection fee is called the *proceeds*.

Finding the Proceeds of a Noninterest-Bearing Draft

Let us assume that S. A. Hipple, Cotton Merchants, Galveston, Texas, sold 700 bales of cotton amounting to \$58,960.50 to the Parker Cotton Mills, Charlotte, N.C. A draft was sent to the Parker Cotton Mills which granted them 30 days after sight in which to pay for this shipment. The draft was dated September 28 and was accepted on October 10. If the draft was discounted on October 16 at 7% and the bank charged a $\frac{1}{4}$% collection fee, the proceeds would be computed as follows.

maturity date (30 days after October 10) = November 9
term of discount (October 16–November 9) = 24 days
discount (7% of \$58,960.50 for 24 days) = \$275.15
collection fee ($\frac{1}{4}$% of \$58,960.50)† = \$147.40

proceeds = \$58,960.50 − (\$275.15 + \$147.40) = \$58,537.95

†A convenient method of calculating a fraction of 1% is to take 1% and divide by the fraction. In this case, 1% of \$58,960.50 = \$589.60; \$589.60 ÷ 4 = \$147.40.

Finding the Proceeds of an Interest-Bearing Draft

J. Robertson and Co. sold furniture to Haverson Stores in the amount of $13,560. A draft dated July 10 was submitted in payment with terms of "90 days after date," bearing interest at 6%. If the draft was discounted on August 30 at $6\frac{1}{2}$% and the bank charged a collection fee of $\frac{1}{10}$%, the proceeds would be computed as follows.

face of note	= $13,560.00
interest (6% of $13,560.00 for 90 days)	= 203.40
maturity value	= $13,763.40
maturity date (90 days after July 10)	= October 8
term of discount (August 30 to October 8)	= 39 days
discount ($6\frac{1}{2}$% of $13,763.40 for 39 days)	= $96.92
collection fee ($\frac{1}{10}$% of $13,763.40)	= $13.76

proceeds = $13,763.40 − ($96.92 + $13.76) = $13,652.72

EXERCISE 52

1. On April 10, Johnson Bros. accepted a 60-day sight draft, amounting to $562.50 drawn on them by Perkins & Co. Perkins & Co. had the draft discounted at the bank at 6% on April 15. What were the proceeds?

2. A 90-day sight draft in the amount of $425.00 was accepted on October 10 and discounted on November 1 at 5%. If the bank charged a $\frac{1}{3}$% collection fee, what were the proceeds?

3. James Owen accepted a draft, dated December 9, drawn by Parker Bros. for $815.50 at 6% payable 4 months after date. Parker Bros. had the draft discounted at the bank on January 9 at $6\frac{1}{2}$%. A charge of $\frac{1}{8}$% for collecting it when due was also charged. What were the proceeds?

4. A draft for $800, due in 3 months and bearing interest at $4\frac{1}{2}$%, was discounted 60 days before it was due. If the discount rate was 5%, what were the proceeds?

5. Find the proceeds for the following drafts.

	Face Value	Date	When Due	Acceptance Date	Discount Date	Discount Rate
(a)	$1500.00	4–15	4 months after date	Apr. 25*	May 10	4%
(b)	750.40	12–18	60 days after date	Jan. 4*	Jan. 22	5%
(c)	3275.00	6–20	60 days after sight	July 1	July 15	$5\frac{1}{2}$%
(d)	544.20	7–11	2 months after sight	July 25	Aug. 1	6%

*Due date is always calculated from the date of the draft for these problems, regardless of the date of acceptance.

6. A 90-days sight draft in the amount of $850 was accepted on May 10 and discounted on July 1 at 5%. If the bank charged a $\frac{1}{4}$% collection fee, what were the proceeds?

7. Parker Bros. accepted a draft for $3500 bearing interest at 6%, dated April 5 and due in 6 months after date. It was discounted on August 20 at $6\frac{1}{2}$%. If the bank charged a $\frac{1}{3}$% collection fee, what were the proceeds?

8. Owens Company accepted a draft, 90-day sight, amounting to $560 drawn on them by Watts Bros. Watts Bros. had the draft discounted 34 days before the due date at 6%. What were the proceeds?

9. A 60-day sight draft dated May 2 in the amount of $255 was accepted by Jackson Bros. on June 1. If it was discounted at 5% on July 1, what were the proceeds?

10. H. Mofit accepted a draft for $620 on October 25 that was dated October 10 and due 3 months after sight. It was discounted at $4\frac{1}{2}$% on November 16. If the bank charges a $\frac{1}{8}$% collection fee, what were the proceeds?

EXERCISE 53 Chapter Summary and Review of Interest

1. Mr. Howard needs \$3500 to remodel his home, to be repaid at the end of 180 days. He can borrow this sum from the local bank at 8% simple interest to be repaid at the end of 180 days. How much must he pay the bank at maturity?

2. Mr. Jacobs obtained a loan of \$1500 from his bank on January 3 to be repaid 3 months later (not a leap year). The bank discounted the loan at 6%. (a) How much did Mr. Jacobs receive? (b) What was the true interest rate?

3. Find the (a) ordinary, (b) banker's, and (c) accurate interest on \$2400 at 6% for a loan dated May 16 and due August 16 of the same year.

4. Compute the ordinary simple interest by the 60-day, 6% method for the following: (a) \$900 at 6% for 90 days, (b) \$2600 at 6% for 60 days, (c) \$450 at 2% for 26 days, (d) \$720 at 3% for 30 days.

5. Mr. Harper borrowed \$600 that was discounted at 8% to be repaid in 90 days. (a) What were the proceeds of this loan and (b) the true interest rate to the nearest tenth of 1%?

6. Mrs. Kelly held a 6-month note for \$500 at $6\frac{1}{2}$% interest. This note was dated June 10 and was due December 10. If this note was discounted at 6% on September 9, what were the proceeds?

7. If a bank charges 5% for discounting notes, how much would a borrower receive now if he had to repay \$1500 in 6 months?

8. Mr. Brown needs \$2000 for 1 year. If the bank discounts the loan at 6%, how much must he borrow in order to have \$2000. Prove your answer.

9. How many conversion periods are there and what is the interest rate per conversion period? (a) 8% converted annually, (b) 6% converted quarterly, (c) 4% converted monthly, (d) 7% converted semiannually.

10. Harold White deposited \$500 in a savings account that paid 6% compounded monthly. If he left it there for 4 years, (a) How much would he have? (b) How much interest did he earn (use 5 decimals from the table)?

11. How much would $500 at 6% simple interest earn in 4 years? Compare your answer to the amount earned in problem 10.

12. How long would it take for money to double itself (approximately) at (a) 5% compounded semiannually, (b) 8% compounded quarterly?

13. Record the maturity date for a draft dated (a) November 12 and payable 60 days after date, (b) June 10 and payable 3 months after date, (c) March 3 and payable 30 days after sight if it is accepted March 15, and (d) October 4 and payable 6 months after sight if it is accepted October 25.

14. Holmes Bros. accepted a draft dated July 10 for $326.80 at 5%, payable 90 days after date. (a) How much was paid if Holmes Bros. paid it when due? (b) If this draft had been discounted at $5\frac{1}{2}$% 30 days before maturity and a $\frac{1}{4}$% collection fee was charged, what were the proceeds?

15. A draft for $3000, dated March 15 and due in 2 months after sight, was accepted on March 30. If it was discounted on April 15 at 6%, find the proceeds.

Objectives

Installment purchases and periodic loan payment plans require credit on the installment plan. An explanation of what the cost of credit is on these bases and how it is used is the main purpose of this chapter. There are several ways in which credit costs may be computed. These are all discussed and illustrated with many applications provided for practice and study.

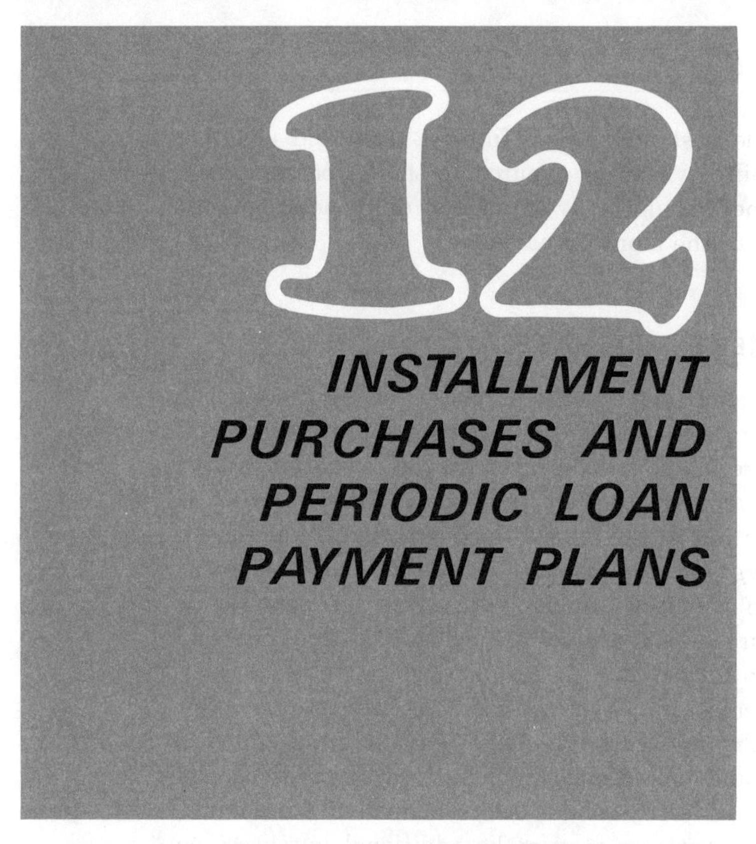

12 INSTALLMENT PURCHASES AND PERIODIC LOAN PAYMENT PLANS

According to figures released by the Federal Reserve Board, installment credit accounts for approximately 81% of the total consumer credit in the United States and is now more than twice that of 15 years ago. In June 1972 this amounted to 115 billion dollars or approximately 10 billion dollars per month. Regular charge accounts at the same time were a little over 8 billion dollars.

Buying or borrowing on the installment plan has become a way of life. Credit on the installment plan makes it possible to enjoy more of the desired things of life while paying for them. Advertising slogans such as "easy payment plan," "liberal terms, no downpayment with low monthly payments," "just one payment per month takes care of all your troubles," etc., have been effective. Also, the increased use of credit cards has encouraged buying to a great extent.

The use of credit is a necessary part of our economy but for the consumer it can be ruinous if not used wisely. A prospective borrower should shop for credit if he is going to use it to the best advantage. Interest rates and all other charges against a loan or a purchase should be compared. The consumer should be aware of cost differences, not just the amount of the monthly payment.

INSTALLMENT PURCHASES

When any article or service is purchased on the installment plan, a charge is added to the purchase price and equal payments are made periodically until the obligation is paid in full. The charge for credit may be a *flat sum* which is added to the cost at the time of the purchase or it may be an *interest charge* on the unpaid balance for each payment period. The contract invariably provides for penalties if payments are late and gives the seller title to goods sold until paid in full.

Department stores favor the monthly interest charge for their installment accounts. It is advantageous for the accounting and billing departments. Such accounts usually are, or can be easily converted to, "revolving accounts" whereby the customer may make additional purchases on the account from month to month so long as he does not exceed his credit limit and pays his installments regularly. This again is an advantage to the store.

For small accounts, credit can be costly. In this case the customer would do better to use a regular 30-day account which is payable in full each month and on which no service charge is made. Many stores also grant 60-day and 90-day accounts with no charge made for some purchases if paid within that time.

CREDIT CHARGES (DOLLAR AMOUNT)

Credit charges are the difference between the cash price and the price paid on a time plan. These charges cover such items as checking the credit of the customer, bookkeeping services, insurance, bad debts, and other protection against losses.

The installment payment schedules illustrated here are representative of those in use today.

SCHEDULE A—EASY PAYMENT CHART

Unpaid Balance	Carrying Charge	Up to 12 Monthly Payments*	Unpaid Balance	Carrying Charge	Up to 12 Monthly Payments*	Up to 18 Months: Carrying Charge	Up to 18 Months: Monthly Payment*
Up to $20.00	$ 2.00	$ 5.00	$250.01–260	$26.00	$24.00		
$ 20.01–30	3.00	5.00	260.01–270	27.00	25.00		
30.01–40	4.00	5.00	270.01–280	28.00	26.00		
40.01–50	5.00	5.00	280.01–290	29.00	27.00		
50.01–60	6.00	6.00	290.01–300	30.00	28.00		
60.01–70	7.00	7.00	300.01–310	31.00	29.00	$46.50	$20.00
70.01–80	8.00	8.00	310.01–320	32.00	30.00	48.00	21.00
80.01–90	9.00	9.00	320.01–330	33.00	31.00	49.50	22.00
90.01–100	10.00	10.00	330.01–340	34.00	32.00	51.00	22.00
100.01–110	11.00	11.00	340.01–350	35.00	33.00	52.50	23.00
110.01–120	12.00	11.00	350.01–360	36.00	34.00	54.00	23.00
120.01–130	13.00	12.00	360.01–370	37.00	35.00	55.50	24.00
130.01–140	14.00	13.00	370.01–380	38.00	36.00	57.00	25.00
140.01–150	15.00	14.00	380.01–390	39.00	37.00	58.50	25.00
150.01–160	16.00	15.00	390.01–400	40.00	38.00	60.00	26.00
160.01–170	17.00	16.00	400.01–410	41.00	39.00	61.50	27.00
170.01–180	18.00	17.00	410.01–420	42.00	40.00	63.00	28.00
180.01–190	19.00	18.00	420.01–430	43.00	41.00	64.50	29.00
190.01–200	20.00	19.00	430.01–440	44.00	42.00	66.00	29.00
200.01–210	21.00	20.00	440.01–450	45.00	43.00	67.50	30.00
210.01–220	22.00	21.00	450.01–460	46.00	44.00	69.00	30.00
220.01–230	23.00	22.00	460.01–470	47.00	45.00	70.50	31.00
230.01–240	24.00	22.00	470.01–480	48.00	46.00	72.00	31.00
240.01–250	25.00	23.00	480.01–490	49.00	47.00	73.50	32.00
			490.01–500	50.00	48.00	75.00	32.00

Check if balance is over $300 and 18 months desired ☐

*All monthly payments are for the amount shown; last payment is for the odd amount remaining due.

SCHEDULE B—SCHEDULE OF PAYMENTS

Consolidated Amount of Purchases	Monthly Payment	Consolidated Amount of Purchases	Monthly Payment	Consolidated Amount of Purchases	Monthly Payment
$ 20–100	$ 5.00	$351–400	$20.00	$651–700	$35.00
101–150	8.00	401–450	23.00	701–750	38.00
151–200	10.00	451–500	25.00	751–800	40.00
201–250	13.00	501–550	28.00	801–850	43.00
251–300	15.00	551–600	30.00	851–900	45.00
301–350	18.00	601–650	33.00	901–1000	50.00

Over $1000 arranged in Credit Office.

Schedule A lists the: (1) carrying charge and monthly payment for 12 months and (2) carrying charge and monthly payment for 18 months for purchases of $300 or more.

Schedule B is representative of those used by many large department stores. Under this plan, "a service charge of $1\frac{1}{2}\%$ of the previous month's ending balance of $1000 or under, rounded to the nearest $10 figure, plus 1% of such balance over $1000, will be charged." For this store, when this regulation is applied, the charges each month are: (1) for a balance up to $24 inclusive, $1\frac{1}{2}\%$ of $20 or 30¢; (2) for a balance of $25–$34 inclusive, $1\frac{1}{2}\%$ of $30 or 45¢; (3) for a balance of $35–$44 inclusive, $1\frac{1}{2}\%$ of $40 or 60¢; etc.

If a purchase is made and Schedule A is used, it is an easy matter to determine the difference between the cash price and the installment price. A purchase of $315 with 12 months in which to pay for it would cost $32 more than the cash price. But if an advertisement states the downpayment, if any, and the monthly payment for a given number of months, it is then necessary to determine what the charges for credit are in order to compare the installment price with the cash price.

example: A tractor is advertised at $675 cash or $25 down and the balance in easy payments of $30 each for 24 months. What is the carrying or installment charge?

solution:

downpayment	$ 25
+ payments ($30 × 24)	720
= installment price	$745
− cash price	675
= installment charge	$ 70

example: A coat may be purchased for $80 cash or over a period of 6 months with a charge of $1\frac{1}{2}\%$ on the unpaid balance per month. If a downpayment of $20 is made, what are the: (a) monthly payment schedule, (b) total cost on the installment plan, and (c) difference between the cash and installment price?

solution:

cash price	\$80
downpayment	\$20
balance due	\$60

monthly payment
on the balance = \$60 ÷ 6 = \$10

(a)

Month	Balance	Installment Payment	Service Charge ($1\frac{1}{2}\%$)	Total Payment
1	\$60	\$10	\$0.90	\$10.90
2	50	10	0.75	10.75
3	40	10	0.60	10.60
4	30	10	0.45	10.45
5	20	10	0.30	10.30
6	10	10	0.15	10.15
Total	—	\$60	\$3.15	\$63.15

(b) total cost of installment plan
= \$63.15 + \$20.00 = \$83.15

(c) installment charge
= installment price − cash price
= \$83.15 − \$80.00 = \$3.15

example: Mrs. Page bought a piano, priced at \$1250 plus $4\frac{1}{2}\%$ tax on the following terms: 10% of cash price down and 18 months in which to pay the balance. A service charge of 12% was added. What was the monthly installment? What was the total cost of the piano?

solution:

cash price	\$1250.00
sales tax ($4\frac{1}{2}\%$)	56.25
total price	\$1306.25
downpayment	125.00
balance (paid on time)	\$1181.25
plus 12% service charge	141.75
balance due on installment	\$1323.00

amount of each installment
= $1323.00 ÷ 18 = $73.50†
total cost of piano
= cash price + sales tax + credit charges
= $1250.00 + $56.25 + $141.75
= $1448.00

EXERCISE 54

1. A TV set can be purchased for $320 cash or $32 down and 12 payments of $25 each. How large is the carrying charge?

2. H. Barnes bought merchandise amounting to $118.25 from the Howe Department Store on the installment plan with no down-payment. Service charges amounted to 1% on the unpaid balance each month. (a) If 5 payments of $20 each plus service charges were made, how much was the last payment (sixth payment) including the service charge? (b) What was the total credit charge?

3. Miss Orr can purchase a coat for $110 cash or on a 90-day account, to be paid in 3 equal installments, that includes a service charge of $7. If the time plan is used, what would be the amount of each installment?

4. An outboard motor was advertised for sale as follows: "Used 35 horsepower outboard motor, good condition. $275 cash or $50 down and 6 payments of $42.50 each. Phone ________." Find the installment charge.

5. Under a budget plan, the Ames Department Store charges $1\frac{1}{2}$% on the unpaid balance to the nearest $10 at the end of each month. If Mr. Dodge's balances for 3 successive months were $293.13, $128.60, and $165.20, what were the service charges for each month? (The balances would be thought of as $290, $130, and $170 to which the $1\frac{1}{2}$% charge would apply.)

6. Baines Co. offered desk chairs for $17.50 cash plus 4.2% sales tax or $5.00 down plus a carrying charge of $1.25 for each chair on a

†If there had been a fraction of $1.00 remaining, it would have been added to the last payment.

3-month account. What would be the total cost of 6 chairs purchased by the time plan? How much are the monthly payments?

7. Find the installment charge for articles listed as follows: (a) cash price $457.68, downpayment $50.00, 12 installments of $35.70 each, and (b) cash price $98.65, downpayment 15%, 6 installments of $16.00 each.

8. Answer the following questions, using the time payment plan. (a) An article could be bought for a cash price of $345 less 15% discount, sales tax $3\frac{1}{2}$%, or a downpayment of 10% of gross cost and the balance to be paid in 6 equal installments with a total carrying charge of $13. Calculate the monthly payment amount. (b) An article costing $78 plus 3% excise tax and a 2% sales tax may be purchased for $10 down and 12 payments of $8 each. What were the total cash price, the total price on the installment plan, and the installment charge?

INTEREST RATES

Credit costs on installment purchases vary and can be deceiving. They are usually stated as a percentage but often are expressed in some other manner such as "so many dollars a month," "low, low daily cost," and "easy monthly payments to meet your budget." To compare costs, these charges must be reduced to a common base. Carrying charges are not interest. Nevertheless, if a customer is to have a choice, he cannot compare these costs with the cost of a loan to finance the purchase of merchandise unless they are reduced to a true annual interest rate (*effective rate of interest*).

When the installment contract calls for $1\frac{1}{2}$% a month on the unpaid balance, the true annual interest is equal to 18%. Since the interest for 1 month is $1\frac{1}{2}$%, then for a year it must equal $12 \times 1\frac{1}{2}$% or 18% (effective interest rate). Accordingly, the effective or true annual interest rate for 1% per month is 12%, for $\frac{3}{4}$% per month is 9%, etc.

When credit charges are expressed as a flat fee which is added to the purchase price and then repaid in a number of equal installments, the problem of determining the *effective* or *true annual interest rate* can become involved, especially for accounts for less or more than a year. There are several methods which may be used. However, the *constant ratio formula* used by the Federal Reserve System and recommended by Consumers Union will be used in this text, as well as a system called *the average balance method.*

Constant Ratio Formula

$$R = \frac{2mI}{P(n + 1)}$$

where

R = effective interest rate
m = number of installments per year (12 if monthly, 52 if weekly)
I = finance charges (dollar amount)
P = net amount of credit after downpayment or actual cash received on an installment loan
n = number of payments required

example: A heater may be purchased for \$39.90 cash or on an easy time plan that requires a downpayment of \$5.00 and \$3.10 per month for a year. If the time payment plan is used, what is the annual interest rate?

solution: cash price = \$39.90
downpayment = 5.00
balance due = \$34.90

installment cost =
(\$3.10 × 12) + \$5.00 = \$42.20

installment charges =
(\$42.20 − \$39.90) = \$2.30

Substituting in the formula,

$$R = \frac{(2)(12)(2.30)}{(34.90)(12 + 1)} = .122 \text{ or } 12.2\%$$

Average Balance Method

When this method is used, interest is found on the average of the principal balance at the beginning of the time plan and the principal balance due on the last payment.

$$r = I \div \frac{(P_1 + P_2)T}{2}$$

where

I = interest or installment charge
r = interest rate
P_1 = beginning principal balance
P_2 = ending principal balance
T = time in years (term of the contract)

Notice that this formula is actually the basic interest formula $I = Prt$. Solving for r, $r = I \div Pt$ in which case P = the average principal balance. In the problem above, discussed under the constant ratio formula:

beginning principal balance	= \$34.90	
ending principal balance	= \$ 2.91	(\$34.90 ÷ 12)
average principal balance	= \$18.91	(\$34.90 + \$2.91) ÷ 2
installment charge	= \$ 2.30	(see first solution)
time	= 1 year	

Substituting in the formula,

$$R = 2.30 \div (18.91 \times 1) = .122 \text{ or } 12.2\%$$

The Consumer Credit Protection Act of July 1, 1969 requires that lending institutions indicate the true interest rate on loans and the basis on which it is calculated. Legislation in recent years also requires that installment purchase contracts indicate the true interest rate the consumer is paying.

EXERCISE 55

Record all interest rates correct to the nearest tenth of 1%.

1. What is the annual interest rate on an installment contract for the following monthly charges on the unpaid balance: 1%, $1\frac{1}{4}$%, 2%, $\frac{1}{2}$%, $\frac{3}{4}$%, and $1\frac{1}{2}$%?

2. An article costing \$280 was purchased on the following terms: \$20 downpayment, the balance in 18 monthly installments, and a carrying charge of \$30. What was the annual interest rate? (In this problem, $m = 12$ if the constant ratio formula is used.)

3. The cash price of a desk is \$65.00 or, if purchased on the installment plan, 10% down and the balance in 6 monthly payments of \$10.50. If the time plan is used, what is the annual interest rate?
4. Mrs. Rappen purchased a carpet priced at \$450 cash from the J. C. Furniture Co. Terms were \$70 down and the balance in 12 monthly payments of \$35 each. If Mrs. Rappen bought the carpet on time, what was the total cost? What was the annual interest rate?
5. A rifle may be purchased for \$10.00 down and \$3.50 a month for 9 months. What are the total cost and the annual interest rate on this basis if the cash price is \$37.50 ($m = 12$)?
6. A stove sells for \$330.00 cash or \$40.00 down and \$10.50 a month for 36 months. Find the annual interest rate charged.

PERIODIC LOAN PAYMENT PLANS

In the preceding section, some aspects of installment credit as it applies to the purchase of goods and services were considered. We shall now consider installment credit as it applies to the borrowing of money.

Personal loans may be obtained from various lending agencies—insurance companies, commercial banks, savings and loan associations, credit unions, specialized finance companies such as those dealing in automobiles or real estate, small loan companies, and individuals. Again, as previously stated, when applying for credit it is important to check and compare dollar cost and interest rates before signing a loan agreement. Company policies and legal regulations differ between lending agencies, all of which should be taken into consideration. Some of the smaller finance companies loan on an installment basis only and sometimes attach a penalty to early settlement of a loan. On the other hand, with a few exceptions, banks do not charge for early settlement of a loan. However, many banks charge a minimum fee for any loan regardless of size or time involved, in which case the interest on a small loan for a short period of time could be excessive.

Small loan or finance companies are permitted to charge a higher rate of interest than that allowed ordinarily in order to offer credit to individuals who would be unable to obtain it otherwise. Like all lending or finance companies, certain regulations pertaining to loans are established by law and vary from state to state. These change from time to time. However, at the time that this book goes to press, the following example is illustrative of how interest charges are made in at least one state by small state-controlled finance companies. A charge of $2\frac{1}{2}\%$ a month (30% annual interest rate) is made on the unpaid principal balance for loans up to $300, 2% a month (24% annual interest rate) on the next $200, and $\frac{5}{6}\%$ a month (10% annual interest rate) on any sum over $500. In other words, if a person borrows $1000, each month he would pay $2\frac{1}{2}\%$ on the unpaid balance for $300, 2% on the next $200, and $\frac{5}{6}\%$ on $500.

It is also common practice among lending agencies to charge a fee which is a small percent ($1\frac{1}{2}\%$ to 5%) of the loan for closing costs and other expenses incidental to handling the loan, in addition to the interest charge on the unpaid balance.

Other than loans on real estate purchases, installment loans are usually short term—3 years or less. However, many loans for home repairs are being made up to 5 years. Personal loans are made for many reasons such as the purchase of a home, home improvements, purchase of an automobile, consolidation of debts, purchase of durable goods, emergencies such as unexpected medical expenses, refinancing, a trip, or a vacation.

Credit unions are able to offer loans to members on liberal terms and at the same time encourage them to save through the purchase of shares. A member who borrows from a credit union not only pays a low rate of interest but is the recipient of dividends or earnings from the lending activities of the union since he is a part owner. Conditions for membership and borrowing are regulated by law.

The most favorable loan is one in which the only cost is the interest on the unpaid balance. Many loans, in addition to the interest, include in the payments additional charges such as insurance to protect the lender against loss and other costs pertinent to the transaction. Some of these additional charges would, of course, have to be met by the borrower even if he paid cash. This is especially true of real estate or automobile transactions.

On loans in which the interest charge is added to the loan and then divided into 12 equal payments, the actual interest rate is about double that quoted. For example, if $4\frac{1}{2}\%$ is added, the true interest rate is 9%; if 8%, the true rate is 14.8%; if 6%, the true rate is 11.1%. Such charges are sometimes called flat rates.

FINDING INSTALLMENT PAYMENTS AND THE TRUE INTEREST RATE WHEN A FLAT RATE OF INTEREST IS CHARGED

example: Mrs. Holmes obtained a loan of $300 under the 6% plan (6% of the total loan for each year) to be repaid in 24 equal monthly installments. Find the carrying charge, the monthly installment, and the interest rate charged.

solution: carrying charge (6% of $300 for 2 years) = $ 36
principal + carrying charge = $336
monthly installment ($336 ÷ 24) = $ 14

Substituting in the constant ratio formula,

$$r = \frac{2(12)(36)}{(300)(24 + 1)}$$

$$= .115 \text{ } or \text{ } 11.5\% \text{ interest rate}$$

FINDING INSTALLMENT PAYMENTS WHEN INTEREST IS CALCULATED ON THE UNPAID BALANCE

A loan may be liquidated through payments in which the same amount is applied to the principal each period plus the interest on the unpaid balance. Or the loan may be *amortized* in which case all payments are equal (including principal and interest).

When the Same Amount is Applied to the Principal with Each Payment Plus the Interest on the Balance

For example, Mr. Dodge obtained a loan of $180 from his credit union which he was to repay in monthly installments of $30 plus 1% on the unpaid balance until liquidated.

The following is a schedule of his payments.

Payment Number	Principal Balance	Payment On Principal	Payment Interest (1%)	Total Payment
1	$180.00	$ 30.00	$1.80	$ 31.80
2	150.00	30.00	1.50	31.50
3	120.00	30.00	1.20	31.20
4	90.00	30.00	.90	30.90
5	60.00	30.00	.60	30.60
6	30.00	30.00	.30	30.30
Total	—	$180.00	$6.30	$186.30

At the end of 6 months the loan has been paid in full plus interest in the amount of $6.30.

When a Loan is Amortized

Amortized loans (all payments are equal) are usually for large amounts and spread over a period of years. For example, Mr. and Mrs. Warren purchased a home for $30,000. They made a downpayment of 20% and borrowed the balance at 6% for 20 years. (a) How much is their monthly payment? (b) How much of the payment for the first and second months is applied to the principal and how much to the interest?

price of home	$30,000	
downpayment	6,000	(20% of $30,000)
balance	$24,000	(sum borrowed)

(a) The total interest on this loan may be calculated by the average principal balance method (pages 274–275).

beginning principal balance = $24,000
ending principal balance = 100
  ($24,000 ÷ 240†)
average principal balance = $12,050
  ($24,000 + $100) ÷ 2

†Number of months in 20 years.

interest on average principal balance ($12,050 × .06 × 20)	= $14,460
total amount to be repaid ($24,000 + $14,460)	= $38,460
monthly payment ($38,460 ÷ 240)	= $160.25

This method results in an answer which is a little less than the true amount except for very short periods of time.

(b) With each succeeding month, the amount applied to the principal increases while the interest decreases since the balance of the loan becomes increasingly smaller.

			Amount Applied to	
Payment	Balance	Payment	Interest ($\frac{1}{2}$%)	Principal
1–first month	$24,000.00	$160.25	$120.00	$40.25
2–second month	23,959.75	160.25	$119.80	$40.45

Comparison of Methods

When these methods of repayment are compared, it can readily be seen that if the second loan is paid in the same manner as the first, the first payment would equal $220 ($100 principal + $120 interest) instead of $160.25. By amortizing the loan, the payment schedule becomes less burdensome since the principal and interest added together are spread throughout the life of the loan on an equalized basis.

If the first loan is amortized by the average principal balance method, then the monthly payments are determined as follows:

average principal balance ($180.00 + $30.00) ÷ 2	= $105.00
interest at 12% (1% per month) on $105.00 for 6 months)	= $ 6.30
total amount due ($180.00 + $6.30)	= $186.30
monthly payment ($186.30 ÷ 6)	= $ 31.05

EXERCISE 56

1. Mr. and Mrs. Rogers obtained a loan for home improvements in the amount of $2000 which was to be repaid in 3 years. The annual interest rate was 10%. On an amortized basis, what were the monthly installment payments?

2. Mr. Hupp borrowed $200 from the Local Finance Co. at $2\frac{1}{2}$% per month on the unpaid balance. If he repaid the loan in 4 monthly payments of $50 plus interest, what was (a) the amount of each payment and what was (b) the total interest cost?

3. The Lindsays purchased a home for $28,500. They paid 10% down and financed the balance at 8% over a period of 15 years on the installment plan. Find the monthly payment and the amount applied to principal for the first month if the loan was amortized.

4. Mr. C. Matteson holds a second deed of trust for $2600 at 7%. If this obligation is paid in 15 years, how much are the monthly payments on an amortized basis and how much did Mr. Matteson receive in interest during the life of the loan?

5. A bank offers unsecured personal loans at a flat rate of 8%. If a man borrowed $350 for a year and repaid it in 12 equal installments, how much were the payments? What was the true annual interest rate (use the constant ratio method)?

6. A couple bought 12 acres of orchard land at $2500 an acre from the owner. They made a downpayment of 10%. The owner agreed to carry the balance at $6\frac{1}{2}$% to be paid in equal monthly installments for 10 years. Find the amount of the monthly payment. Amortization

7. Mr. Guthrie borrowed $200 which he repaid in 12 monthly installments of $19.50. What was the annual interest rate?

8. For problem 6, set up an amortization table for the first 4 months showing the amount applied to principal and interest.

9. Miss Cole wanted a fur coat worth $720. She could buy it from the store for 10% down and 12 monthly payments of $65. Would it be more economical for her to borrow $720 at 8% for the same period of time to be repaid in 12 installments? If so, how much would she save?

10. Assume that the following charges (flat rates) are made on $400 for a year and that the loan in each case is to be repaid in 12 equal

monthly installments: 3%, 5%, $7\frac{1}{2}$%, and 9%. What true annual interest is the borrower paying?

11. A loan was obtained for $600 at 6% to be repaid every 6 months for 5 years on an amortized basis. What was the amount of each installment?

12. A store advertised cameras for $98, 10% down and the balance in weekly payments of $4 for 24 weeks. What was the interest rate?

EXERCISE 57 Chapter Summary

1. Mr. Graham purchased a camera selling for $425.00. He made a downpayment of $25.00 and agreed to pay the balance in monthly installments of $24.25 for 18 months. (a) Find the carrying or installment charge. (b) What is the annual interest rate (nearest tenth of 1%)?

2. In problem 1, suppose Mr. Graham made a downpayment of $25.00 as stated and paid the balance plus a service charge of 12% in 18 equal installments. (a) Find the amount of each installment and (b) the total cost of the camera.

3. Mrs. Arns made purchases amounting to $140.96 which she paid at the rate of $35.24 a month for 4 months. After the first payment (no interest) she paid $1\frac{1}{2}$% on the unpaid balance and $35.24 a month until the total amount was paid. How much interest or carrying charge did she pay?

4. A chair selling for $150 plus 5% sales tax may be purchased over a period of 6 months with a charge of $1\frac{1}{2}$% on the unpaid balance per month. If a downpayment of $19.50 is made, what are the (a) monthly payments, (b) total cost on the installment plan, and (c) difference between the cash and installment cost? Prepare a schedule.

5. What is the annual interest rate on an installment contract for the following charges on the unpaid balance? $2\frac{1}{2}$%, 3%, $1\frac{1}{2}$%, $\frac{3}{5}$%?

6. A cultivator may be purchased for $450, $50 down and the balance plus a 12% flat fee to be paid in equal installments for 16 months. What is the annual interest rate (nearest tenth of 1%)?

7. Mr. and Mrs. Adam purchased a home for $35,000 with the following terms: $5000 downpayment, balance at 7% for 20 years. How much is the monthly payment?

8. In problem 7, how much of the first and second monthly payments are applied to interest and how much to principal?

Objectives

This chapter is an introduction to computer number systems, and it concerns itself only with binary numbers. Its purpose is to explain what binary numbers are, how they compare to numbers in the decimal system, and how they are used, as well as some practice in addition, subtraction, and multiplication.

13 COMPUTER NUMBER SYSTEMS

In the course of human history, men have developed various languages in order to communicate with each other. Similarly, systems have also been developed to denote numbers and their relationships.

The system which we use is called the decimal system. It is discussed in detail on page 4 of the first chapter. The base for the decimal system is 10 with 10 discrete symbols, 0 to 9 inclusive. However, a number system may be based on any number of discrete symbols.

With the development of the electronic computer there was a need for a system that was easier to use than the decimal system. The binary system, because of its adaptability and simplicity, is the basic code used. The base is 2. There are two discrete symbols, 0 and 1. Each circuit in an electronic computer has two positions, "on" and "off." 0 represents the *off* position and 1 represents the *on* position. These 2 digits are used in various combinations to denote language, symbols, decimal numbers, and binary numbers or values.

In our own system, values are determined by position. This is also true of the binary system. In the decimal system, we move one place to the left to denote 10 (the base); in the binary system, we also move one place to the left to denote 2 (the base). The highest number that can represent the units position in the binary system is 1. Then to make the number 2, we must carry 1 to the next position to the *left*. This becomes the 2 position and is written as 10.

In the same way, using the decimal system, we move to the left for the next position for 100 or 10^2. In the binary system it is 4 or 2^2. Then the next position would be 10^3 or 1000 and 2^3 or 8, respectively.

Position values for the decimal system are 10^0, 10^1, 10^2, 10^3, 10^4, etc., that is, 1, 10, 100, 1000, 10,000, etc. Corresponding values for the binary system are 2^0, 2^1, 2^2, 2^3, 2^4, etc. or 1, 2, 4, 8, 16, etc.

POSITION VALUES FOR THE TWO SYSTEMS

Powers of digits*	5	4	3	2	1	0
Base 10	100,000	10,000	1,000	100	10	1*
Base 2**	32	16	8	4	2	1*

*Any number raised to the zero power equals 1, i.e., $2^0 = 1$, $5^0 = 1$, etc.
**In the language of the computer, 0 and 1 are referred to as *bits*. The 0 is the *no* bit and represents the off condition, and 1 is called a bit and represents the on condition. As you move from right to left, values are referred to as the 2-bit, the 4-bit, etc.

To compare the numbers in the binary system to those in the decimal system, we see from the chart that the values of 2, 4, etc. correspond to 10, 100, etc. From these we can easily determine how odd numbers are written in binary. For example, if $1 = 1$, $2 = 10$, then $3 = 1 + 10 = 11$. The number 4 in the decimal system is written as 100 in the binary system. The number 1 is the same in both systems. Then 5, i.e., $4 + 1$, in the decimal system, when written in the binary system is $100 + 1 = 101$.

POSITION VALUES*

Decimal number	Binary notation
1	1
2	10
3	11 (2 + 1)
4	100
5	101 (4 + 1)
6	110 (4 + 2)
7	111 (4 + 3)
8	1000
9	1001 (8 + 1)
10	1010 (8 + 2)
11	1011 (8 + 3)
12	1100 (8 + 4)
13	1101 (12 + 1)
14	1110 (12 + 2)
15	1111 (14 + 1)
16	10000
etc.	etc.

*For converting decimal numbers to binary numbers.

CONVERTING A DECIMAL NUMBER TO ITS BINARY EQUIVALENT

To convert any decimal system number to the binary number, proceed as follows: divide the decimal number by 2, repeating this process until the quotient reaches zero. Record remainders to the right as illustrated in the examples shown here.

example: Convert 51 to its binary equivalent.

solution:

2)51
2)25 with the remainder 1
2)12 with the remainder 1
2) 6 with the remainder 0
2) 3 with the remainder 0
2) 1 with the remainder 1
0 with the remainder 1

The remainders, read in reverse, represent the binary equivalent, 110011.

example: Convert 16 to its binary equivalent.

solution:

$$\begin{array}{rl} 2\overline{)16} & \\ 2\overline{)\ 8} & \text{with the remainder } 0 \\ 2\overline{)\ 4} & \text{with the remainder } 0 \\ 2\overline{)\ 2} & \text{with the remainder } 0 \\ 2\overline{)\ 1} & \text{with the remainder } 0 \\ 0 & \text{with the remainder } 1 \end{array}$$

Answer = 10000 (see previous chart)

CONVERTING A BINARY NUMBER TO ITS DECIMAL EQUIVALENT

Keep in mind the fact that the first position (units) is 2^0, the second position is 2^1, etc.

example: Convert 1001100 to its decimal equivalent.

solution: (1) Beginning with the first position, move to the left and record the value of each digit position as follows:

$$\begin{array}{lcl} \text{1st position} & = 0 \times 2^0 = 0 \times 1 = & 0 \\ \text{2nd position} & = 0 \times 2^1 = 0 \times 2 = & 0 \\ \text{3rd position} & = 1 \times 2^2 = 1 \times 4 = & 4 \\ \text{4th position} & = 1 \times 2^3 = 1 \times 8 = & 8 \\ \text{5th position} & = 0 \times 2^4 = 0 \times 16 = & 0 \\ \text{6th position} & = 0 \times 2^5 = 0 \times 32 = & 0 \\ \text{7th position} & = 1 \times 2^6 = 1 \times 64 = & 64 \end{array}$$

(2) Add right-hand column. Total = 76 (answer)

check:

$$\begin{array}{rl} 2\overline{)76} & \\ 2\overline{)38} & \text{with a remainder } 0 \\ 2\overline{)19} & \text{with a remainder } 0 \\ 2\overline{)\ 9} & \text{with a remainder } 1 \\ 2\overline{)\ 4} & \text{with a remainder } 1 \\ 2\overline{)\ 2} & \text{with a remainder } 0 \\ 2\overline{)\ 1} & \text{with a remainder } 0 \\ 0 & \text{with a remainder } 1 \end{array}$$

Answer = 1001100

example: Convert 11001 to its decimal equivalent.

solution: position of digits from right to left

position	
1	$1 \times 2^0 = 1 \times 1 = 1$
2	$0 \times 2^1 = 0 \times 2 = 0$
3	$0 \times 2^2 = 0 \times 4 = 0$
4	$1 \times 2^3 = 1 \times 8 = 8$
5	$1 \times 2^4 = 1 \times 16 = 16$

$16 + 8 + 1 = 25$ (answer)

check:

2)25
2)12 with a remainder 1
2) 6 with a remainder 0
2) 3 with a remainder 0
2) 1 with a remainder 1
0 with a remainder 1

Answer = 11001

ADDING BINARY NUMBERS

Rules for addition:

1. $0 + 0 = 0$
2. $0 + 1 = 1$
3. $1 + 1 = 0$ with 1 to carry to next position on left

example: A:

decimal	binary
8	1000
+ 5	+0101†
13	1101

proof:

$1 \times 2^0 = 1$
$0 \times 2^1 = 0$
$1 \times 2^2 = 4$
$1 \times 2^3 = 8$
13

Or

2)13
2) 6 1
2) 3 0 = 1101
2) 1 1
0 1

†Zeros preceding the digit 1 on the left do not change its value but may be used for convenience.

example B:

```
  10
 +11
 ---
 101
```

example C:

```
   110
  +111
  ----
 +1101
```

Check B and C answers against the chart shown on page 287.

These examples illustrate rule 3. The 1 is written over the next column to left as an aid in adding that column, or columns.

example D:

decimal	binary
41	101001
+29	+ 11101
70	1000110

(This example also illustrates rule 3 for addition. This is explained later.)

1 + 1 = 0 with 1 to carry to second position and to be to 0 + 0. Then 1 + 0 + 0 = 1 in second column.

3rd position: 1 + 0 = 1
4th position: 1 + 1 = 0 with 1 to carry over to next position
5th position: add 1 that was carried over, then 1 + 1 + 0 = 0 with 1 to carry to the sixth position
6th position: 1 + 1 = 0 with 1 to carry to seventh position
7th position: bring down the 1 that was carried over

The same rules apply for adding binary numbers with "binary points" as for whole numbers. Values to the right of the binary point are fractional values (minus powers of 2) of 2, that is $\frac{1}{2}$, $\frac{1}{4}$, $\frac{1}{8}$, and so forth.

example A:

```
   11.11
 + 11.01
  ------
  111.00
```

example B:

```
   101.10
 +  10.10
  1000.00
```

example C:

```
   100.10
 +111.00
  1011.10
```

SUBTRACTING BINARY NUMBERS

Two methods of subtracting binary numbers are illustrated here.

I. Rules for subtraction:
1. $0 - 0 = 0$
2. $1 - 0 = 1$
3. $1 - 1 = 0$
4. $0 - 1 = 1$ with 1 borrowed from the left

example A:

decimal		binary
14		1110
− 6		−0110
8	=	1000

example B:

decimal		binary	
17		10001	
− 9		−1001	
8	=	01000	or 1000

For example B, the fourth position $= 0 - 1 = 1$. The fifth position $= 0 - 0 = 0$.

II. A second method is to change the subtrahend to its complement and proceed as in addition illustrated as follows. This is an easier method that I and one which is used by computers. The complement of a binary number is easily written by changing each 0 to 1, and each 1 to 0.

example: The complement of 1000 = 0111

0101 = 1010

01010 = 10101

example A:

decimal number		binary number		
8		1000		1000
−5		−0101	or	− 101
3	=	11		11

procedure: 1. Change lower number to its complement. Our problem then becomes

```
 1000        1000
 1010   or    010
10010        1010
```

2. Add

3. Shift the first digit on left to units position and add, that is,

```
 1000        1000
 1010   or    010
10010        1010
    ↘1+         ↘1+
   11          11
```

example B:

```
 10001
 −1001
```

procedure:

```
Step 1.   10001
          +0110   complement of 1001
Step 2.   10111   added sum
Step 3.       ↘1  shift 1 in left hand
                  column and add
           1000   answer (0111 + 1)
```

Compare this answer with that shown for example B under rule I.

MULTIPLYING BINARY NUMBERS

I. Rules for multiplication:
1. $0 \times 0 = 0$
2. $0 \times 1 = 0$
3. $1 \times 1 = 1$

In multiplying binary numbers, the partial products are positioned and added just as they are when multiplying decimal numbers.

example A:

```
  100
   11
 ----
  100
 100
 ----
 1100
```

example B:

```
   111
   101
  ----
   111
 1110
------
100011
```

II. When multiplying large numbers, it is easier to add the first two partial products and in turn add their sum to the next partial product, etc.

example:

```
  39 =     100111
  46 =     101110
 ----     -------
 234      1001110
156      100111        11101010, sum of
----                   first 2 partial products
1794    100111        1000100010, sum of
                       first 3 partial products
      100110
      -----------
      11100000010      sum of all
                       partial products
```

proof:

```
2)1794
2) 897  0
2) 448  1
2) 224  0
2) 112  0
2)  56  0
2)  28  0
2)  14  0
2)   7  0
2)   3  1
2)   1  1
     0  1
```

The numbers in column to right when copied from bottom to top read 11100000010 which agrees with our problem; that is, 1794 expressed in binary numbers is equal to the product obtained earlier.

When multiplying numbers with binary points, follow the same rules as those used for whole numbers.

DIVIDING BINARY NUMBERS

I. Rules for division:
The same procedure is used as in the decimal system except that the rules for binary multiplication and subtraction must be used. This can become very cumbersome and complicated with pencil and paper. For that reason, only the simplest of problems are illustrated here. Since computers subtract by adding complements, it would be awkward to illustrate rule II here.

example A:

decimal number	binary number
4	100
9)36	1001)100100
	1001
	000000

example B:

decimal number	binary number
5	100
12)53	1100)110101
48	1100
5	101 remainder

EXERCISE 58

1. Write down the binary values that correspond to the following decimal numbers: 10^0, 10^1, 10^2, 10^3, 10^4 or 1, 10, 100, 1000, 10000.

2. Convert the following decimal numbers to binary numbers: 236, 79.

3. Convert the following binary numbers to the decimal number equivalent and check: 1011, 111011.

4. Add:

(a)	(b)	(c)
11011	1000	1100
1100	1001	1111

5. Check answers to (a) in problem 4 by converting 11011 and 1100 to decimal numbers and checking result as illustrated on page 289, example A.

6. Change 36 and 12 to binary numbers and add. Compare your answer with the decimal number equivalent.

7. Subtract the following binary numbers using either method.

(a) $\begin{array}{r} 1011 \\ \underline{0111} \end{array}$ (b) $\begin{array}{r} 10010 \\ \underline{1101} \end{array}$ (c) $\begin{array}{r} 11100 \\ \underline{1011} \end{array}$

8. Convert the following to binary numbers and check your answers with those shown in the decimal system.

(a) $\begin{array}{r} 17 \\ \underline{-\ 8} \\ 9 \end{array}$ (b) $\begin{array}{r} 26 \\ \underline{-14} \\ 12 \end{array}$

9. Multiply the following binary numbers.

(a) $\begin{array}{r} 101 \\ \underline{110} \end{array}$ (b) $\begin{array}{r} 1101 \\ \underline{111} \end{array}$ (c) $\begin{array}{r} 11011 \\ \underline{11101} \end{array}$

10. Prove your answer for 9(a) by converting both factors and answer to decimal numbers.

Objectives

This chapter is devoted to a study of the more common units of metric measurements and their relation to the English system. Although problems include converting from one system to the other, the goal is to give the student some exposure to thinking in terms of metric units and working with them in preparation for the day when they will replace our present system.

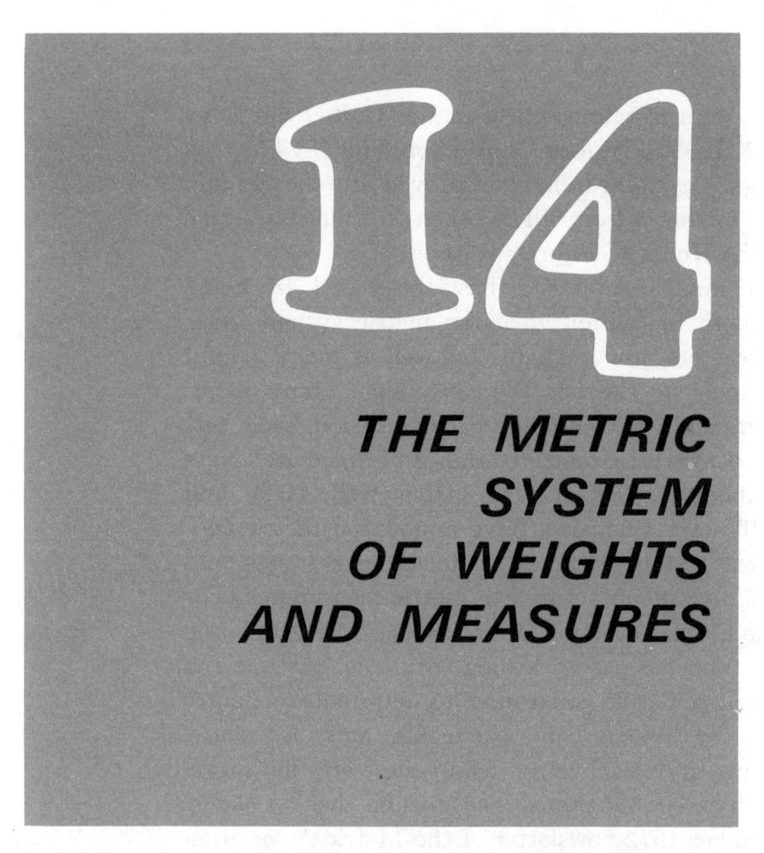

14 THE METRIC SYSTEM OF WEIGHTS AND MEASURES

Modern communication and transportation have brought the nations of the world closer, and since trade is a vital factor in their economy, a common international system of units is essential. The metric system was adopted in 1875 as the international system of weights and measures.

The United States is the only leading industrial country of the world that does not require the use of the metric system universally, and it has not initiated a conversion program to date. It does an import-export business with metric nations that is approximately $50 billion annually. These countries are increasingly demanding metric measurements. Consequently, if the United States adopts the metric system it will have a decided trade advantage in the world market. Not only will business benefit but every American will gain from the simplicity of using metric units.

Great Britain, Australia, and New Zealand have initiated conversion schedules that will extend over a period of several years. Many developments in this area are also taking place in Canada. A great deal of activity concerning the metric system is taking place in the United States as well. In 1972 the Pell Metric Conversion Bill, S 2483, was introduced and overwhelmingly approved by the Senate. However, no further action will be taken on it until hearings are held in the House of Representatives.

Meanwhile many conferences have been and are being held to promote the adoption of the metric system. Consultant services are now available (as well as many helpful educational tools) to assist in the conversion. Many manufacturing companies have already decided it is in their best interests to convert to metric units, and they are doing something about it. For example, Honeywell, IBM, and Caterpillar Tractor Company have announced metric conversion plans.† Many articles such as socket sets and wrenches are now manufactured in metric measurements as well as in the English system. Metric measurements are used for industries and businesses dealing with lens, cameras and optics, drugs, scientific instruments, automobiles, automobile tools for foreign cars, chemicals, and electronics. Some canning and packaging companies are indicating metric as well as United States measures on their products.

The November 1972 newsletter of the Metric Association announced that the Ford Motor Company has built a large plant at Lima, Ohio to build all metric engineered engines for its Pinto automobile. Also in that same bulletin was the announcement that the Sequoia Hospital in Redwood City, California has completed the conversion of all patient clinical records and measurements to the metric system. These include new scales for weighing patients, new thermometers, and metric tapes.

The metric system has many advantages over the cumbersome English system of measurement which we use. All units are based on multiples or powers of 10. Therefore, one is not burdened with facts such as 12 inches = 1 foot, 16 ounces = 1 pound, etc.

The greatest difficulty one has in converting from one system to another is in acquiring a "mental image" of the new unit of measure. For example, we all know or have a mental image of the length of a yard but not of a meter. However, if we know that a meter equals 39.37 inches or

†From *Metric Association Newsletter.*

1.094 yards, we can immediately envision the length of a meter in relation to our experience in working with yards. Similarly, a kilometer for the unitiated has little meaning unless we know that a kilometer is approximately $\frac{6}{10}$ths of a mile. In that case then, 10 kilometers is approximately 6 miles.

In adopting the metric system, one has to learn to think in terms of metric measurements and forget others. This is not easy to do unless we are exposed to or have experience in the use of metric units. This takes time. Since we will all be (and many are now) using metric units of measure, it is important that we learn to use them. Anyone interested in pursuing this subject beyond the scope of this text is invited to write to the Metric Association, Inc., 2004 Ash Street, Waukegan, Illinois, 60085.

CONVERTING ENGLISH MEASUREMENTS TO THE METRIC SYSTEM, AND CONVERSELY

LINEAR, SQUARE, AND CUBIC MEASUREMENTS

Before we solve any problems using the metric system exclusively, we shall first solve some problems that will involve converting English measurements to metric measurements and conversely. This will help us appreciate (have a mental image of) the metric measurements in terms of actual length, area, or volume when working with metric measurements alone. Tables 1, 2, and 3 are for this purpose.

TABLE 1. METRIC AND ENGLISH EQUIVALENTS OF COMMON UNITS OF MEASURE (LINEAR)

Metric	English	English	Metric
millimeter (mm)	.0394 in.	inch (in.)	25.4 mm
centimeter (cm)	.3937 in.	foot (ft)	.3048 m
decimeter (dm)	3.937 in.	yard (yd)	.9144 m
meter (m)	39.37 in.	rod (rd)	5.0292 m
	or 3.2808 ft	furlong (fur)	201.1684 m
dekameter (dkm)	32.8083 ft	mile (mi)	1.6093 k
hectometer (hm)	328.083 ft		
kilometer (km)	.62137 m		

TABLE 2. UNITS OF CAPACITY (LIQUID)

Metric	English	English	Metric
liter	1.0567 qt liquid or .9081 qt dry	fluid ounce (fl oz)	29.5729 m
		fluid pint (fl pt) (16 fl oz)	.4732 liters
dekaliter (dki) (10 liters)	2.6418 gal	quart (qt) or 2 pints (pt)	.9463 liters
hectoliter (hi) (100 liters)	26.418 gal	gallon (gal)	3.7853 liters
		Dry Measure	
		pint (pt)	.5506 liters
		quart (qt)	1.1012 liters
		bushel (bu)	35.2383 ft

TABLE 3. WEIGHT

Metric	English	English	Metric
1 gram (g)	.0353 oz	ounce (oz)	28.3495 g
dekagram (dkg) (10 g)	.3527 oz	pound (lb) (16 oz)	.4536 kg
hectogram (hg) (100 g)	3.5274 oz	hundredweight (cwt)	45.3592 kg
kilogram (kg) (1000 g)	2.2046 lb	ton	.9072 t
metric ton (t) (1000 kg)	1.1023 tons*		

*Short ton

Linear Measurements (Table 1)

Before solving any problems, consider the following relationships:

1. Both the millimeter (mm) and the centimeter (cm) are smaller than an inch. This is an advantage of the metric system in that it provides units of measure that are very small compared to an inch—measurements that are necessary in many fields of science and industry.

2. The metric system has a basic unit to which all other measurements are a power of 10, that is, 10, 100, 1000,

etc. or 1/10, 1/100, 1/1000, etc. Notice that a centimeter is 10 times a millimeter in value, a decimeter equals 10 cm, etc. We do not have this advantage in the English system; we have to remember that there are 12 in. in a foot, 3 ft in a yard, etc.

3. There is no metric measurement that corresponds to or is close to 1 in. or 12 in. (1 ft), common measurements in the English system. However, notice that a meter, which is the basic metric linear measurement, equals 39 + in. or a little more than a yard or 3 ft.

EXERCISE 59

This exercise is to test your knowledge of the relationship between the English and metric systems. Study Table I, then see if you can answer the following questions correctly without referring to the table. Write down your answers and then compare with the table when you have answered them all.

1. The basic unit of linear measure in the metric system is the ____?

2. All units of measure are a ______ __ ______ of the basic unit.

3. An inch is *more* than or *less* than a centimeter.

4. Approximately how many inches are there in a meter?

5. Is a meter *more* than or *less* than a yard? By approximately how much?

6. Is a mile *more* than or *less* than a kilometer?

7. If the distance between two points is 160 mi, will it contain more or less kilometers? Why?

Examples in converting from the English system to the metric system:

example A: A lot is 120 ft deep. How many meters does this represent?

solution: If 1 ft = .3048 m (from the table), then in 120 ft there will be 120 × .3048 or 26.576 m.

example B: How many feet are there in 3 m?

solution: 1 m = 3.2808 ft. Therefore, in 3 m there will be 3 × 3.2808 or 9.8424 ft.

example C: If 1 mi = 1.6093 km how many kilometers are there in 100 mi?

solution: 100 × 1.6093 = 160.93 k.

example D: A race track is 1 mi around. Convert this distance to meters.

solution: 1 mi = 1.6093 km; 1 km = 1000 m. In 1.6093 km there will be 1000 × 1.6093 or 1609.3 m.

Examples in converting from the metric system to the English system:

example A: 400 m = how many feet?

solution: 1 m = 3.2808 ft (from the table)
In 400 m there will be 400 × 3.2808 or 1312.32 ft.

example B: 400 m = how many yards?

solution: From problem A, 400 m = 1312.32 ft. Since we know that there are 3 ft in a yard, then 1312.32 ÷ 3 = 437.44 yd.

example C: A highway sign indicates a speed of 110 km/ph. How many miles per hour is this?

solution: 1 km = .62137 mi
110 km = 110 × .62137 = 68.3507 mi

example D: The Ace Garage Company carries spark plugs in sizes 14 to 18 mm in diameter. Convert sizes 14, 16, and 18 mm to the English equivalents in inches.

solution: 1 mm = .0394 in.

then, 14 mm = 14 × .0394 = .5516 in.
16 mm = 16 × .0394 = .6304 in.
18 mm = 18 × .0394 = .7092 in.

This problem illustrates the simplicity of using metric measurements instead of the English measurements where one would be dealing with whole units of measure instead of decimal fractions for objects with small measurements.

SQUARE AND CUBIC MEASUREMENTS

All square and cubic measurements are obtained in the metric system in the same manner in which they are obtained in the English system. Tables of units of area and volume are included in metric units only for the purposes of this chapter.

Units of Capacity (Table 2)

This is not a complete table but it does furnish a comparison between the two systems for common units of measure. Conversion from one system to the other is similar to that discussed under linear measurements.

example A: How many liquid pints are there in a liter?

solution: 1 liter = 1.0567 qt (liquid)
2 pt = 1 qt
then 1.0567 × 2 = 2.1134 pt in a liter

example B: Convert 4 qt to liters.

solution: 1 qt = .9463 liters
4 qt = 4 × .9463 = 3.7852 liters

example C: How many liters are there in 10 gallons?

solution: 1 gal = 3.7853 liters
10 gal = 10 × 3.7853 = 37.853 liters

Units of Weight (Table 3)

As in linear measurement, notice that the metric system has an advantage in having units of measure that are much smaller than the English ounce. For example, the gram is universally used as the basic unit for medications. It is used in most fields of science where weight is involved. The ounce, on the other hand, equals 28.3495 g which is much too heavy or large a measure for many purposes.

example A: How many grams are there in a pound?

solution: 1 oz = 28.3495 g, 1 lb = 16 oz
16 × 28.3495 = 453.592 g in a pound

example B: How many pounds are there in 5 kg?

solution: 1 kg = 2.2046 lb
5 kg = 5 × 2.2046 or 11.023 lb.

example C: If an English ton = .9072 t, how many metric tons are there in 250 English tons?

solution: 250 × .9072 = 226.8 t

EXERCISE 60

1. (a) What is the basic metric unit in linear measure? (b) How does it compare with a yard?
2. How many miles are there in 100 km?
3. If a building is 320 ft high, express it in meters.
4. A cross-country race was held that was 20 mi in length. (a) Convert this distance to kilometers. (b) If 1 km = 1000 m, how many meters are there in 20 mi?
5. The rainfall in a certain area in South Africa is 250 mm. How many inches is this?

6. Express a rainfall of 40 in. in (a) millimeters, (b) centimeters, (c) decimeters.

7. (a) How many quarts (dry measure) are there in a liter?
   (b) How many liters are there in a liquid quart?

8. How many liters are there in 5 gal?

9. In a certain hospital, babies weighing 1.5 kg are not allowed to go home until they weigh $2\frac{1}{4}$ kg. (a) How many pounds is 1.5 kg? (b) What must the babies weigh in pounds before being released?

10. How many ounces are there in a kilogram?

Use of Metric Measurements Without Reference to the English Equivalents

Now that we have an appreciation of the value of metric measurements to some degree—that is a mental image of the metric unit—we shall use them without reference to the English system in the following section. Tables 4, 5, 6, and 7 will be used.

Metric Measurements

Notice that in all the metric tables each unit is $\frac{1}{10}$, $\frac{1}{100}$, $\frac{1}{1000}$, etc. or 10, 100, 1000, etc. times the basic unit. You can also think of them as being some power of 10 as related to the basic unit. The following definitions will help identify each unit as related to another.

milli	= thousandths	kilo	= thousands
centi	= hundredths	hecto	= hundreds
deci	= tenths	deka	= tens

TABLE 4. LENGTH: BASIC UNIT IS THE METER

unit		meter
millimeter (mm)		.001
centimeter (cm)	= 10 mm	.01
decimeter (dm)	= 10 cm	.10
meter (m)	= 10 dm	Basic Unit
dekameter (dkm)		10
hectometer (hm)	= 10 dkm	100
kilometer (km)	= 10 hm	1000

TABLE 5. AREA: BASIC UNIT IS 100 SQUARE METERS (m^2) OR 1 ARE (A), PRONOUNCED AIR*

Unit	Meters
1 are (a) or (1 dkm^2)	100 m^2
100 ares (a)	1 ha or (100^2) m
100 hectares (ha)	1 km^2 or (1000^2) m

*All area measurements are derived from linear measurements just as they are in the English system. For example, 1 square dekameter (dkm^2) equals 10^2 or 100 m^2; an area 15km by 10 km would equal 150 square kilometers (km^2), etc.

VOLUME: BASIC UNIT IS A CUBIC METER (m^3) WHICH IS KNOWN AS A STERE (s). The same rule applies to obtaining cubic measurements as in the English system.

TABLE 6. LIQUID MEASURE: BASIC UNIT IS THE LITER, PRONOUNCED LEETER

Unit	Liters
milliliter (ml)	.001
centiliter (cl) = 10 ml	.01
deciliter (dl) = 10 cl	.10
liter (l)	Basic Unit
dekaliter (dkl)	10
hectoliter (hl) = 10 dkl	100

TABLE 7. WEIGHT: BASIC UNIT IS A GRAM (g)

Unit	Gram
milligram (mg)	.001
centigram (cg) = 10 mg	.01
decigram (dg) = 10 cg	.10
gram	Basic Unit
dekagram (dkg)	10
hectogram (hg) = 10 dkg	100
kilogram (kg) = 10 hg	1000
metric ton (t) = 10 kg	10,000

It is important to keep in mind the definitions of milli, centi, deci, deka, hecto, and kilo as they are the key to the size of the unit in relation to the basic unit. For example, since milli = thousandths, then

a millimeter = $\frac{1}{1000}$ or .001 m
a milliliter = $\frac{1}{1000}$ or .001 liters
a milligram = $\frac{1}{1000}$ or .001 g, etc.

The same identification should be noted for centi, deci, etc. as they readily identify the unit in terms of meters, liters, or grams.

Remembering these prefixes will also help you to quickly determine the answers to questions such as "How many millimeters are there in a meter?" Since milli means thousandths, and from that we know that a millimeter is .001 m, then there must be 1000 mm in a meter, 100 cm in a meter, 100 dm in a meter, etc.

Like the English system, shorter terms are used for some units. For example, it is easier and simpler to use and say "1 are" than "100 square meters." The term "ha" is often used for the same reason.

EXERCISE 61

This is a self-test or for class discussion. Study the tables and then answer the following questions without referring to the tables.

1. Express the following in terms of the basic unit.
   (a) millimeter (mm)
   (b) centiliter (cl)
   (c) decigram (dg)
   (d) kilometer (km)
   (e) deciliter (dl)
   (f) centigram (cg)
   (g) milligram (mg)
   (h) hectogram (hg)

2. (a) How many centimeters are there in a meter? (b) A kilogram equals how many grams?

3. How much larger is a centimeter than a millimeter? A kilometer than a hectometer? A kilometer than a dekameter?

4. How many meters are there in 1 hm? In 3 hm? In 100 cm?

Check your answers with the tables.

example A: A millimeter is what part of a centimeter?

solution: Without referring to the tables, we should have the answer immediately if we remember that milli means thousandths ($\frac{1}{1000}$) and centi means (hundredths ($\frac{1}{100}$). That is, a centimeter is 10 times a millimeter or a millimeter = $\frac{1}{10}$ of a centimeter.

example B: How many hectometers are there in a kilometer?

solution: Again, if we remember that hecto means hundreds and kilo means thousands, we know at once that there are 10 hm in a kilometer.

example C: How many square meters are there in 6 a?

solution: 1 a = 100 m^2. Then in 6 a, there are 6 × 100 or 600 m^2.

example D: A hectogram is how much more than a dekagram?

solution: If we know the definition of hecto and deka, the answer is simple. Referring to the table, we know that a hectogram is 10 times more than a dekagram.

example E: Change 6 kg to grams.

solution: 1 kg = 1000 g by definition. In 6 kg there will be 6 × 1000 = 6000 g.

EXERCISE 62

1. How many dekaliters are there in 5 liters? 20 liters, 500 liters?
2. A housewife purchased the following: 5 hg of meat, 250 g of dried beans, and 3 kg of potatoes. What was the total weight of her purchase in grams?

3. Mr. Brown bought a lot that measured 20 m wide and 100 m deep. (a) How many square meters in this? (b) Express area in nearest ares.

4. Mary bought 2 liters of punch, 1 cl of cream, and 1 dkl of milk. Express her total purchases in the basic metric unit.

5. A game preserve of 1500 ha equals how many ares?

6. Express the following in the basic unit? (a) decimeter, (b) kilometer, (c) dekaliter, (d) centigram, and (e) milligram.

EXERCISE 63

1. The A & B Construction Company erected a building that was 165 m in height. What does that represent in feet (nearest whole number)?

2. (a) What is meant by deci, kilo, milli, hecto? (b) How many meters are there in a hectometer?

3. (a) How many centimeters are there in a meter? (b) How many centimeters equal a decimeter? (c) One kilometer equals how many meters?

4. If the posted speed on a highway is 120 km, what is it in miles?

5. A lens supplied by a local manufacturer is advertised as having a focal length of 19 mm, length of the lens as 2 in and the barrel diameter $\frac{3}{4}$ in. Express all measurements in the metric system.

6. The Haven Optical Company supplies simple lenses with the following measurements:

lens diameter (mm)	focal length (mm)
12	25.4
25.4	76
22	25.4

Convert these measurements to the English equivalent to 2 decimals.

7. The Brown Trucking Company had a maximum limit of 8 tons on their trucks when loaded. Using a short ton (2000 lb) how many metric tons is this (nearest whole number)?

8. The Hampton Canning Company includes both metric and English measurements on their labels. (a) Find the metric equivalent in grams

for the following (nearest whole number): 9 oz, 24 oz or $1\frac{1}{2}$ lb. (b) Find the English equivalent for the following: 305 g and 400 g. Express answers in ounces to 2 decimals.

9. The Arne Bottling Company put both metric and English measurements on its bottles. Find the missing equivalents to 3 decimals. (a) 12 fl oz = ________ liters, (b) .473 liters = ________ oz.

10. (a) How many liquid pints are there in a liter? (b) If an automobile had a capacity of 19 gal, how many liters could it hold?

11. Mr. Parker lived on a hill 1500 m above sea level. Convert this distance to feet (nearest whole number).

12. If the rainfall in a certain area was 42 in, determine the amount in millimeters.

13. A piece of land measures 80 by 120 m. How many ares is this?

14. Convert 326 mg to grams.

Objectives

Review and practice are necessary if one is to maintain the required skill and understanding needed to be successful in any area of study. The problems and exercises offered here are intended to help the student with additional applications or to offer more testing material for the instructor. Whatever the need, the many problems and exercises in this section should help to meet it.

CHAPTER I BASIC OPERATIONS OF ARITHMETIC

1. A grocer sold apples in bags of 3 dozen each. If he purchased them in the bulk, how many bags could he fill from 50 boxes containing 66 apples to the box? (Express remainder, if any, as a fraction.)

2. What is the average daily payroll per employee for the Ace Box Co. if the total daily payroll for 25 employees equaled the following amounts: $675, $650, $700, $630, and $625?

3. A salesman for the Hemp Motor Co. traveled 315, 265, 179, 250, 310, 276, 410, 395, and 160 miles during 9 consecutive days. (a) What was the average daily mileage? (b) If he was allowed 9¢ a mile for travel and $12 a day for other expenses, how much did the company reimburse him for the trip?

4. Miss Brown typed, on an average, $2\frac{1}{2}$ stencils per hour for 8 hours. If she was paid $3.50 per hour, what was the cost of typing per stencil?

5. A typist was given a manuscript of 324 pages to type. She received $2.75 per hour for her work. If it took her 72 hours to complete the job, how many pages on an average did she type per hour? What were her total gross earnings?

6. The cost of manufacturing an article increased during the first year from $8.95 to $11.50, the second year to $15.26, and the third year to $17.55. What was the dollar increase in (a) 1 year, (b) 2 years, and (c) three years?

7. Smith owned $\frac{1}{5}$ of a piece of property. If the total value of the property was $19,266, (a) what was his share worth? (b) If he sold $\frac{1}{3}$ of his share for $1500, what was his profit?

8. Hayes spent $\frac{1}{4}$ of his salary for housing, $\frac{1}{3}$ for food, $\frac{1}{6}$ for entertainment, and $\frac{2}{9}$ for clothing. What fractional part did he have left for other needs?

9. The taxes on a piece of property last year were $675.16. This year they amounted to $\frac{7}{5}$ of that figure. How much were the taxes this year?

10. A house was sold for $22,528 which was $\frac{3}{8}$ more than it cost the owner. What did the owner pay for the house?

11. What number must be added to 4.2168 + 0.01762 to get 5.3?

12. Complete the following invoice, and total.

(a) $8\frac{1}{3}$ yards cotton @ $4.80 yard
(b) $\frac{5}{6}$ yards padding @ $2.56 yard
(c) $32\frac{1}{2}$ yards cotton @ $3.20 yard
(d) 25 yards cotton @ $4.15 yard
(e) $15\frac{3}{8}$ yards cotton @ $.90 yard
(f) $26\frac{2}{5}$ yards lining @ $.16$\frac{2}{3}$ yard

13. Complete the following invoice and total.

	Item	Price
(a)	5 dozen pencils	$4.50 a gross
(b)	26 folders	0.15 each
(c)	$1\frac{1}{3}$ dozen erasers	0.10 each
(d)	6 bottles black ink	4.20 a dozen
(e)	$3\frac{1}{3}$ dozen blotters	0.40 each
(f)	$\frac{1}{4}$ dozen bottles red ink	25.20 a dozen

14. Complete the following invoice and total.

(a) 2450 pounds @ \$2.02 cwt.
(b) 316 pounds @ \$1.25 cwt.
(c) 64 pounds @ \$6.25 M
(d) 625 pounds @ \$1.95 M
(e) 2367 pounds @ \$2.13 M
(f) 54 pounds @ \$.87½ cwt.

15. The Analy Market has 3 cash registers. Sales for each on a particular day during a typical hour were as follows: Reg. No. 1: \$3.04, \$0.79, \$15.67, \$32.50, \$1.00, \$22.66, \$19.65, \$0.91, and \$10.16. Reg. No. 2: \$37.89, \$41.20, \$11.99, \$5.60, \$11.27, \$8.86, and \$10.01. Reg. No. 3: \$20.06, \$52.00, \$17.89, \$5.89, \$6.03, \$14.56, \$27.16, and \$30.09. Find the total sales for each register and the total sales for all 3 registers.

16. Deliveries of milk to the Green Haven Cheese Factory from 5 local dairies were as follows: 2470 gallons, 3240 gallons, 2210 gallons, 1570 gallons, and 3765 gallons. How many gallons were delivered to the factory? What was the approximate number of gallons delivered?

17. Mrs. Yule purchased the following items: gasoline, \$4.65; groceries, \$16.26; garden plants, \$25.65; and stamps, \$5.00; she also spent \$5.50 at the hair dresser. What was her total expenditure for the day?

18. Monte Highland District conducted a drive to raise funds for the local orphanage. How much was collected during the first week of the drive if donations were received in the following amounts: \$16.50, \$2.00, \$5.00, \$12.50, \$0.50, \$35.00, \$5.00, and \$10.00?

19. During the week, R. J. Hopland worked $8\frac{1}{4}$, 7, 10, $7\frac{3}{4}$, $6\frac{3}{4}$, and 4 hours. If he received \$6.50 per hour, how much did he earn?

20. An inventory count in the yardage department of a retail store indicated that there were 6 rolls of #216 gingham containing $16\frac{1}{8}$, $27\frac{1}{2}$, $36\frac{2}{8}$, $45\frac{3}{8}$, $50\frac{7}{8}$, and $42\frac{5}{8}$ yards, respectively. At \$1.76 a yard, what was this stock worth?

21. A grocer bought $255\frac{3}{4}$ pounds of bacon which he divided into $\frac{1}{2}$-pound packages. How many packages did he have for sale?

22. A carpenter used $31\frac{1}{3}$ feet of molding which was cut into $\frac{3}{4}$-foot lengths. Allowing $\frac{7}{12}$ feet for waste, how many $\frac{3}{4}$-foot lengths was he able to cut?

23. The monthly electric bills for the Moffits during last year were: January, \$27.08; February, \$25.01; March, \$28.02; April, \$16.57; May, \$16.77; June, \$16.26; July, \$16.01; August, \$14.83; September, \$15.97; October, \$15.47; November, \$16.77; and December, \$21.05. How much more was the average electric bill for the months of November through April as compared to the average for the rest of the year?

24. Sales for the Rockwell Mfg. Co. amounted to \$376,000 last year. During the first 6 months of this year the monthly sales were \$37,000, \$15,000, \$26,500, \$37,500, \$28,750, and \$32,500, respectively. How much must the sales total for the remainder of the year to equal last year's figure?

25. Complete the following report. Find the: (1) net sales (gross sales less returns and allowances), (2) total gross sales, (3) total returns and allowances, and (4) total net sales. Verify all totals.

WOHLERS HARDWARE STORE
HAINSBROOK, COLORADO

Sales Report for the Month of June, 19___

Department	Gross Sales	Returns & Allowances	Net Sales
A	\$ 7,350.10	\$252.25	\$
B	9,072.64	572.48	
C	14,106.61	604.90	
D	11,593.48	747.56	
Total	\$	\$	\$

26. (a) What number must be subtracted from 10.459 + 9.034 to get 6.352? (b) What number must be added to 6.034 + .0069 to obtain 11.06843?

27. Complete the following invoice.

JEFFERS GENERAL MERCHANDISE
Jeffers, Montana

Sold to: Dave Shipley
Portland, Oregon

Invoice No. 2–0678–A
Date, May 2, 19____

Item No.	Quantity	Price	Extension
16	100 lb	$2.16 lb	$
217	1.5 yds	.25 yd	
63A	15.25 gal	.10 gal	
92	75 bbls	3.26 bbl	
31B	126 yds	.99 yd	
17	125 pcs	1.15 each	
		Total	$

28. (a) What is the approximate difference between $3168.50 and $1129.55?
(b) What is the approximate decrease in attendance at the theater last week of 3800 as compared to 2925 this week?

29. (a) Mr. Allen drove his car 2200 miles during the past year and used 122 gallons of gasoline. What was the approximate number of miles per gallon?
(b) Peter Sams paid 32¢ a gallon for gasoline. If he traveled 398 miles during the month, what was the approximate cost?

30. The monthly telephone bills for the Banes Optical Co. for 6 months were: $47.16, $59.25, $63.75, $43.50, $59.25, and $87.16. What was the average monthly telephone bill during this time?

31. An automobile traveled $868\frac{1}{2}$ miles during which time 38.6 gallons of gasoline had been used at a cost of $12.31. (a) How many miles did the car travel on a gallon of gasoline? (b) What was the average cost per gallon of gasoline (3 decimals)? (c) What was the average cost per mile for the trip (3 decimals)?

32. Tickets for the senior play at the local high school were sold at \$1.25 for general admission, \$0.75 for student body members, and \$0.50 for children under 10 years of age. Before opening night, tickets had been sold as follows: Team A—76 general admissions, 225 student body memberships, and 83 children. Team B—67 general admissions, 175 student body memberships, 105 children. How much was collected by each team and what was the total?

33. Hays bought 3 items at $\$2.17\frac{1}{2}$ each and $6\frac{1}{2}$ articles at $\$0.78\frac{1}{4}$ each. How much change should he receive from a \$20 bill?

CHAPTER 2 PERCENTAGE

Record all percents to the nearest tenth of 1% unless the remainder can be expressed as a common fraction such as $12\frac{3}{8}\%$, $15\frac{1}{9}\%$, and $14\frac{1}{6}\%$.

1. What is $12\frac{1}{2}\%$ of 326, 0.176, $3\frac{3}{4}$, $\frac{5}{8}$, and 19?

2. What percent is $\frac{3}{4}$ of 26, $\frac{1}{4}$, 16, .012, and $\frac{5}{9}$?

3. 216 is 6% of what? 3% of what? $\frac{1}{2}\%$ of what? 110% of what?

4. (a) 182 increased by 12% of itself is how much? (b) 842 decreased by 30% is how much?

5. If $26\frac{1}{2}\%$ of a number is 622.75, what is 130% of the number?

6. Mr. Hope purchased a lot for \$1500 which he sold 4 years later for \$3500. What was his profit expressed in dollars and in percent?

7. Mitchell received a commission of \$77 on sales of \$550. What was his commission rate?

8. James earned \$510 in commissions during December. His commission rate was 15%. What were his total sales?

9. Howard, Thomas, and Braden owned a business jointly which was worth \$376,000. Their interests in the business were 32%, 14%, and

54%, respectively. How much did their respective interests represent in dollars and cents?

10. Holt, Lang, Smith, and Shibley divided their profits in proportion to their shares in the business. Holt's share was $17,600, Lang's share was $25,000, Smith's share was $19,500, and Shibley's share was $32,600. On this basis, how much did each receive if profits for the year amounted to $15,152?

11. Sales for the E & P Stores increased from $96,500 last year to $127,540 this year. What percent increase does this represent?

12. Due to unseasonable weather this year, tomatoes were selling at 69¢ a pound in July which was an increase of 41% as compared to May prices and represented an increase of $91\frac{2}{3}$% over the previous year. What was the price of tomatoes in May of this year and July of the previous year?

13. If Mr. Opal spent 11% or $594 of his income for entertainment last year, what was his annual income?

14. The floor space occupied by the 5 departments of Havens & Co. (merchants) was as follows: Dept. A, 37,000 square feet; Dept. B, 29,600 square feet; Dept. C, 14,800 square feet; Dept. D, 55,500 square feet; and Dept. E, 48,100 square feet. What percentage did each share of the total floor space?

15. A house worth $18,500 was mortgaged for 70% of its value. What was the downpayment?

16. An article sold for $125.05 which was 22% more than the cost. What was the cost?

17. A certain item sold for $4.00. Of 600 of these items, 55 were still on hand at the end of the year. To clear the stock, they were reduced in price to $3.50 each. If these items cost $1.90 each, what were the: (a) total profit, (b) average selling price, and (c) percent reduction in selling price of the last 55 items?

18. A company had assets worth $630,000,000 and liabilities amounting to $320,000,000. Express the relationship of assets to liabilities as a (a) fraction and (b) percent?

19. If $3\frac{2}{5}$ percent of a certain number is subtracted from 98.046, the result is 41.946. What is the number?

20. If a yearly income of $1300 represents $6\frac{1}{2}\%$ of the investment, how much was invested?

CHAPTER 3 TRADE AND CASH DISCOUNTS

1. A local store sold chairs to clear at $57.50 each less 15%. What was the net price?

2. If the net price of a book is $2.72 after a 20% discount, what is the list price?

3. Mr. Axtel received a discount of 14% amounting to $28 on a suit. What were the list and net prices?

4. The Cox Mfg. Co. sold a piece of equipment for $291.50 that was originally priced at $318.00. What discount rate did this represent?

5. Find the net decimal equivalent for the following series: $10\text{–}5\text{–}1\frac{1}{2}\%$, 12–7–5%, $50\text{–}40\text{–}12\frac{1}{2}\%$, $33\frac{1}{3}\text{–}5\text{–}2\%$, and $65\text{–}12\frac{1}{2}\text{–}3\%$.

6. In problem 5, what are the single decimal equivalents for the series shown? What are the single discount equivalents?

7. Find the net price of an article listed at $312 less 10% and 5%.

8. Which is the best deal, a chair listed at $75 less 33% and 10% or $75 less 43%? By how much?

9. A desk was sold for $76.95 after discounts of 10% and 5% had been taken. What was the list price?

10. Trade discounts of 20%, 10%, and $7\frac{1}{2}\%$, which amounted to $197.08, were allowed on a refrigerator. What were the list and net prices?

11. Barker Bros. purchased a lathe listed at $1050 less 5%. They were billed October 1 with terms of 2/10, n/30. They paid this obligation on October 19. What was the amount of their remittance?

12. A local retailer was able to purchase some TV sets from Melody Stores for $325 less 15%. He was granted a 1% cash discount for payment within 10 days. What did these TV sets cost if he took advantage of both the trade and cash discounts?

13. Mr. Dodge received an invoice for a shipment of goods which amounted to $752 with terms of 2%–10th E.O.M. The invoice was dated May 25. On June 9, a partial payment of $352 was made. Find the amount of the discount and the balance due on the invoice.

14. The Bryan Bicycle Co. offered a special discount of 15% and 10% on their No. 206 bikes that normally sold for $65 net. An additional 1% discount was allowed for cash. A dealer bought six on this basis. If he paid cash and there was a 4% sales tax as well, what was the total amount of the purchase?

15. Mays Equipment Co. received a bill for $525.75 dated June 10 with terms of 3/10, 1/15, n/30. If it was paid on July 15, what was the amount of the remittance?

16. The Welk Hardware Store listed their best garden rakes at $5.50 less 10% and 5%. To meet competition, the net price was dropped to $4.47. What additional discount was added to meet this drop in price?

17. Find the last date of payment for the following problems that will allow the buyer to receive the best discount.

Invoice	Terms	Date of Invoice
a	2/10, n/30	February 15
b	1/10, E.O.M.	May 25
c	2% 15 Prox.	October 20
d	3/10, 1/30, n/90	November 22

18. Find the amount of discount and balance due for the following invoices.

In-voice	Amount of Invoice	Date of Invoice	Terms	Partial Date	Payment Amount
a	$396.50	3/1	2/10, n/30	3/11	$145.00
b	198.00	5/26	1/10 E.O.M.	6/8	120.00
c	1056.75	9/15	3/30, n/60	10/12	725.00

19. The Bluebell Paper Co. shipped the following items on 2 invoices to the Ames Publishing Co. with terms of 1%–10th E.O.M.

(a) Invoice No. 6–0972

70 reams, $8\frac{1}{2} \times 11$—16# white mimeo paper @ $2.76 ream
10 reams, $8\frac{1}{4} \times 11$—16# pink mimeo paper @ $2.88 ream
Trade discount of 50% and 10%

(b) Invoice No. 6–0973
    4 rolls, $\frac{3}{4}''$, #810 mending tape @ $1.25 a roll
    2 rolls, 1″, #898 filament tape @ $3.49 a roll
    1 #H–10 filament tape dispenser @ $4.50 each
    No trade discount but a sales tax of 4% was added.

If both of these invoices were dated June 22 and paid July 10, how much should be remitted for payment in full?

20. Compare $35.00 less 15–10–1% with $35.00 less 26%. Which is the best price? What is the difference?

CHAPTER 4 MERCHANDISING

1. An article costs $3.75 and sells for $4.70. What are the markup and the rate (a) based on cost and (b) based on selling price?

2. The Roden Electric Supply House buys portable heaters for $47.50 each. How much is the selling price if a 15% markup is based on (a) the selling price or (b) the cost price?

3. An electric saw is advertised by a dealer for $67.88 less 25% and 5%. If a markup of 22% on cost is realized, how much did the dealer pay for the saw?

4. What was the cost to a dealer on an article with a markup of $35.75 which was 15% of the selling price?

5. The Green Brae Shop reduced one of its Christmas speciality items $33\frac{1}{3}$% after the holidays. If the original markup was 40% of the selling price and the cost of the items was $3.50 each, what was the reduced selling price?

6. Haines Dress Shop marked down all of their velvet skirts $33\frac{1}{3}$% in a January sale to sell at $4.50 each. What were the amount of the markdown and the original selling price?

7. Wooden & Co. marked up their shirts 45% on cost. Some were later reduced 20% for clearance. What was the markup on cost for these sale shirts?

8. What is the markdown rate on a rug that normally sells for $175 if the price is reduced to $120?

9. A loss of $12.60 was incurred on an article which amounted to 4% of the selling price. What were the selling price and the cost price?

10. An article sells for $55. Net profit is 10%; overhead is 18%; selling expense is 22%. Gross profit is based on the selling price. Find the cost.

11. Corrick & Sons prices its china to realize a net profit of 15% on the selling price. If the cost of a set of china is $112.50, overhead amounts to 15%, and the selling expense is 9%, what is the selling price? How much is the net profit? The gross profit? (Markup is based on the selling price.)

12. Gamble Rug Co. sold a rug for $315 that included a markup of 35% on the selling price. It was purchased from N. & E. Wann & Co. (wholesaler). At what price was the rug listed by the wholesaler if trade discounts of 10% and 5% were granted?

13. Mainland's Department Store sold a table at a loss of $13\frac{1}{2}$% or $19.71, based on the retail price. What were the (a) cost and (b) retail price?

14. The gross profit on an article sold by James Bros. was $12.80 or 8%, based on cost. Buying expenses amounted to $17.50. What is the cost exclusive of buying expenses?

15. The Manner Clothiers purchased men's jackets at $35.55 less a trade discount of 20%. If the markup is $16\frac{2}{3}$% of the cost, what is the selling price?

16. A merchant sold a piece of electrical equipment for $845. After he allowed a discount of $12\frac{1}{2}$%, he realized a gross profit of $33\frac{1}{3}$% of the selling price. What was the cost price to the merchant?

17. The Sterling Furniture Co. bought merchandise for $346.20 less 15% and $2\frac{1}{2}$%. Buying expenses were $34.60. If the merchandise was sold at a markup of 20% of cost, what was the selling price?

18. Neward Gift Shop wishes to earn a 10% profit on a shipment of 135 fancy wastebaskets which they purchased for $2.65 each. Shipping and other handling charges amounted to $26.55. Overhead expenses were 20%; selling expenses were 8%. (a) How much must the company sell each basket for if a 10% profit is to be realized? (b) What is the net profit for the total shipment? (Base = selling price.)

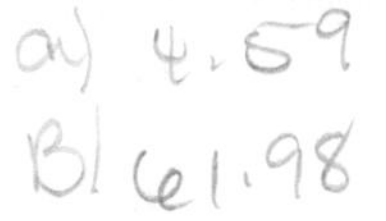

19. A hardware store buys cabinet showers at \$62 less $12\frac{1}{2}\%$. They are sold at a profit of 20% on the net selling price after allowing the customer a discount of 15%. Find the list price and net retail price.

20. Find the retail price on a coat that cost \$65 if 20% of retail is allowed for overhead, 8% is allowed for selling expense, and a net profit of 15% is desired?

CHAPTER 5 DEPRECIATION

1. A building worth \$35,000 is to be written off in 30 years with no scrap value. If the straight-line method of depreciation is used, what are the annual depreciation and the rate of depreciation? What is the book value at the end of 15 years?

2. A typewriter that cost \$420 is considered to have a useful life of 5 years. At the end of that time its scrap value is estimated at \$35. Prepare a depreciation schedule by the straight-line method.

3. A desk worth \$197.50 is to be written off in 10 years with a scrap value of \$20.00. Find the annual depreciation rate and the book value at the end of 5 years. (Use the straight-line method.)

4. The yearly depreciation on office furniture costing \$2420 was \$158. Scrap value equaled \$50. What was the estimated life of the furniture?

5. A planing mill and equipment worth \$67,500 is to be depreciated over 30 years by the straight-line method. Its scrap value is estimated to be \$2000. An additional first-year depreciation is taken. On this basis, find the book value at the end of 5 years.

6. Prepare a depreciation schedule by the sum-of-the-years-digits method for problem 2. Compare the book value at the end of 3 years with the value at the same time by the straight-line method. How much more is the book value by the straight-line method?

7. A piece of furniture costing \$820 has an estimated life of 10 years and a salvage value of \$30. Find the depreciation for (a) the third year and (b) the sixth year by the sum-of-the-years-digits method.

8. Find the book value at the end of the third and eighth years for problem 7.

9. A man purchases \$900 worth of laundry and bathroom equipment which he estimates will last 10 years. At the end of that time this equipment will be worthless. What is the depreciation charge at the end of the first and second years by the sum-of-the-years-digits method of depreciation?

10. Find the book value of the laundry and bathroom equipment in problem 9 at the end of the eighth year.

11. If an additional depreciation charge was taken for the purchases in problem 9, what would be the depreciation charge for the first year?

12. A truck purchased for $3200 has an estimated life of 5 years with a scrap value of $675. If the truck was sold for $1900 after 3 years, how much did the owner gain or lose by the (a) straight-line method of depreciation and (b) sum-of-the-years-digits method of depreciation?

13. In problem 12, what is the accumulated depreciation by the declining balance method at the end of 3 years?

14. A cultivator purchased for $289.50 had an estimated life of 10 years. Its scrap value is 20% of the cost. What is the book value at the end of 5 years by the sum-of-the-years-digits method of depreciation?

15. If the cultivator in problem 14 was sold at the end of 6 years for $80, did the owner gain or lose on the transaction? By how much? Use the sum-of-the-years-digits method.

16. An asset worth $360 is written off in 5 years. Scrap value is estimated at 15% of cost price. Prepare a depreciation schedule using the declining-balance method.

17. An automobile worth $3200 is written off in 5 years. Scrap value is estimated at $325. What is the depreciation for the first year by the (a) straight-line method, (b) declining-balance method, and (c) sum-of-the-years-digits method of depreciation?

18. If the declining-balance method of depreciation is used, what rate should be used for assets that are written off in 12 years? In 20 years? In 30 years? In 50 years?

19. A machine costs $540 and has an estimated life of 8 years. It has no salvage value. However, an additional first-year depreciation is taken. What is the depreciation for the first year by the three methods of depreciation discussed in this text?

20. In problem 19, find the book value at the end of 4 years by the declining-balance method of depreciation.

CHAPTER 6 PROPERTY AND SALES TAXES

1. The budget for a town is \$5,650,000. Of this amount, 85% must be raised by property taxes. What is the tax rate per \$100 if the total assessed valuation of property in the town is \$25,360,000? (Carry answer to 3 decimals.)

2. Mrs. Smith's home is assessed at 24% of its market value. If the market value is \$25,650, what is the assessed valuation?

3. In problem 2, if the tax rate is \$9.067 per \$100, how much did Mrs. Smith pay in property taxes?

4. What is the tax on a piece of property if the assessed valuation is \$660 and the tax rate is \$8.33 per \$100?

5. Find the amount of tax for the following: (a) assessed valuation \$4500, tax rate $8\frac{1}{3}\%$, (b) assessed valuation \$32,000, tax rate \$10.91667 per \$100 and (c) assessed valuation \$16,500, tax rate \$0.0756 per \$1.00.

6. Find the assessed value and amount of tax paid for the following:

	Market Value	Assessment Rate	Tax Rate
(a)	\$30,000	25%	\$11.50 per \$100
(b)	15,750	22.4%	\$0.096 per \$1
(c)	52,500	27%	$9\frac{1}{2}\%$

7. Mrs. Hale owned a home which was assessed at \$5500. If the assessment rate was 22%, what was the market value?

8. Taxes amounting to \$1794 were paid on a piece of property in a district where the tax rate was $11\frac{1}{2}\%$. At this rate, what would be the expected market value of the property if the assessment rate was 30%?

9. David Shipley purchased a home for \$28,500 in a city with an assessment rate of 25%. The tax rate is \$9.35 per \$100. How much of his total tax is spent on education if 65% of the city taxes are spent for this purpose?

10. Mrs. Hays owns a rental unit from which she receives a gross income of \$175.00 per month. The tax rate is \$10.9667 per \$100. The market value of this property is \$18,000; the assessment rate is 26%.

During the year, repairs amounted to $15.63. Also, a new roof was laid at a cost of $350.00. If the cost of the new roof was depreciated over 10 years by the straight-line method, how much was Mrs. Hays's net income for the year?

11. The tax rate in a city is $8.12 per $100. If the annual budget requires $5,317,000 to be raised through property taxes, what is the total assessed value of the property in the city (nearest dollar)?

12. Mr. Perkins purchased a chair for $79.50, a coffee table for $56.20, and a lamp for $45.00. As a preferred customer of the store, he was given a 5% discount. If 1% local and 3% state sales taxes were added, what was the total amount of Mr. Perkins's purchase?

13. J. Hawks ordered 3 tires for his car at a cost of $17.95 each. The excise tax on each tire was $2.86. What was the total amount of this purchase if a 4.2% sales tax was added?

14. A camera was sold for $125.15 including sales tax of 3%. What was the list price?

15. H. Kemp purchased 2 dresses at $27.50 each. A $\frac{1}{2}$% city tax and a 5% state tax was charged on 1 dress. The other dress was mailed out of the state (no tax) with a postage charge of 75¢. What was the total cost of these purchases?

16. The total sales for the Hand-i-mans Shop was $17,556.25 for the current month. A 2% city sales tax and a 3% state sales tax were added on all purchases made at this store. How much was collected for each purpose and what was the total tax?

17. Mom's Speciality Shop sold household kitchen gadgets. The cost of 1 item was $3.50 and was sold at a markup of 30% of cost. The Alta View Apts. purchased 45 of these articles and was given a discount of 10% and 5%. Sales tax amounted to $2\frac{1}{2}$%. What was the total of this purchase?

18. Locust City had a 1% city tax. Mr. Orr purchased an article for $136.50 which was shipped to a point within the state. How much did the article cost if the state tax rate was $3\frac{1}{2}$% and shipping costs were $10.65?

19. The sales tax on an electrical appliance was $6.06 or 3.2%. What was the billing price?

CHAPTER 7 PAYROLL

1. Mr. Smith worked for the local laundry at $2.15 an hour. Overtime was figured at time and a half. If he worked 8 hours on Monday, 10 hours on Tuesday, $8\frac{1}{2}$ hours on Wednesday, 8 hours on Thursday, and 9 hours on Friday, what was his gross pay for the week?

2. Miss Taylor earns $130 per week plus overtime at $1\frac{1}{2}$ times the regular rate for all hours over 40. (a) What are her regular and overtime hourly rates? (b) How much did she earn in a week in which she worked 45 hours?

3. Nash Products pays its typists $425 a month. Find their weekly, hourly, and overtime rates (overtime is time and a half). Use a 40-hour week.

4. Miss Smith works for the Amber Co. as a stenographer for $550 a month. She was absent without leave for 5 days. What was the amount of her gross earnings for the month if she worked 40 hours per week and an 8-hour day?

5. Find the F.I.C.A. tax on gross earnings amounting to (a) $345.50 and (b) $560.00.

6. Find the income tax withheld on a monthly salary of $620 with 2 exemptions for a married person.

7. Find the income tax withheld on a weekly salary of $98.50 for a single person with 1 exemption.

8. Mr. Adam earns $8.75 an hour for a 40-hour week. During a week he worked 8 hours overtime at time and a half the regular rate. What were his gross earnings, F.I.C.A. tax, and income tax withheld for the week? If he also had additional deductions of $3.50, what was his net income? (He has 1 exemption and is single.)

9. In how many weeks will Mr. Adam, in problem 8, pay in the maximum allowed for F.I.C.A. taxes? (Use a 40-hour week and no overtime.)

10. Mr. Ames is a salesman for the Manning Mercantile Co. In addition to a regular monthly salary of $675, he receives a commission of 5% on all sales in excess of $1500 during a month. If his sales amounted to $1875 during a certain month, what were his: (a)

gross earnings, (b) F.I.C.A. taxes, (c) income taxes withheld, and (d) net earnings? (Mr. Ames claims 2 exemptions and is married.)

11. Mr. Remick, a married man with 3 exemptions, sells real estate and receives a commission of $3\frac{1}{2}\%$ on all sales. If his sales amounted to $256,525 for the month, find his gross earnings, standard deductions, and net earnings. Assume he has paid no F.I.C.A. taxes for the year.

12. Mrs. Harvey assembles, on an average, 217 items per day at 12¢ each. How much does she earn in a week if she maintains this average for 6 days?

13. The A & B Co. pays $1\frac{1}{2}$ times the regular rate for all overtime in excess of 40 hours during the week except for Sunday when it pays twice the regular rate. How much would an employee earn who had worked $46\frac{1}{2}$ hours during the week and 6 hours on Sunday if his regular rate was $2.80 per hour?

14. The Holdrite Farm pays its help in cash. Prepare a cash sheet and memorandum for employees whose net earnings during the week were as follows: Employee #26, $80.00; #27, $95.00; #28, $125.00; #29, $62.50; and #30, $78.50.

15. J. Jenkins works on a commission. He earns 5% on sales up to $3000, 6% from $3001 to $3500, and 8% for all sales above $3500. How much did he earn on sales of (a) $3567, (b) $2650, and (c) $4200?

16. Using the income tax table, determine the amount of tax withheld on a weekly basis for a married person with 3 exemptions who earns (a) $95.00, (b) $102.60, and (c) $85.02.

17. If Mr. Watson earns $520 a month how much will he pay in F.I.C.A. taxes during the year?

18. Mr. Haverson works in a shoe factory and receives $0.019 for each pair of shoes which pass over his bench. How much does he earn if he handles 1590 pairs on Monday, 1625 on Tuesday, 1670 on Wednesday, 1650 on Thursday, and 1640 on Friday?

19. Mr. Owens earns $12,000 a year. How much does his employer withhold from his salary during the year for (a) F.I.C.A. taxes? (b) Federal Income Taxes if he is married and claims 3 deductions?

20. J. Roberts earned $12,600 last year. In what month would he have paid all F.I.C.A. taxes required for the year 1973? How much did he pay in the last month in which it was collected?

CHAPTER 8 FEDERAL INCOME TAXES

1. Mr. Owens, a single man, earns $8600 as a wage earner. This is his only income. (a) If he does not itemize deductions, which form should he use? (b) Would he use the tax tables or the Tax Rate Schedule X, Y, or Z?

2. Mr. and Mrs. Howard had an adjusted gross income of $9800. Itemized deductions totaled $1750. (a) Which form should they use if they are going to use itemized deductions? (b) How much is their tax if they have one dependent child and file a joint return?

3. (a) If Mr. and Mrs. Howard in problem 2 did not itemize their deductions, how much would their tax equal? (b) Are they better off to itemize deductions?

4. Mary Williams earns $7600 in wages and $750 in rentals. She itemized her deductions. She cannot use the short form 1040A. For what reasons?

5. A single taxpayer (1 exemption) had an adjusted gross income of $13,250. Use the short form and compute his tax as outlined on the form. Deductions were not itemized.

6. Determine whether the standard or itemized deductions should be taken for the following. Also, determine the amount of the deductions to be used in each case.

	Adjusted Gross Income	Total of Itemized Deductions
(a)	$10,500	$2160
(b)	9,800	1300
(c)	6,450	986
(d)	16,560	1962
(e)	8,250	620

7. Mr. Allen's income was from a pension (taxable), dividends, and interest. (a) Which form would he use? (b) What schedules would he need to record this income for the tax return? (c) Where would this income be reported on Form 1040?

8. J. Albright earned $6850 during the year as a self-employed person. (a) Where is this information recorded? (b) Where does the self-employment tax appear on Form 1040?

9. How many exemptions would be allowed for the following?

Status	Number of dependents
married (joint return)	2
married (separate return)	1
married (joint return) (both over 65)	none
single (head of household)	1

10. Complete Form 1040 for Mr. and Mrs. John Smith who wish to file a joint return. They have one dependent child, Jennie. Mrs. Smith's name is Mary. Social Security numbers are 568-26-4483 (H), and 568-26-4483 A (W). Their income is: wages $11,260.00, interest $426.86, dividends $522.00, rentals $528.50, and self-employment $956.20. (Use 1972 rate of .075 for self-employment.) Itemized deductions equal $2168.05. Use $750.00 for each exemption. No payments were made on the estimated income. However, Mr. Smith took only *two* tax withholding allowances instead of three for Federal Income Tax withholding. (Forms may be obtained from the nearest IRS office without cost.)

CHAPTER 9 INSURANCE

1. A house worth $45,000 is insured against fire for 80% of its value. At $0.36 per $100, find the annual premium.

2. If a building is insured against fire for $12,000 at $2.86 per $1000, find the annual premium.

3. Calculate the fire insurance premium for 1 year and the premium for the full term on the following, using annual rate factors: (a) $1.562 per $1000 on $15,000 for 3 years, (b) $0.45 per $100 on $22,000 for 2 years, and (c) $2.57 per $1000 on $56,500 for 3 years.

4. The local school district insured one of its school buildings against fire for $725,000 at $1.54 per $100 for 3 years. What are the (a) premium for 3 years and (b) amount saved on a 3-year policy compared to 3 1-year policies?

5. What is the amount of insurance paid on the following losses? (Coinsurance claims are not involved.)

No.	Amount of Insurance	Amount of Loss
a	$15,000	$ 2,600
b	17,500	21,000
c	5,000	5,000

6. An apartment building is insured against fire for $325,000 which is divided between 3 companies. Insurance is carried by Company A for $120,000 at $0.76 per $100, by Company B for $175,000 at $0.61 per $100, and by Company C for the balance at $0.89 per $100. Find the annual premium paid to each company and the total premium.

7. In problem 6, if fire caused damages in the amount of $76,500, how much should each company pay the insured?

8. Property valued at $46,500 was insured for 80% of this amount. Fire caused damages amounting to $13,200. If the policy carried an 80% coinsurance clause, how much should the carrier pay?

9. A house valued at $32,000 was insured for $26,000 with an 80% coinsurance clause. Damages, due to fire, were $10,500. What was the amount of the indemnity?

10. Mr. Root insured 2 adjoining rental properties with the same company: property A, worth $25,000, for $17,000; property B, worth $32,000, for $26,000. A fire completely destroyed property B and caused $12,000 damages to property A. If the insurance policies contained an 85% coinsurance clause, what should be the indemnity paid on each property and the total payment?

11. In problem 10, what part of the fire loss would the owner bear?

12. Mr. Opal insured some personal articles for $10,000 for a period of 6 months. What was the cost of this insurance if the rate was $0.216 per $100?

13. Mr. and Mrs. Smith insured their furnishings from May 20 to August 10 for $7500 at a rate of 64¢ a hundred. What was the insurance cost?

14. Find the cost of insuring a piano for $950 at $0.56 a $100 for 126 days.

15. A home owner cancelled a 1-year insurance policy, costing $52.70, 76 days after it was purchased. What was the amount of the refund?

16. Mr. and Mrs. Babcock insured their home, worth $28,000, at $7.50 a $1000 and the furnishings, worth $15,000, at $5.40 a $1000 on May 21 for a year. On September 5, they cancelled the policy. (a) What was the amount of the refund? (b) How much did the insurance cost Mr. and Mrs. Babcock?

17. In problem 15, what would the refund equal if the insurance company had cancelled the policy?

18. Mr. Manning owns a home for which he carries $18,500 insurance at a cost of $0.85 per $100. What premium does he pay on (a) a 1-year policy, (b) a 2-year policy, and (c) a 3-year policy?

19. Mrs. Smith, who is 36, drives her car to work—a distance of 10 miles each way. She purchased insurance for the following coverage: comprehensive, $100-deductible collision, 50/100 thousand dollars bodily injury, $1000 medical payments, and $5,000 for property damage. If her car is classified as M1, what are the premium charges for each coverage and the total cost of the insurance? (Mrs. Smith is the only driver in the household.)

20. John Driver is 28 and single. He owns a car that he uses for pleasure. He carries comprehensive, bodily injury for 15/30, and property damage for $5000. His car classification is G4. What is his premium for the year?

21. In problem 20, how much would Mr. Driver's insurance cost if he is married?

22. Mr. Truex paid $127.50 for insurance on his car for the year and cancelled it 196 days later when he sold the car. How much did the insurance company retain and what was his refund?

23. Mr. Noble's insurance on his automobile was cancelled by the insurance company due to an excessive number of minor accidents. He purchased the insurance on April 10 in the amount of $188.20. If it was cancelled on June 5, how much was his refund?

24. Mr. Lindsay, age 45 and married, uses his car to drive to work, a distance of 5 miles. He carries comprehensive, no collision, 50/100 bodily injury, \$1000 medical, and \$5000 property damage. His car is in class H4. What is the cost of his insurance?

25. How much more per year would a \$5000 straight life insurance policy cost at age 50 than at age 30?

26. Find the cost of a \$4000 life insurance policy for a person 35 years of age under a (a) 10-year term policy and (b) straight life insurance policy if the premiums are paid annually.

27. How much less would it cost annually to buy a 20-payment life insurance policy for \$6000 than a 20-year endowment at age 25?

28. A man 50 years of age wants insurance for \$5000 for 10 years. If he lives that long, which would be the best buy: term or straight life? Compare the total amount paid in premiums for both types.

CHAPTER 10 INTEREST

Use the interest formula for the first four problems.

1. Find the interest on \$560 at 4% for 45 days on a 360-day basis and on a 365-day basis.

2. Find the interest on \$900 at $3\frac{1}{2}$% for 5 months.

3. What is the interest on \$6500 at 5% for 4 years, 2 months?

4. Find (a) the ordinary simple interest (30-day month) and (b) the commercial interest (exact time) on \$450 from June 5 to August 10 at $4\frac{1}{2}$%.

5. Find the exact number of days for the following periods of time: (a) March 23 to November 1, (b) May 15 to July 10, and (c) January 1, 1972 to April 11, 1972.

6. What is the accurate interest on \$600 at $4\frac{1}{2}$% from July 15 to October 21?

7. How many days are there for the following time periods if a 30-day month is used: (a) March 6 to July 15, (b) May 4, 1966 to January 2, 1967, and (c) October 10, 1967 to February 18, 1968?

8. Find the ordinary, banker's, and accurate interest on a loan for \$3000 at 6%, dated June 12 and due December 20 of the same year.

9. Find the interest on $420 for 17 days at $4\frac{1}{2}\%$ on a 360-day basis. (Use Simple Interest Tables.)

10. Find the interest on $920 for 95 days at 2% on a 360-day basis. (Use tables.)

11. Find the interest on $500 at $10\frac{1}{2}\%$ for 16 days on a 365-day basis. (Use tables.)

12. Find the interest on $2300 at $1\frac{1}{2}\%$ for 35 days on a 360-day basis. (Use tables.)

13. Find the ordinary simple interest on $1250 at $5\frac{1}{2}\%$ for 68 days (use formula) and verify by use of the table.

14. Compute the accurate interest on a $200 loan, dated May 16 and due October 16 of the same year, if the interest rate is 10%.

15. A man borrows $95. He repays the loan (principal plus interest) 6 months later with a payment of $100. What interest rate did he pay?

16. Mr. Kempton borrowed $300 for 3 months at $5\frac{1}{2}\%$. How much was the payment when the loan was due?

17. What is the maturity value of a loan for $750 for 120 days at $6\frac{1}{2}\%$ based on a 360-day year?

18. Find the principal necessary to earn $11.25 in 3 months at 3%.

19. At what simple ordinary interest rate must $950 be loaned to earn $8.97 in 40 days?

20. How long will it take $2500 to earn $83.33 at 5% (simple ordinary interest)? Express time in terms of days.

21. How much interest will $3000 earn at 7% compounded annually for 5 years? Compounded semiannually for 5 years?

22. If a person deposited $250 at 4% compounded quarterly, how much would he have at the end of 10 years?

23. An investment of $5200 at 5%, compounded semiannually was made 6 years ago. Find the amount of the investment today.

24. What is the effective rate of interest equivalent to 10% converted quarterly? (Record answer to nearest hundredth of 1 percent.)

25. How long will it take any sum of money to double itself at 4% compounded semiannually? (Use Compound Interest Tables and record closest approximate time period.)

26. The sum of $4000 was invested for 10 years. During the first 5 years the interest rate was 5% converted semiannually. The rate then dropped to 4% converted semiannually for the remainder of the time. What was the final amount?

CHAPTER 11 NOTES AND DRAFTS

1. Find the due date of a note dated May 5, 19____ if the time is (a) 30 days, (b) 3 months.

2. Find the term of discount for the notes in problem 1 if they were discounted May 11.

3. If each of the notes listed in problem 1 and 2 was discounted at 6% on May 11, find the proceeds if note (a) is $650 and note (b) is $2300. Neither of these notes is interest bearing.

4. A note for $800 is due in 5 years with interest at 4%. At the end of 3 years it is discounted at 5%. What are the proceeds at the time it is discounted?

5. Find the (a) due date, (b) maturity value, (c) term of discount, (d) bank discount, and (e) proceeds of a noninterest-bearing note dated March 2 for $400 and due in 3 months if it was discounted April 15 at 7%.

6. A manufacturer "sold" a 90-day noninterest-bearing note to his bank for $1575. It was discounted 15 days after the date of origin at 4%. Find (a) the bank discount, (b) the proceeds, and (c) the interest rate equivalent to the discount rate of 4%.

7. The following note, bearing ordinary interest, was discounted by the local bank: date of note, July 10; face of note, $465; time, 180 days; interest rate, 6%; discount rate, 5%; and discount date, October 10. Find the (a) due date, (b) number of days at interest, (c) interest, (d) maturity value, (e) discount term, (f) bank discount, and (g) proceeds.

8. James Bros. accepted a note as payment on a debt from Hiffin & Son amounting to $850, bearing interest at 6% for 3 months. A month later James Bros. took the note to its bank where it was discounted at 6%. How much should James Bros. receive?

9. A man wants $5000 for 3 months. What size loan should he obtain if the bank charges a discount rate of 5%?

10. Jones obtained a loan of $4500 for a year and received $4275 as the proceeds. At what rate was the loan discounted? What simple interest rate was charged?

11. A 1-year note with a face value of $5000 bears interest at 3%. It is purchased by a bank 120 days before it is due. If the discount is $60, what is the discount rate?

12. What are the proceeds on a draft for $1750, dated June 20, due 3 months after date, accepted July 7, and discounted August 10 at $5\frac{1}{2}$%, with a collection fee of $\frac{1}{5}$%?

13. Find the date of maturity, the term of discount, the bank discount, the collection fee, and the proceeds of the following draft: face of draft, $1788; date of draft, June 10; due date, 60 days after sight; acceptance date, June 13; discount date, July 9; rate of discount, 6%, and collection fee, $\frac{1}{4}$%.

14. Find the proceeds of a 60-day sight draft amounting to $457.40, accepted November 18, and discounted November 24 at $6\frac{1}{2}$%.

15. A 60-day sight draft was drawn on Harper Bros. by Vichrome Co. for $350. It was drawn on October 4 and accepted October 15. On November 12 it was discounted at the bank at 5%. The collection fee was $\frac{1}{5}$%. Find the proceeds.

16. A draft for $625 on Smith was drawn by Perkins on April 12 and due 4 months after date. Smith accepted the note on April 20. On July 1 it was discounted at the bank at 6% and the proceeds were deposited to Perkins's account. What was the amount of the deposit?

CHAPTER 12 INSTALLMENT PURCHASES AND PERIODIC LOAN PAYMENT PLANS

1. Mrs. Smith bought a piece of furniture for $780. She made a down-payment of $60 and agreed to pay the balance in equal installments of $120 per month plus 1% interest per month on the unpaid balance.

Prepare a payment schedule showing the amount of each payment and the interest for each. What was the total interest paid during the 6-month period?

2. Mr. and Mrs. Ames turned in 2 older cars as a downpayment on a newer car. The dealer offered them \$675 for the 2 cars on a trade-in for a car priced to sell at \$1995. They were able to finance this purchase through a local bank at \$60 a month for 30 months or through their credit union at \$55 a month for 30 months. What was the installment charge in each case? Find the interest rate for the 2 payment programs by the average principal balance method and the constant ratio formula.

3. A radio may be purchased for \$75.00 cash or 10% down and 12 monthly payments of \$6.50. Find: (a) the total cost on the installment plan, (b) the interest or carrying charges, and (c) the interest rate.

4. In problem 3, how much would the buyer save if he borrowed \$75.00 from the bank at 8% for a year and paid cash for the radio?

5. A sofa may be purchased for \$375 plus 4% sales tax for cash. Or it can be paid for over a 10-month period with a downpayment equal to 10% of the price plus the tax and the balance in equal installments, plus a service charge of $1\frac{1}{2}$% on the unpaid balance (rounded to the nearest \$10 figure). What are: (a) the amount of the downpayment, (b) the monthly payment for the first 2 months including the service charge, and (c) the balance after 2 months?

6. A radio may be purchased for \$76.50 cash or \$10.00 down and \$12.00 per month for 6 months. How large is the carrying charge?

7. In problem 6, would it be better to borrow the money at 6% and pay cash for the radio? If so, how much would be saved?

8. The Starr Furniture Co. advertised floor lamps at \$47.50 less 15% discount with no downpayment on approval of credit and 6 monthly payments of \$8.00 each, or 10% down and 6 payments of \$7.00 each. What are the carrying charges for each plan?

9. A piece of machinery may be purchased for \$670 cash or \$70 down and 12 monthly payments of \$52. What is the interest rate on the time plan?

Chapter 3

1. \$48.88
2. \$3.40
3. \$200, \$172
5. .842175, .77748
6. .15783, .22252
   15.783%, 22.252%

Chapter 4

1. 25 1/3%, 20.2%
2. \$54.63, \$55.88
6. \$6.75
11. S.P. = 184.43
    G.P. = 71.93
    N.P. = 27.66

Chapter 5

17. a. \$575 b. \$1280 c. \$958.35

Chapter 6

1. \$18.937 per \$100
2. \$6156
3. \$558.16

Chapter 9

9. \$10,500
15. \$36.36 Returned
17. \$41.73 Returned

Chapter 11

5. a. June 2 b. \$400 c. 48 days d. \$3.73 e. \$396.27
7. a. Jan. 6 b. 180 days c. \$13.95 d. 478.95 e. 88 days
   f. \$5.85 g. \$473.10

Chapter 12

3. a. \$85.50 b. \$10.50 c. 28.7%
6. \$5.50
7. Yes, Save \$3.20
14. \$257.05 \$133 interest \$124.05 Principal

ANSWERS

Chapter 4 p. 322

1,2,6,11

Chapter 5 p.325

17

Chapter 6 p. 326

1,2,3

Chapter 9 p. 332

9,15,17

Chapter 11 p. 336

5,7

Chapter 12 p. 338

3,6,7,14

Chapter 14 Metric (Use quiz for review)

10. Mr. John belongs to a credit union. He borrows \$150 and pays back \$15 a month plus 1% carrying charge on the unpaid balance. How much does he pay in total carrying charges and what is the true rate of interest?

11. Mr. Crocker borrowed \$200 which he repaid at \$10 a month plus 1% interest on the unpaid balance. How much was the total of the carrying charges?

12. A. Deardon obtained a bank loan for \$400. The bank charges were $\frac{3}{4}$% on the unpaid balance with each payment. What monthly installment would pay off this loan in a year (amortized loan)?

13. Mr. C. Bryant obtained a loan of \$14,000 at $6\frac{1}{2}$% to finance the building of a home. The bank added a charge of $1\frac{1}{2}$% of the loan to carry it. What amortized installment payment would liquidate this obligation in 5 years?

14. Mary Redding bought a home for \$28,500. She made a downpayment of 20%. The remainder was amortized at 7% for 10 years. What was the installment payment? How much of the first month's payment was applied to principal and how much to interest?

15. A car selling for \$4200 was purchased with a trade-in of \$1000 and 36 monthly payments of \$130. What was the interest rate? The rate per month?

16. Mr. Hammer borrowed \$2000 for 5 years at 8%. How much were the payments if they were amortized quarterly?

17. A watch was advertised in the window of a jewelry shop as follows: "\$72, nothing down, \$12.50 a month for six months." Find the interest rate.

18. A house was sold for \$25,000 with a downpayment of 10% and the balance amortized over 25 years. If the interest rate was 7%, find the monthly payment.

19. John bought a bike for \$65 which he paid for in 5 equal monthly payments including interest. If the interest rate was 24%, what was the installment payment?

20. A man buys an \$1800 automobile, pays \$600 down, and wishes to pay the balance by means of monthly payments for $2\frac{1}{2}$ years.

If the interest rate is 6%, how large is each monthly payment if the loan is amortized?

CHAPTER 13 COMPUTER NUMBER SYSTEMS

1. What value in the binary system would correspond to 10^3, 10^5, 10^8?

2. Convert (a) 23 and (b) 56 to binary numbers.

3. Prove your answers in problem 2 by converting them to decimal numbers.

4. (a) Add the numbers 16 and 29. (b) Convert each number to the binary equivalent and add. (c) Prove by comparing answers. (Convert answer in decimal system to binary or conversely.)

5. (a) Subtract 18 from 56. (b) Convert to binary numbers and subtract. (c) Prove your answer.

6. (a) Subtract the following 101110 − 10001. (b) Convert to the decimal system and prove your answer by converting the decimal system answer and compare it to the answer in binary.

7. Convert the following problems to binary numbers, multiply, and prove your answer by converting them to the decimal system. (a) $12 \times 9 = 108$. (b) $36 \times 10 = 360$.

8. (a) Add the following binary numbers. (b) Convert these numbers to decimal system numbers and add. (c) Prove by comparing answers in both systems. Problem: $10101 + 10001 + 1100 = ?$ (Arrange vertically.)

CHAPTER 14 THE METRIC SYSTEM OF WEIGHTS AND MEASUREMENTS

1. How many liters are there in 5 dkl?

2. The normal rainfall in a certain area is 40 in. Convert this to millimeters.

3. How many feet high is a building of 300 m?

4. Convert a speed of 70 mph to kilometers (nearest whole number).

5. A speed of 150 km equals how many miles per hour (nearest whole number)?

6. If 1 liter equals 1.06 qt, how many quarts are there in 5 liters?

7. If 1 g equals .04 oz approximately, how many grams approximately equals 2 oz?

8. A box weighs 100 lb. How many kilograms is this?

9. How many milligrams are there in a hectogram?

10. Mr. Jacobs bought a piece of land that was 120 m by 300 m? Convert this area to ares.

ANSWERS TO ODD-NUMBERED PROBLEMS

CHAPTER 1

Exercise 1

1. 12,840
3. 13,383
5. 150.25
7. 11,397
9. 1.8970
11. 135,355

Exercise 2

(1) Columns

A	127.57
B	165.59
C	371.73
D	326.89
E	226.48
Total	1218.26

(2) Lines

A	391.99
B	181.33
C	153.93
D	186.52
E	304.49
Total	1218.26

Exercise 3

1. 182
3. 2.61
5. 64.87
7. .00192
9. 3217
11. 1858
13. 19.27
15. 16.51
17. 12.689
19. 286.12

Exercise 4

1. 2150
3. 3787.50
5. 48,800
7. 100,000
9. 730
11. 2660
13. 4608.292
15. 1.650
17. 237,000

Exercise 5

I-1. 20.64
3. 14.70
5. .375
7. .210
9. .018

II-1. .09
3. 1.17
5. .44

III-1. .281
3. 300.000
5. .067

Exercise 6

1. $2348.20 and $2300.00
3. $1000 or $1100
5. $1360 or $1400
7. $8+

Exercise 7

1. $28\frac{1}{3}$
3. $24\frac{3}{8}$
5. $45\frac{3}{4}$
7. $29\frac{5}{8}$
9. $4\frac{7}{12}$
11. $161\frac{7}{10}$
13. $65\frac{1}{4}$
15. $5\frac{5}{12}$
17. 196
19. 2

Exercise 8

I-1. $2.11\frac{1}{9}$
3. $.08\frac{1}{3}$
5. $.18\frac{3}{4}$
7. $15.16\frac{2}{3}$
9. $6.37\frac{1}{2}$

II-1. $\frac{1}{6}$
3. $1\frac{7}{8}$
5. $12\frac{5}{12}$
7. $3\frac{5}{8}$
9. $4\frac{3}{8}$

III-1. $3
3. $24
5. $27
7. $22
9. $30
11. $25
13. $6
15. $20
17. $28
19. $120

Exercise 9

1. (a) $4.00, $48.92, $52.05 $180.74, $73.84 (b) $7.82
3. $0.95
5. $2253.10, $751.03, and $1502.07
7. (a) $30,000 (b) $45,000
9. (a) This year $8798.30
   Last year $8583.00
   (b) January $ 38.50 inc
   February 217.25 inc
   March 47.00 dec
   April 332.25 dec
   May 106.00 dec
   June 444.80 inc
   Total $215.30 inc
   ($700.55–$485.25)
   (c) This year $1466.38
   Last year $1430.50
11. $780, $2080, $260

13. $118.45
6.00
2.50
75.00
199.50
.14
8.00
$409.59

15. Gross 138,783
Tare 2,310
Net 136,473

Net 13,158
26,855
32,350
35,152
28,958
136,473

17. $182,000

19. $1603.80
318.60
279.20
357.21
90.40
18.13
$2667.34

21. $24,000 or $20,000
23. $2800
25. $1840

CHAPTER 2

Exercise 10

1. 36%
3. 264%
5. .6%
7. 207%
9. 450%
11. 205%
13. 2000%
15. 225%
17. $\frac{1}{4}$% or .25%
19. $2\frac{1}{2}$%
21. .03
23. .006
25. .026
27. .17
29. 3.00
31. .0005
33. .11
35. 6.52
37. .134
39. .017
41. $\frac{1}{4}$
43. $\frac{33}{400}$
45. $\frac{13}{12}$ or $1\frac{1}{12}$
47. $\frac{3}{50}$
49. $\frac{2}{5}$
51. $\frac{3}{20}$
53. $\frac{4}{5}$
55. $\frac{11}{10,000}$
57. $\frac{1}{250}$
59. $\frac{23}{50}$
61. $37\frac{1}{2}$%
63. $8\frac{1}{3}$%
65. 40%
67. $3\frac{1}{3}$%
69. 50%
71. $187\frac{1}{2}$%
73. $16\frac{2}{3}$%
75. $12\frac{1}{2}$%
77. 140%
79. $133\frac{1}{3}$%

Exercise 11

I-1. $123.75
3. 23.15
5. 7.8
7. .1368

9. $303\frac{1}{3}$

II-1. $33\frac{1}{3}\%$
3. 3%
5. $3\frac{1}{3}\%$
7. 3.5%
9. .05%

III-1. 140
3. 105
5. 1500
7. 6240
9. $52

Exercise 12

1. 3955
3. $5200
5. 15.8%
7. $7\frac{1}{2}\%$
9. 150%
11. $187.50
13. 3000
15. $21,675
17. 5.2%
19. $92.81
21. rug 35%
fabrics 27%
lineo 23%
del ser 15%
100%
23. $4424.37

Review of Chapters 1 and 2

1. (a) $8304.00 (b) $4047.17
(c) $6297.00
3. 15.7%
5. Rose, $3304; Stewart, $2,065; Lynch, $2891
7. $38,330
9. $\frac{41}{120}$
11. $.16\frac{2}{3}\%$ or $\frac{1}{6}\%$
13. $27,216
15. $27,400

Inventory Test

I-1. (a) 11,654
(b) 253.0716
(c) 2351.0767
(d) 546.6345
(e) 2472.1814
3. .325526
5. .0005001
7. 239.8
9. 100.216
11. 931.47 + or 931 16/34
13. $\frac{3}{5}$
15. $\frac{7}{20}$
17. $12\frac{8}{9}$
19. $\frac{2}{7}$
21. 6000
23. $1\frac{1}{2}$
25. $\frac{1}{85}$
27. $\frac{9}{77}$
29. .1136
31. 60%
33. 96.6
35. 6448
37. $16\frac{2}{3}\%$
39. 150%
41. 79.2

II-1. $18,111.60, $6037.20, $6037.20
3. $1,461,000
5. 3.6%
7. $225,000
9. $27\frac{3}{8}$ feet

CHAPTER 3

Exercise 13

1. Net price, $1275.00
discount, $225.00
3. List price, $378.33
net price, $321.58
5. $105.63 or $105.62
7. $22.88

Exercise 14

	Net Price	Discount
1.	$303.75	$ 71.25
3.	364.87	204.13
5.	162.76	351.24
7.	11.90	3.56
9.	207.00	253.00

Exercise 15

	Net Price	Discount
1.	$ 19.24	$ 7.76
3.	230.54	69.71
5.	98.09	78.41
7.	27.90	34.10
9.	3.07	1.88

Exercise 16

	N.D.E.	S.D.E.
1.	.75208	24.792%
	.66500	33.5%
	.50625	49.375%
	.22500	77.5%

3. 25%
5. (a) 40 and 10%, $80.64
   (b) same
7. 21.025%
9. List price, $44.65;
   net price, $32.15

Exercise 17

I-1. March 25
3. January 23
5. August 1
7. December 11
9. June 10

	Cash Discount	Payment
II-1.	$ 0.46	$ 22.70
3.	0.62	61.38
5.	0.20	19.55
7.	3.47	343.43
9.	10.08	493.88

	Discount	Balance Due
III-1.	$4.64	$157.96
3.	2.32	317.68
5.	1.77	148.23
7.	1.39	49.01
9.	0.76	40.74

Exercise 18

1. (a) February 10 (b) $5.47
3. $483.42
5. $851.56
7. $8.50
9. (a) $87.10 (b) $225.47
11. 30%
13. $60.86
15. $248.24

CHAPTER 4

Exercise 19

	Markup	% on Cost	% on Selling Price
I-1.	$35.00	58.3	36.8
3.	22.50	88.2	46.9
5.	7.05	28.3	22.0

II-1. $30.00, 200%
3. (a) $30.00, (b) $66\frac{2}{3}$%
5. $51.75, $28\frac{3}{4}$%

Exercise 20

	Selling Price Based on Cost	Selling Price Based on Selling Price
I-1.	$ 11.75	$ 12.53
3.	$480.00	$500.00
5.	$ 51.75	$ 52.94

II-1. $617.50
3. $5.25
5. $25.33

Exercise 21

	Cost Based on	
	Selling Price	Cost
I-1.	$112.50	$113.64
3.	6.17	6.95
5.	264.42	268.62

II-1. $18.95
3. $1.75
5. (a) $139.20 (b) $208.80

Exercise 22

I-1. $100, 20%
3. $11.00, $8\frac{1}{3}$%
5. $80.59, $12.09

II-1. 22%
3. $3.00
5. $22\frac{2}{9}$%

Exercise 23

1. $8.70
3. $33.00
5. $221.68

Exercise 24

1. (a) $23.40 (b) $24.38
3. 25%
5. 13.6%
7. (a) $31.67 (b) $190.00
9. $1.12
11. (a) $256.50, (b) $228.00
13. (a) $110.00, (b) $11.00 (c) $35.20

CHAPTER 5

Exercise 25

1. $160
3. (a) $420, (b) $2400
5. (a) $640, (b) $12,800
7. (a) $610 (b) $1870
9. $315.00

Exercise 26

1. (a) $592 (b) 296
5. $714.29
7. (a) $950, (b) $1583.33
9. (a) $\frac{1}{3}$, (b) $\frac{2}{13}$, (c) $\frac{2}{31}$

3. (Ex. 26)

End of Year	Annual Depreciation	Depreciation To Date	Book Value End of Year
0	$	$	$900.00
1	330.00	330.00	570.00
2	247.50	577.50	322.50
3	165.00	742.50	157.50
4	82.50	825.00	75.00
Total	$825.00	XXX	XXX

Exercise 27

1. $391.68
3. $243.00
5. 25%, $13\frac{1}{3}$%, $6\frac{2}{3}$%, $16\frac{2}{3}$%
7. (a) $1250, $1187.50, $1128.13
   (b) $21,434.37
9. (a) $120 and $80
   (b) $25 and $20

Exercise 28

1. (a) $1325.00, (b) $13,250.00
   (c) $31,875.00
3. $1000
5. (a) $600, (b) $1200,
   (c) $1090.91
7. (a) $\frac{1}{3}$, (b) $\frac{1}{4}$, (c) $\frac{1}{15}$
9. $789.23

CHAPTER 6

Exercise 29

I-1. 6.7%
3. $0.10667 per $
5. $11.916 per $100
7. $0.25 per $100
9. $0.0302

II-1. $195.00
3. $10,817.10
5. $273.02

	Assessed Valuation	Taxes Paid
III-1.	$5,200.00	$650.00
3.	7,150.00	719.29
5.	31,400.00	2616.67

IV-1. $3137.50
3. $160,000
5. $933,034,091

7. $28,000
9. $13.81

Exercise 30

1. (a) $1.93 (b) $1.44
   (c) $0.12 (d) $28.82
   (e) $0.81 (f) $0.04
3. $0.13
5. $65.00
7. $77.63

Review

1. $22.50, 81.8%
3. 1.9%
5. $303.33
7. $57.50
9. $8.90 per $100, $88.93 per $1000 and 8.893%
11. $3990.00
13. $200.00
15. $6.63
17. $212.50
19. (a) $95 (b) $63
    (c) $750
21. $364.93
23. $3000, $2900
25. (a) 20% (b) $12\frac{1}{2}$%
    (c) 10% (d) $3\frac{1}{3}$%
    (e) 5%

CHAPTER 7

Exercise 31

1. $104.58
3. $99.34
5. $254.80
7. $372
9.

	Weekly Rate	Monthly Rate
(a)	$166.40	$721.07
(b)	82.20	356.20
(c)	72.00	312.00
(d)	130.00	563.33
(e)	200.00	866.67
(f)	113.00	489.67

Exercise 32

1. (a) \$15.80 (b) \$25.51
   (c) \$3.54 (d) \$6.58
   (e) \$13.70
3. (a) \$9.65 (b) \$26.60
   (c) \$128.75
5. (a) \$1050 (b) \$61.43
   (c) \$190.00 (d) \$798.57

Exercise 33

1. \$136.00
3. \$3.40, \$2.205
5. \$3.51

7. (Ex. 32)

Employee No.	Total Hours	Earnings Regular	Overtime	Total
32	$41\frac{1}{2}$	\$ 86.00	\$ 4.84	\$ 90.84
33	$39\frac{1}{2}$	78.21	—	78.21
34	$41\frac{1}{2}$	140.00	7.88	147.88
35	$41\frac{1}{2}$	130.00	7.31	137.31
36	$39\frac{1}{2}$	108.63	—	108.63
37	43	112.00	12.60	124.60
Total	$246\frac{1}{2}$	\$654.84	\$32.63	\$687.47

9. \$310.38

Exercise 34

1. Payroll A:

Employee	Total Hours	Earnings Regular	Overtime	Total
Ames	$40\frac{1}{2}$	\$ 88.00	\$ 1.65	\$ 89.65
Bell	$43\frac{1}{2}$	78.00	10.24	88.24
Ellis	40	128.00	—	128.00
King	$42\frac{1}{2}$	180.00	16.88	196.88
Moore	39	142.35	—	142.35
Wells	41	208.00	7.80	215.80
Total	$246\frac{1}{2}$	\$824.35	\$36.57	\$860.92

Employee	Deductions F.I.C.A.	Tax With.	Other	Total	Net Earnings
Ames	\$ 5.24	\$ 7.40	\$ 2.50	\$ 15.14	\$ 74.51
Bell	5.16	5.00	5.00	15.16	73.08
Ellis	7.49	18.70	5.00	31.19	96.81
King	11.52	22.00	2.50	36.02	160.86
Moore	8.33	15.90	2.50	26.73	115.62
Wells	12.62	34.00	2.50	49.12	166.68
Total	\$50.36	\$103.00	\$20.00	\$173.36	\$687.56

3. Cash Sheet:

	Net	Bills				Coins				
Employee	Earnings	20	10	5	1	.50	.25	.10	.05	.01
Ames	$ 74.51	3	1		4	1				1
Bell	73.08	3	1		3				1	3
Ellis	96.81	4	1	1	1	1	1		1	1
King	160.86	8				1	1	1		1
Moore	115.62	5	1	1		1		1		2
Wells	166.68	8		1	1	1		1	1	3
Total	$687.56	31	4	3	9	5	2	3	3	11

Change Memorandum:

Number	Amount		
31 twenties	$620.00	2 quarters	.50
4 tens	40.00	3 dimes	.30
3 fives	15.00	3 nickels	.15
9 ones	9.00	11 pennies	.11
5 halves	2.50	Total	$687.56

CHAPTER 8

Exercise 35

I-1. Line 14 (gross income) was less than $10,000
3. (a) $741
(b) no, forced saving

II-1. $2000

III. answers listed in text

IV-1. $1301.72
2. $1126.96

Exercise 36

1. (a) $1230.66
(b) Reduced her taxable income; saved $34.34
3. (a) $1317.50
(b) $1346.75
(c) $2214.74
(d) $1894.00
(e) $1777.50
5. $1185.63
7. $1280.16 or $1286.52
9. (a) Form 1040A
(b) $1200

Exercise 37

1. $5.34, $7.275, $7.875
3. $960.34
5. (a) $127.20, (b) $3.18
7. (a) No, because they are less than 15% of adjusted gross income
(b) 1040A
(c) $1748.50

CHAPTER 9

Exercise 38

I-1. $22.50
3. $40.40
5. $39.68

II-3. $3.28
5. $3.22

Exercise 39

1. (a) $16,500 (b) $24,000
3. A $10,400.00
B 6,933.33
C 8,666.67
D 26,000.00
$52,000.00

Exercise 40

I-1. $10,526.32
3. $24,000.00
5. $3174.60

II-1. $76,000
3. (a) $135,416.67
(b) $24,583.33
5. $34.375

Exercise 41

1. 21%, 13%, 43%, 34%
3. Net premium, $21.13; refund, $11.37
5. $439.57

Exercise 42

1. $7.00 less
3. $20.00
5. (a) $100.63 (b) $118.84
7. $25.00

Exercise 43

1. Term (a) $79.50 (b) $151.60
3. $12.79, $2.50
5. (a) 20 years, $495.00;
40 years, $1158.00
(b) 20 years, $795.00;
40 years, $2311.00
7. $3065.00, $1935.00
9. $397.50 annual basis,
$405.40 semiannual basis

Exercise 44

1. (a) $118.80 (b) $47.52
(c) $1483.20 (d) $417.36
(e) $2770.20
3. (A) $8750 (B) $5250
(C) $21,000
5. $8.78
7. $7.66
9. $10,000
11. $164
13. $46.82
15. Comprehensive = $ 21
50/100 bodily injury = 51
2000 medical = 15
5000 property damage = 19
Uninsured liability = 8
Total = $114

17.

Years	(a)	(b)	(c)
25	$24.95	$67.00	$208.20
40	35.35	115.55	219.85
60	144.90	268.80	none offered

19. $198.00 paid premium

CHAPTER 10

Exercise 45

	Approximate Time	Exact Time
1.	89 days	91 days
3.	330 days	335 days
5.	168 days	171 days
7.	371 days	377 days
9.	118 days	120 days

Exercise 46

I-1. $12.50
3. $63.00
5. $16.00
7. $13.78
9. $122.50

II-1. $7.49
3. $11.73
5. $3.33
7. $5.00
9. $16.90

III-7. $5.10
9. $17.54

IV-1. $22.19
3. $59.07
5. $19.92

Exercise 47

1. $6.75
3. $18.00
5. $2.06
7. $21.60
9. $1.58
11. $9.17
13. $49.50
15. $7.92
17. $2.90
19. $1.86

Exercise 48

I(a)-1. $2.56
3. $1.35
5. $3.17
(b)-7. $14.04
9. $28.61

II-1. $13.76
3. $21.60
5. $43.73

III-1. Bank A, $1.37 less interest
3. (a) November 1
(b) $16.63
5. Ordinary interest, $13.50; accurate interest, $13.32

Exercise 49

	Principal	Maturity Value
I-1.	$600.00	$603.75
3.	350.00	376.25
5.	543.75	558.25

II-1. 9%
3. 2%
5. $4\frac{1}{2}$%

III-1. 90 days
3. 45 days
5. 450 days or 1 year, 90 days

IV-1. 45 days
3. $7\frac{1}{2}$%
5. $680.40
7. (a) $0.80 (b) $7.50
(c) $1.35
9. 4%
11. $1\frac{1}{3}$ years or
1 year, 120 days

Exercise 50

1. $17\frac{1}{2}$ years
3. Semiannually 3%, conversion periods 10;
quarterly $1\frac{1}{2}$%, conversion periods 20;
monthly $\frac{1}{2}$%, conversion periods 60;
annually 6%, conversion periods 5
5. $62.89
7. $1434.74
9. $1766.10
11. $386.42

CHAPTER 11

Exercise 51

1. Maker, Mr. Hope;
   payee, Mann Bros.;
   assignee, First National Bank;
   maturity date, February 11;
   maturity value, $507.50;
   term of discount, 40 days;
   discount, $3.10;
   proceeds, $504.40
3. $4217.35
5. $1525.77
7. $1237.81
9. 6.4%, 8.7%

Exercise 52

1. $557.34
3. $817.25
5. (a) $1483.83 (b) $747.79
   (c) $3251.98 (d) $539.21
7. $3563.04
9. $253.94

Exercise 53

1. $3640
3. (a) $36.00 (b) $36.80
   (c) $36.30
5. (a) $588 (b) 8.2%
7. $1462.50
9. (a) 1, 8%
   (b) 4, $1\frac{1}{2}$%
   (c) 12, $\frac{1}{3}$% or .$33\frac{1}{3}$%
   (d) 2, $3\frac{1}{2}$%
11. $15.25 less
13. (a) January 11
    (b) September 10
    (c) April 14
    (d) April 25
15. $2977.50

CHAPTER 12

Exercise 54

1. $12.00
3. $39.00
5. $4.35, $1.95, $2.55
7. (a) $20.72 (b) $12.15

Exercise 55

1. 12%, 15%, 24%, 6%, 9%, 18%
3. 26.4%
5. $41.50 total cost,
   34.9% interest rate

Exercise 56

1. $64.12
3. Installment, $228.48; amount applied to principal, first month, $57.48
5. Payments, $31.50;
   14.8% interest rate
7. 31.4%
9. Yes, a saving of $74.40
11. $69.90

Exercise 57

1. (a) $36.50 (b) 11.5%
3. $3.18
5. 30%, 36%, 18%, 7.2%
7. $212.86

CHAPTER 13

Exercise 58

1. $2^0, 2^1, 2^2, 2^3, 2^4$
   or 1, 2, 4, 8, 16
3. (a) 11 (b) 59

5. $$\begin{array}{r} 11011 = 16+8+0+2+1 \quad = 27 \\ +1100 = \ 8+4 \quad = 12 \\ \hline 100111 = 32+0+0+4+2+1 = 39 \end{array}$$
7. (a) 100 (b) 101 (c) 10001
9. (a) 11110 (b) 1011011
   (c) 1100001111

Exercise 59

Self testing

CHAPTER 14

Exercise 60

1. (a) meter (b) 1.0936 yd
3. 97.536 m
5. 9.85
7. (a) .9081 (b) .9463
9. (a) 3.3069 (b) 4.96035

Exercise 61

Self testing

Exercise 62

1. .5, 2, 50
3. (a) 2000 (b) 20
5. 150,000

Exercise 63

1. 541
3. (a) 100 (b) 10 (c) 1000
5. 19 mm, 50.8 mm,
   19.05 mm
7. 7
9. (a) .355 (b) 16 oz
11. 1500 m = 4921 ft
13. 96 a

CHAPTER 15

Chapter 1

1. $91\frac{2}{3}$ bags
3. (a) $284\frac{4}{9}$ (b) $338.40
5. $4\frac{1}{2}$ page average per hour,
   $198.00 gross earnings
7. (a) $3853.20 (b) $215.60
9. $945.22
11. 1.06558
13. (a) $ 1.88
    (b) 3.90
    (c) 1.60
    (d) 2.10
    (e) 16.00
    (f) 6.30
    Total $31.78
15. $106.38, $126.82, $173.68,
    Total: $406.88
17. $57.06
19. $284.38
21. $511\frac{1}{2}$ packages
23. $6.53
25.

	Net Sales
A	$7,097.85
B	8,500.16
C	13,501.71
D	10,845.92

    Total gross sales, $42,122.83;
    returns and allowances,
    $2177.19; net sales,
    $39,945.64
27. $216.00
    0.38
    1.53
    244.50
    124.74
    143.75
    $730.90
29. (a) 20 miles
    (b) $120

31. (a) $22\frac{1}{2}$ mi
    (b) $0.319 or 32¢
    (c) $0.014
33. $8.38

Chapter 2

1. $40\frac{3}{4}$, 0.022, $.46\frac{7}{8}$, $.078\frac{1}{8}$, $2\frac{3}{8}$
3. 3600, 7200, 43,200, 196.36
5. 3055
7. 14%
9. $120,320.00, $52,640.00, $203,040.00
11. 32.2%
13. $5400
15. $5550
17. (a) $1232.50, (b) $3.95, (c) $12\frac{1}{2}$%
19. 1650

Chapter 3

1. $48.88
3. List price, $200.00; net price, $172.00
5. .84218, .77748, .26250, .62067, .29706
7. $266.76
9. $90.00
11. $997.50
13. Amount of discount, $7.18; balance due, $392.82
15. $525.75
17. (a) February 25
    (b) June 10
    (c) November 15
    (d) December 2
19. $115.88

Chapter 4

1. Markup $0.95, on cost $25\frac{1}{3}$%, on selling price 20.2%
3. $39.64
5. $3.89
7. 16%
9. $315.00, $302.40
11. Selling price, $184.42; net profit, $27.66; gross profit, $71.92
13. (a) $165.71 (b) $146.00
15. $33.18
17. $385.81
19. List price, $79.78; net price, $67.81

Chapter 5

1. Annual depreciation, $1166.67; rate of depreciation, $3\frac{1}{3}$%; book value, $17,500
3. Depreciation rate, 9%; book value, $108.75
5. $45,333.35
7. $114.91, $71.82
9. $163.64, $147.27
11. $310.91
13. $2508.80
15. Lost $20.01
17. (a) $575.00 (b) $1280.00 (c) $958.33
19. Straight line, $162.00; sum of digits, $204.00; declining balance, $216.00

Chapter 6

1. $18.937 per $100
3. $558.16
5. (a) $375.00 (b) $3493.33 (c) $1247.40
7. $25,000
9. $433.02
11. $65,480,296.00
13. $65.05
15. $57.26
17. $179.44
19. $189.38

Chapter 7

1. $97.29
3. Weekly rate, $98.08; hourly rate, $2.452; overtime rate, $3.678
5. (a) $20.21 (b) $32.76
7. $12.70
9. 31
11. Gross earnings = $8978.38
    Deductions = $3404.16
    Net earnings = $5574.22
13. $172.90
15. (a) $185.36
    (b) $132.50
    (c) $296.00
17. $365.04
19. (a) $631.80
    (b) $1532.40

Chapter 8

1. (a) Form 1040A
   (b) Tax tables
3. (a) $1019.00
   (b) yes
5. $2228.38
7. (a) Form 1040
   (b) Schedule E
   (c) Schedule B
9. 4, 2, 4, 2

Chapter 9

1. $129.60
3. (a) $23.43, $63.26
   (b) $99.00, $183.15
   (c) $145.21, $392.07
5. (a) $2600 (b) $17,500
   (c) $5000
7. A $28,246.15
   B $41,192.31
   C $ 7,061.54
   $76,500.00
9. $10,500
11. $8400.00
13. $15.84
15. $36.36
17. $41.73
19. Comprehensive = $ 46.00
    $100 deductible collision = $ 96.00
    50/100 bodily injury = $ 66.00
    $1000 medical = $ 15.00
    $5000 property damage = $ 25.00
    Total = $248.00
21. $76.00
23. $159.33
25. $90.70
27. $118.14

Chapter 10

1. $2.80, $2.76
3. $1354.17
5. (a) 223, (b) 56, (c) 101
7. (a) 129, (b) 238, (c) 128
9. $0.89
11. $2.30
13. $12.99
15. 10.52%
17. $766.25
19. $8\frac{1}{2}$%
21. $1207.66, $1231.81
23. $6993.42
25. $17\frac{1}{2}$ years

Chapter 11

1. (a) June 14
   (b) August 5
3. (a) $647.40
   (b) $2267.03

5. (a) June 2 (b) $400.00
   (c) 48 days (d) $3.73
   (e) $396.27
7. (a) January 6 (b) 180 days
   (c) $13.95 (d) $478.95
   (e) 88 days (f) $5.85
   (g) $473.10
9. $5063.29
11. 3.5%
13. maturity date, August 12
    term of discount, 34 days
    bank discount, $10.13
    collection fee, $4.47
    proceeds, $1773.40
15. $347.74

Chapter 12

1.

End of Month	Balance	Payment	Charges 1%	Total Payment
1	$720	$120	$7.20	$127.20
2	600	120	6.00	126.00
3	480	120	4.80	124.80
4	360	120	3.60	123.60
5	240	120	2.40	122.40
6	120	120	1.20	121.20
Total	—	$720	$25,20	$745.20

3. (a) $85.50 (b) $10.50
   (c) 28.7%
5. (a) $52.50 (b) $38.85
   (c) $38.25 (c) $270.00
7. Yes, save $3.20
9. 7.4%
11. $21.00
13. $275.96
15. Interest rate, 30%;
    monthly interest charge,
    2.5%
17. 14.3%
19. $13.78

Chapter 13

1. 2^3, 2^5, 2^8, or 8, 32, 256
3. (a)

1	$1 \times 2^0 =$	1
2	$1 \times 1^1 =$	2
3	$1 \times 2^2 =$	4
4	$0 \times 2^3 =$	0
5	$1 \times 2^4 =$	16
		23

   (b)

1	$0 \times 2^0 =$	0
2	$0 \times 2^1 =$	0
3	$0 \times 2^2 =$	0
4	$1 \times 2^3 =$	8
5	$1 \times 2^4 =$	16
6	$1 \times 2^5 =$	32
		56

5.
```
  111000
 -10010
 ------
  100110
```

proof:
```
2)38
2)19  0
2) 9  1
2) 4  1
2) 2  0
2) 1  0
2  0  1
```

7. (a)
```
      1100
      1001
      ----
      1100
   1100
   -------
   1101100
```

proof: $0 \times 2^0 = 0$
$0 \times 2^1 = 0$
$1 \times 2^2 = 4$
$1 \times 2^3 = 8$
$0 \times 2^4 = 0$
$1 \times 2^5 = 32$
$1 \times 2^6 = 64$
108

(b)
```
     100100
       1010
    -------
    1001000
  100100
  ---------
  101101000
```

proof: $0 \times 2^0 = 0$
$0 \times 2^1 = 0$
$0 \times 2^2 = 0$
$1 \times 2^3 = 8$
$0 \times 2^4 = 0$
$1 \times 2^5 = 32$
$1 \times 2^6 = 64$
$0 \times 2^7 = 0$
$1 \times 2^8 = 256$
360

Chapter 14

1. 50 liters
3. 984.24 ft
5. 93 mph
7. 50 g
9. 100,000 milligrams

INDEX